GLOBALIZING TOBACCO CONTROL

TRACKING GLOBALIZATION

Robert J. Foster, editor

GLOBALIZING TOBACCO CONTROL

Anti-smoking Campaigns in California, France, and Japan

Roddey Reid

Indiana
University
Press
BLOOMINGTON AND INDIANAPOLIS

This book is a publication of

Indiana University Press
601 North Morton Street
Bloomington, IN 47404–3797 USA

http://iupress.indiana.edu

Telephone orders 800-842-6796
Fax orders 812-855-7931
Orders by e-mail iuporder@indiana.edu

The paper used in this publication meets the minimum requirements of American National Standard for Information Sciences—Permanence of Paper for Printed Library Materials, ANSI Z39.48-1984.

Manufactured in the United States of America

Library of Congress Cataloging-in-Publication Data

Reid, Roddey, date
Globalizing tobacco control : anti-smoking campaigns in California, France, and Japan / Roddey Reid.
p; cm. — (Tracking globalization)
Includes bibliographical references and index.
ISBN 0-253-34667-3 (cloth : alk. paper) — ISBN 0-253-21809-8 (pbk. : alk. paper)
1. Smoking—Government policy—Cross-cultural studies. 2. Smoking—Government policy—California. 3. Smoking—Government policy—France. 4. Smoking—Government policy—Japan. 5. Smoking—California—Prevention. 6. Smoking—France—Prevention. 7. Smoking—Japan—Prevention. I. Title. II. Series.
[DNLM: 1. Smoking Cessation—California. 2. Smoking Cessation—France. 3. Smoking Cessation—Japan. 4. Cross-Cultural Comparison—California. 5. Cross-Cultural Comparison—France. 6. Cross-Cultural Comparison—Japan. 7. Health Knowledge, Attitudes, Practice—California. 8. Health Knowledge, Attitudes, Practice—France. 9. Health Knowledge, Attitudes, Practice—Japan. 10. Health Promotion—California. 11. Health Promotion—France. 12. Health Promotion—Japan. 13. Social Change—California. 14. Social Change—France. 15. Social Change—Japan. WM 290 R357g 2005]

RA1242.T6R435 2005
362.29'67—dc22 2005011500

1 2 3 4 5 10 09 08 07 06 05

For L.B.

Contents

Acknowledgments

A lengthy interdisciplinary enterprise such as this one has benefited from conversations with many colleagues near and far.

The University of California, San Diego (UCSD), is particularly rich in resources for students of public health and tobacco control. I have been fortunate to have as colleagues senior epidemiologists who, early on, granted key interviews that shaped my research project. I want to thank in the School of Medicine David Burns and especially John Pierce, who helped broker early interviews in the tobacco control community and read portions of the manuscript. Both Joseph Gusfield (Sociology) and Michael Schudson (Communication) gave crucial interviews at the beginning of my research for which I am grateful. Colleagues in the UCSD Science Studies Program, the Ethnic Studies Department, and the Center for Iberian and Latin American Studies provided support and venues for testing out some of my findings. The Department of Literature has furnished me with a unique environment that nourishes interdisciplinary studies of culture and society without which my career as a scholar would have taken a very different course.

State health officials, public health researchers, anti-smoking advocates, and advertising executives took time away from busy schedules and pressing matters to grant interviews to a researcher from well outside their respective professional fields. In the U.S., a special debt of gratitude goes to Robert Robinson of the Office on Smoking and Health (CDC), who also read portions of the manuscript; Stanton Glantz of the University of California, San Francisco; Colleen Stevens of the California Department of Health Services; Paul Keye of Paul Keye and Partners; Lisa Unsworth of Arnold Communications; John Elder of San Diego State University; Peggy Toy of the University of California, Los Angeles (UCLA), Center for Health Policy Research; W. James Popham, formerly of IOX Associates; Lourdes Baezconde-Garbanati of the University of Southern California; Robert Hornik of the Annenberg School of Communication, University of Pennsylvania; Michael Cummings of the Roswald Park Cancer Institute; Julia Carol of Americans for Nonsmokers' Rights; Gregory Connolly of the Massachusetts Department of Public Health; Kathy Harty of the Minnesota Health Department; Mike Pertschuk of the Advocacy Institute; Nancy Kaufmann of the Robert Wood Johnson Foundation; Rose-Mary Romano of the Academy for Educational Development; Elaine Arkin; and Steven Rabin and Diana Fievelson of Isosphere.

Internationally, many served as informants and some also as hosts. In particular I wish to thank Derek Yach, former project manager of the World Health Organization's Tobacco-Free Initiative. In France I enjoyed the opportunity to interview several times many key players in the anti-smoking movement. I wish to thank Albert Hirsch of the Hôpital Saint-Louis; Philippe Boucher, former director of the Comité national contre le tabagisme (CNTC); Pascal Mélihan-Chenein of the Ligue contre le cancer; Serge Karsenty of the Centre national de la recherche scientifique (Nantes); Claude Got, Faculté de Médecine (Paris); Gérard Dubois, Faculté de Médecine (Amiens); Karen Slama of the International Union against Tuberculosis and Lung Disease (Paris); Danielle Grizeau and Béatrice Speisser of the Institut national de prévention et d'éducation pour la santé (INPES); Michel Le Net of the Institut de la communication sociale; Laurence Folléa of *Le Monde*; Jean Tostain; Ruben Israel of GlobaLink; and Christian Péchang. I owe a special debt of gratitude to Jean-Yves Nau of *Le Monde*, who granted several interviews and gave me access to his newspaper's priceless tobacco control archive.

In Japan I wish to thank the executive director, Yoshihiko Wakumoto, and the staff of the Japan Foundation's Center for Global Partnership for their support. Frank Baldwin, Tokyo representative of the Social Science Research Council for the Abe Fellowship Program, helped orient me and broker introductions. I especially want to thank Tadao Shimao of the Japan Anti-Tuberculosis Association; Kenji Makino of the Science University of Tokyo; Bungaku Watanabe of the Tobacco Information Center, Tokyo; Yumiko Mochizuki-Kobayashi of the National Institute of Public Health; Suketami Tominaga of the Aichi Cancer Center; Nobuko Nakano of Women's Action on Smoking; Tadao Hozumi of Nagashima and Ohno, Tokyo; Yoshio Isayama of the Lawyers' Organization for Nonsmokers' Rights; Tsuneo Matsumoto of Hitotsubashi University; Kiroku Hanai of Shumei University; the late Yayori Matsui of Asia-Japan Women's Resource Center; Masakazu Nakamura of the Osaka Cancer Prevention and Detection Center; Kyoichi Miyazaki of Can Do Harajuku; Masumi Minowa of the National Institute of Public Health; Yoneatsu Osaki and Tomofumi Sone of the National Institute of Public Health; the late Yuichiro Goto of Tokai University; Masayuki Nishiki of the Hiroshima University School of Medicine; Shigeru Iwamori of Hiroshima City ASA Hospital; Mick Corliss of the *Japan Times*; Mark Levin of the University of Hawai'i; Naohito Yamaguchi of the National Cancer Center; Satoshi Ukai of Hitotsubashi University; Huu-Dung Dao of Josai International University; Yoshihiko Ikegami of *Gendai shisō*; and Naohiro Masuda of Tokyo Broadcasting System. Noriko Mizuta, president of Josai International University (Chiba Prefecture), deserves special thanks for hosting me as a visiting scholar for two years.

I am indebted to staffs at various libraries and archives who provided invaluable assistance: the Tobacco Control Archives, University of California, San

Francisco; the Tobacco Education Clearinghouse of California Resource Center; the Institut national de l'audiovisuel (Paris); the archives of Dentsu Advertising (Tokyo); The Shelden Margen Public Health Library, University of California, Berkeley; and the Biomedical Library and Social Science and Humanities Library, UCSD.

I had the opportunity to present findings at scholarly meetings, including the annual conferences of the Society of Social Studies of Science, the Society for the Study of Literature and Science, the International Communication Association, the American Anthropological Association, and the Twentieth-Century French Studies Association. I have also delivered papers on my research at the Chinese Academy of Social Sciences (Beijing), Josai International University (Chiba Prefecture), and the Gender and Image Research Group (Tokyo).

Earlier versions of some of this material have appeared in *Communication Theory* ("Fractured Media Sphere and U.S. Health Promotion: Cigarettes as Icon of Flawed Modernity," 15, no. 3 [2005]), *Science as Culture* ("Healthy Families, Healthy Citizens: The Politics of Speech and Knowledge in the California Anti–Secondhand Smoke Media Campaign," 29 [1997]: 541–581; and "Tensions within California Tobacco Control in the 1990s: Health Movements, State Initiatives, and Community Mobilization," 13, no. 4 [2004]: 515–537); *Film Quarterly* ("Unsafe at Any Speed: Todd Haynes Visual Culture of Health," 50, no. 3 [1998]: 32–44), and *Gendai shisō* ("Doko ni ite mo anzen de wa nai: Toddo Heizu ni yoru kenkō to risuku no shikaku bunka" [Sept. 2000]: 161–177); as well as the anthology, *Doing Science + Culture* ("Researcher or Smoker? or, When the Other Isn't Other Enough in Studying 'across' Tobacco Control," ed. Roddey Reid and Sharon Traweek [New York: Routledge], 119–150). Permission to reprint portions of these publications is gratefully acknowledged.

Colleagues have encouraged my interdisciplinary study of science and medicine over the years. It was never a simple undertaking. Special thanks go to participants in the "Postdisciplinary Approaches to the Technosciences" resident research group (January–June 1996) at the University of California Humanities Research Institute for which I served as convener; they include Adele Clarke, Joan Fujimura, Emily Martin, Jackie Orr, Molly Rhodes, and Sharon Traweek. This project owes its existence to other colleagues who have provided criticism and suggestions including Michael Bernstein, Geof Bowker, Charles Briggs, Linda Brodkey, Sandra Buckley, Patrick Carroll, Steve Fagin, Michelle Fine, Mike Fortun, Tak Fujitani, Luce Giard, Jonathan Mark Hall, David Halperin, Marcel Hénaff, Lisa Lowe, Mike Lynch, Constance Penley, Paul Rabinow, Steve Shaviro, Leigh Star, Lesley Stern, Paula Treichler, Carole S. Vance, and Linda Williams. Lisa Bloom gave constant intellectual and moral support throughout the long career of this project.

I thank Lisa Cartwright, who first brought my manuscript to the attention of

Indiana University Press; Robert Foster, the series editor and astute reader of texts; Rebecca Tolen, acquisitions editor, whose suggestions helped make the text more readable; and Andrea Zuercher, who performed expert copyediting. David Hess and Toby Miller gave crucial advice on revising the manuscript. Kurt Fowler of the Tobacco Control Section (California Department of Health Services) and Cynthia Newcomer of the Media Campaign Resource Center (Office on Smoking and Health, U.S. Centers for Disease Control and Prevention) provided invaluable service in securing reproduction rights for illustrations of U.S. anti-smoking ads and spots.

Finally, this project has enjoyed support in the form of a Japan Foundation Abe Fellowship awarded by the Social Science Research Council and the American Council of Learned Societies, which afforded me the invaluable opportunity to include Japan in this study by making it possible to live and work for two years in Japan (June 1998–March 2000), during which I spent time preparing for and conducting interviews, attending anti-smoking advocates' meetings, and learning Japanese. Throughout duration of my research, the UCSD Academic Senate supported my work through three research grants, and a subvention from the Office of Graduate Studies and Research helped offset costs of publication.

R. R.

GLOBALIZING TOBACCO CONTROL

Introduction

Globalization and Liberal Governance in Tobacco Control

It is a long way from Geneva to Kobe [Japan]. But the cause that brings us together from far corners of the earth shrinks any physical distance that might separate us. The tobacco epidemic, ladies and gentlemen, spares no nation and no people. Wherever we come from and whatever we do, we are never truly safe from the long arms of the tobacco industry as they search the world for new markets and victims.

—Gro Harlem Brundtland, director-general, World Health Organization, keynote address to the international conference, "Making a Difference to Tobacco and Health: Avoiding the Tobacco Epidemic in Women and Youth" (1999)

One may be labeled the "World Cup" fallacy, the notion that cross-national learning is like picking the best soccer team. The task is to find the best model . . . from around the world and transplant it elsewhere. This approach is, of course, foolish. No institution, policy, or program is transplantable in this simplistic way. . . . The other danger is the opposite fallacy, the notion that since nations (cities, neighborhoods, families) always differ in some respects, there is no way they can learn policy lessons from each other.

—Theodore R. Marmor, "Global Health Policy Reform" (1997)

The individual must realize that perpetuating the present system of high-cost, after-the-fact medicine will only result in higher costs and greater frustration. The next major advances in health . . . will be determined by what the individual is willing to do for himself and for society at large. If he is willing to follow reasonable rules for healthy living, he can extend his life and enhance his own and the nation's productivity. . . . If he participates fully in private and public efforts to reduce the hazards of the environment he can reduce the causes of premature death and disability. He can either remain the problem or become the solution to it; beneficent government cannot.

—John H. Knowles, President, Rockefeller Foundation, "Responsibility for Health," *Science* (1977)

Smoking at the End of the Twentieth Century

As many readers can attest, one of the tangible aspects of traveling today between countries and in the U.S. between California and many other states is the commonsense experience of moving between smoke-filled and smoke-free environments. Not long ago, it began to crop up in conversations and media reports in the form of accounts of putting up with tobacco smoke in waiting lounges and restaurants, travelers' complaints about the inability to smoke upon deplaning, or the chore of having to send even unworn articles of clothing to the dry cleaners after a trip abroad to remove the smell of stale cigarette smoke. Nonsmoking seats on Japanese bullet trains were difficult to obtain, cigarettes disappeared in electronic and print advertising in France but not in cafés, and in some California cities smokers had to step into the street before lighting up. The subject was endlessly fascinating, for it brought together in terms of national and regional differences everything from questions of health risks, bodily pleasures, and personal conduct to issues of government regulation and transnational corporate interests. It also could be seen as a crisis of consumption, or even a crisis of *consumer citizenship*—a moment in which, in the name of healthy populations and good governance, public and private entities are compelled to step in and attempt to reverse and redirect expanding consumption by populations for whom access to consumer goods many of these same agencies had previously defined as an integral part of belonging to the nation-state (T. Miller 1993, xvi, 170; Lash and Ury 1994, 309; Larner 1997; R. Foster 2002, 93, 111). By the end of the twentieth century it seemed that there was nothing more personally and socially entailing than one's smoking and the smoking of others. U.S. historian Allan Brandt went so far as to claim in 1990 that the cigarette, one of the very icons of American modernity, individualism, and pleasure, threatened to disappear (Brandt 1990; Gusfield 1993). Throughout the 1990s news accounts in the U.S., France, and Japan were full of stories about new regulations restricting smoking in public places. At times French television and newspapers portrayed French measures against secondhand smoke as the importation of puritanical intolerance from America, and some Japanese commentators viewed the mild restrictions implemented by their government as the intrusion of U.S. utilitarian thinking in Japanese cultural life, while many American journalists invoked the old modernization narrative (which always placed the U.S. in the lead) and mocked the "backwardness" of French and Japanese anti-smoking policies.

In the pages that follow, closer scrutiny of anti-smoking policies and health promotion in these three nations will unsettle many of the easy assumptions of these discourses that confidently attach behavior and policy to cultural differences or stages of development. By the same token, it will lead us to reconsider global tobacco control in terms of contingent histories of tobacco consumption and of public health practices of citizen mobilization, epidemiological research, and the communication of scientific findings to citizen-consumers. In particular, this book looks at some of the recurring tensions in contemporary tobacco control at it attempts to counter the global expansion of cigarette smoking. I take as my point of departure the World Health Organization's (WHO's) Tobacco Free Initiative, launched in 1999, which proposed a new approach in health promotion coupled with a significant treaty (the Framework Convention on Tobacco Control) based on the model of the Kyoto Protocol on global warming. This initiative appealed to local and transnational community identities in order to recruit populations into the WHO's project of reducing tobacco-related diseases and replacing smoking with nonsmoking as the global norm.

I devote much of the book to investigating the conceptualization and implementation of anti-smoking initiatives that preceded the WHO program in France, Japan, and the United States, especially in California, whose anti-tobacco education program was adopted by the WHO as a global model. Since its inception, the California program has been the subject of extensive public health interest in the U.S. and other countries by virtue of its combination of local ordinances on clean indoor air; tax increases; aggressive social marketing–based media campaigns; and the establishment of ethnic networks, which were unique at the time of their creation in 1990–91, to promote community mobilization and participation.

The interpenetration of transnational and local practices characterizes much of tobacco control worldwide through the uneven circulation of personnel, expertise, funding, epidemiological studies, health promotion materials, and policies. The book traces in these anti-smoking efforts how tensions between local constraints and global imperatives, civil society initiatives and state policy bureaucracies, and self and other shaped the struggle to regulate the production, marketing, and consumption of tobacco products in highly industrialized societies. Special attention is paid to links between these initiatives and market-based neoliberal modes of governance gaining favor at that time in international policy circles that stressed, on the one hand, the benefits of globalization, the reduction and privatization of state services, and the introduction of corporate management and marketing techniques into the public and nonprofit sectors; and, on the other, the recruitment of citizens as responsible members of aggregate and particular populations at local, national, and international levels (Barry, Osborne, and Rose 1996; Dean 1999; N. Rose 1999).

"Globalizing California"

On 1 November 1999 Derek Yach, project manager of the WHO's Tobacco Free Initiative (TFI), addressed a meeting of tobacco control advocates in Lake Tahoe, California.[1] Organized by the California Department of Health Services, the gathering brought together over 400 directors of local tobacco control projects funded by state cigarette taxes together with rising figures in news media from fourteen countries around the world. The latter were flown in by the WHO. The purpose was to introduce them to anti-smoking activists and to tobacco control tactics developed in California and to familiarize the Californian project directors with the perspectives of global tobacco control.

Yach opened his remarks by situating the state of California within global trends, including that of anti-smoking policies:

> We are rapidly approaching the end of this century. It was a century that saw California emerge to become the global leader in the entertainment and information technology fields. It was a century that witnessed the rise of the tobacco epidemic in developed countries and the start of its control in those countries. It was the century that saw globalization become a buzzword, a means of expanding trade and marketing for good and for bad. (Yach 1999a)

After outlining the goals of the TFI, which include countering the tobacco industry's worldwide targeting of women, youth, and ethnic communities, Yach pitched his appeal to the local members of the audience:

> California is a microcosm of the world. With your myriad cultures you perhaps know more than most about what works and what doesn't. You have a great story to tell—a story that saves not just lives but precious resources for the state. . . .
>
> But it is not enough to do good work. Go out and tell your story. Tell the world that tobacco control works in your diverse communities. That you have achieved smoking prevalence rates way below those currently seen in most countries: and that your economy continues to boom and your people continue to gain in health and wealth. . . . And do more than tell your story: act globally as you act locally. Here are a few ways you can do this. The tools of the twenty-first century are being designed in your backyard. The twin advances in entertainment and information technology allow all of you to sit down in your office and serve the world. A few minutes of advice to remote NGOs in Nepal, India, or China via the Internet builds local-global communities with a shared sense of purpose and shows that you care. (Yach 1999a)

Yach was bringing to California the TFI, a highly articulate, self-conscious project of global public health policies with the express purpose of taking California's "best practices" of tobacco control (social marketing–based counteradvertising, community mobilization) and "globalizing" them (Yach's term; Yach 1999b) while having state activists and health officials see themselves and their work as belonging to what he termed the "global health village." California's anti-tobacco campaign had lowered smoking rates from 24 percent to 18 percent and greatly reduced inhabitants' exposure to secondhand smoke (Pierce et al. 1998, 2–13, 2–16). The WHO's goal seemed to be nothing less than the internationalization of a new social movement and local policy initiatives (Kearney 1995, 560). The theoretical and practical point of articulation are California's diverse populations, especially its diasporic communities linked through global electronic media and the Internet. The TFI was to be coupled with a significant treaty (the Framework Convention on Tobacco Control). At stake was the struggle between the tobacco industry and the public health community over the fate of entire populations in terms of questions of smoking, health, and community affiliation.

In fact, already in the preceding month members of state tobacco control ethnic networks had staged a joint press conference with the Department of Health Services to denounce the targeting of their communities by the new Philip Morris Virginia Slims $40 million "Find Your Own Voice" print media campaign. This campaign promoted smoking through testimonial narratives of immigration, community identity, and assimilation. The press conference was a response to this threat of global tobacco marketing that capitalizes on community differences and the dynamics of population movements and acculturation to promote its product (APITEN 1999b).

Local and Global Data, Particular and Universal Claims

Building upon the momentum created by the Tenth World Conference on Tobacco or Health in Beijing in 1997 and the U.S. Master Agreement Settlement of November 1998, Yach's appeal also drew upon the publication of *The Global Burden of Disease* in 1996 by the Harvard School of Public Health and *Curbing the Epidemic* in 1999 by the World Bank. These reports decried industry marketing practices and the growth in worldwide cigarette consumption while forecasting upwards of ten million tobacco-related deaths by 2020. The TFI proposed a striking new departure in health promotion that valorized local and community differences as the means through which the WHO could meet its universal goal of denormalizing smoking and stemming the expanding consumption of cigarettes.

Just how complex this task would be emerged in international gatherings of tobacco control advocates that took place in the months that followed. For example, several weeks after Yach's speech in the United States, at another conference sponsored by the WHO in Kobe, Japan, titled, "Making a Difference to Tobacco and Health: Avoiding the Tobacco Epidemic in Women and Youth," much time was spent sorting out region- and gender-specific epidemiological data and wrestling with the question of how transferable data were from industrialized to industrializing countries. At one point a Johns Hopkins University epidemiologist reminded participants of the effectiveness of universal biological arguments in discussions with skeptical state officials (who insisted on the need for local data), stating, "A lung is a lung in whichever country it happens to be." Conversely, in another session South Asian activists questioned one of the conference's working assumptions that "tobacco" meant only cigarettes, a meaning they claimed was not as pertinent to their local situations, where large numbers of consumers chew tobacco.[2]

The inevitability of negotiating tensions between the specific and general, the particular and the universal, in the struggle against tobacco use was brought home again one year later at the Eleventh World Conference on Tobacco or Health held in Chicago in August 2000. There, African American attendees noted with disappointment that the opening ceremonies did little to acknowledge the international conference's diverse constituencies, dominated as the proceedings were by the all-white representatives of the sponsoring U.S. health nongovernmental organizations (NGOs). And later during a session devoted to developing new strategies for twenty-first-century tobacco control in terms of juridical venues, a leading French epidemiologist, who had trained at Johns Hopkins, stood up and pointed out that the panelists' assumptions driving the discussion were based on Anglo-American common law, which limited the scope of their recommendations. Here, in these instances, one cultural or social particularity threatened to erase all others in the eyes of participants and seemed to confirm the judgment of two writers on international health policy that "international generalizations (some would say 'fixations') almost always ride roughshod over the unique and the particular" (Björkman and Altenstetter 1997, 9).

Questions of Embodiment, Truth, and Ethics in Health

Finally, these tensions have played out not only in terms of regional and cultural differences but also in terms of particular embodied practices. The health management of populations almost always requires articulating subjects ethically in terms of conduct in relation to their bodies and the bodies of others. This became an issue even in my fieldwork. For example, while I was in Paris, my status as a smoker or nonsmoker had to be established with an assistant—who, as it turned out, was American—before an interview was granted with a leading French anti-smoking activist. In early fieldwork performed for this book as well as in scholarly venues where I delivered papers, this occurred again and again, especially in the U.S. among Americans. Scarcely an interview or presentation took place without my being queried whether I was a smoker or non-smoker. Other scholars who have studied volatile medical controversies concerning abortion, contraception, women's health issues, and HIV/AIDS have discovered, too, that researchers' own bodies and subjectivity come under scrutiny. I was also queried about my sources of funding. This will not surprise readers familiar with the U.S. "tobacco wars," which have not only framed smoking as a drug addiction but have also revealed repeated instances of tobacco industry funding of scientific research that questions the results of widely accepted smoking studies (Reid 2000). The question of what I did or had done with my body or what some institution or firm hoped to do with the results of my work suggests a mutual entailment of questions of embodiment, truth, ethics, and identity filled with tension about the origin and production of research that intersected with debates concerning the reliability and validity of tobacco control data and policies across different communities, regions, and nation-states. Here, the ghost of an intractable otherness hovers over public health research and policy tied, on the one hand, to questions of bodily practices and self-interest and, on the other, to cultural or social differences. As a result, modes of access to prospective informants for this book varied across institutions, organizations, and nations. I argue in this book that the tension between embodied subjects, local practices, aggregate data, and general strategies has been central to many anti-smoking campaigns in the late twentieth century, especially as the global campaign against tobacco use sponsored by the WHO began to take off.

The Tobacco Epidemic and the Rise of New Populations

I have termed the decline of smoking as a crisis of consumer-citizenship, for state agencies now find themselves in the business of discouraging consumption of a legal product that they either continue to promote through subsidies, diplomatic initiatives, and so forth, or once did so. This product is virtually like no other, for it is not only identified with early mass consumption practices that helped create the first national markets of consumer goods and by the same token helped define American, French, or Japanese modern identity but is also associated with the state itself by virtue of a long history of state-controlled tobacco monopolies (Brandt 1990; R. Foster 2002). The crisis first began around 1980 in the U.K., the U.S., Canada, and Australia and struck principally older male smokers and not long afterward older women. It arose from the experience of tobacco-related illnesses among family and friends and the published reports of deaths of celebrities; publicity generated by early epidemiological studies conducted in the U.K., the U.S., and Japan; the shift from industrial-based economies to service-based ones with their large white-collar middle classes for whom quality-of-life issues became paramount; the transfer of dirty industries abroad, which in some cases were replaced by new high-tech factories that required pollutant-free shop floors; and the influence of health and environmental movements and their expanded sense of risk and physical vulnerability (Beck 1992; Hacking 1990; Castel 1981, 1991; Armstrong 1993). These developments forced cigarette manufacturers such as Philip Morris (now Altria), British American Tobacco, Japan Tobacco, R. J. Reynolds, SEITA (France; now Altadis, France/Spain), and Rothmans International (South Africa; now British American Tobacco) to seek out new markets at home and abroad. In turn, this has led them to target heavily segments of local and overseas populations: young women, adolescents, and, in some cases, ethnic and racial communities. These are the new populations of not only of tobacco marketing but also tobacco control. Marlboro, Dunhill, Mild Seven, Camels, Gauloises, Fine, and Rothmans have risen to the paradoxical status of "global cigarettes" pitched at populations in particular regions.

The response of anti-tobacco movements has been to attempt to counter these marketing campaigns and the smoking practices they underwrite through social marketing campaigns; health promotion; advertising curbs; tax increases; restrictions on smoking in public places; and the creation of an international protocol governing tobacco production, marketing, and consumption. If one way to understand public health is in terms of making bodies of populations available for public and private interventions (Lupton 1995, 70), the anti-tobacco struggle may be viewed as having contributed to the emergence of population segments as objects of solicitude, study, and health education who were scarcely present in early studies on the health hazards of smoking: women, adolescents, and various social and ethnic groups. For example, the 1964 U.S. surgeon general's report on smoking had relatively little to say about the smoking habits of women or blacks and nothing about adolescents (U.S. Dept. of Health, Education, and Welfare 1964). Later U.S. studies developed segmentation practices that studied behaviors by gender, age, and economic and education status and also by ethnicity and race. These surveys helped articulate anti-smoking campaigns' "outreach" mission to "underserved" populations, and they set the stage for California's ambitious tobacco control program that sought in the 1990s an active collaboration with Chicano/Latino, African American, American Indian, and Asian American communities. However, although California's community mobilization efforts along with its social marketing–based media campaign were what eventually attracted the attention of the WHO, paradoxically they were also what set it apart from most initiatives outside the U.S. With regard to race and ethnicity, U.S. public health segmentation practices are unique and stand as the exception to the rule, especially in highly industrialized nations, where the focus is on the economically disadvantaged, women, and adolescents. In fact, in Western Europe the gathering of population data by race and ethnicity is rarely practiced because of the legacy of the Nazi holocaust, and in France it is actually illegal. And with respect to smoking, in Japan the focus is on the category of young women, who are considered the most at risk. So in each of the three sites of tobacco control in this book, different segmentation practices are in play, and the fate of different populations is at stake.

Secondhand Smoke, the Social Body, and the Normative Nonsmoker

Since the 1980s no other single issue besides tobacco industry marketing practices has galvanized the anti-smoking movement in highly industrialized nations to the extent that secondhand smoke (or environmental tobacco smoke, ETS) has done. Thickly social in its implications, secondhand smoke seems to entail the fate of all citizens, and a revolution in public attitudes began toward the regulation of what was considered private behavior in shared public spaces. In different ways public officials and activists called upon citizens to take actions on their own behalf; intervene in each other's behavior; and generally regulate themselves, smokers, and the tobacco industry through personal acts and government legislation. The succession of anti-smoking campaigns and municipal ordinances, state laws, and national regulation targeting secondhand smoke linked the management of populations to interlocking questions of public health, epidemiological knowledge, social marketing expertise, social identities, and good citizenship. It is here that the elaboration of a normative nonsmoker is most visibly articulated, often in terms of a "flexible body" or "neutral body" conceived as a porous membrane constituted by rates of flow across its boundaries and defended by "zero tolerance" policies toward pathogens emanating from tobacco smoke (E. Martin 1994; Hirsch and Karsenty 1992). This body would be that of the normative nonsmoking (or soon to be nonsmoking) consumer-citizen who realizes his or her potential through healthy strategies of self-care and community policing. Over against this body stand other citizen-subjects less susceptible to the neoliberal appeals to self-governance and healthy lifestyles. It is in campaigns against secondhand smoke, where matters of policy, knowledge, and populations intersect, that the shared tensions of the arts of government and public health are potentially the most explosive. To take just one example, in anti–secondhand smoke advertising, ineluctably TV spots addressing particular groups are viewed by everyone else. As a result, what seems to be a welcome singularizing move that acknowledges specific group needs may also stigmatize the group as irreducibly different from and outside of the aggregate "general population" altogether. Conversely, ads addressing the "general population" alone may likewise erase different communities' or groups' health issues.

Media Campaigns and Social Marketing

Much of this book's focus is on the media component of anti-smoking campaigns, and there are several reasons for this. First, the goal of translating scientific findings into lay terms capable of stimulating public debate, introducing new norms of healthy behavior, and mobilizing citizen-consumers heightened the role of media-based health promotion in many anti-smoking efforts. Health officials and activists also desired to counter the massive marketing campaigns of the tobacco industry and its long-standing cultural presence in cinema by carrying the battle to the public media sphere. Moreover, in the 1970s and 1980s a new field of media expertise arose, especially in the U.S., which claimed to bring issues affecting the general interest to a mass public by drawing on the market-based knowledge of the private sector. Called "social marketing" by its proponents, it promised to forgo the presumption of older forms of commercial and governmental authority to dictate consumption and conduct to consumers and citizens in favor of ascertaining and responding to the needs of the population through market surveys and focus groups. Stressing the ability of visual and aural media to reach illiterate populations, social marketers claimed that they could tailor communication between government agencies and voluntary organizations and the citizen-consumers the former desired to address by availing themselves of the latest techniques developed by advertisers, the entertainment industry, and the behavioral sciences (for suggestive titles, see Kotler and Zaltman 1971; Kotler 1975; Manoff 1985).

Finally, the salience of media campaigns in tobacco control that seek to alter citizen-consumer behavior and delegitimize the culture of smoking confirms the observation of U.S.-based anthropologist Arjun Appadurai, according to which, in an era of globalization and mass population movements, "electronic mediation transforms preexisting worlds of communication and conduct" and offers "resources for the experiments with self-making in all sorts of societies" (Appadurai 1996, 3). I am especially interested in this process of making and unmaking of bodies and selves in tobacco control media campaigns whose market-driven techniques intersect with a radical form of liberal governance at the end of the twentieth century called neoliberalism, which, among other things, advocated the introduction of corporate management and market-based concepts in public-sector and nonprofit organizations (more about which below).

The Discourse of Cultural Difference and Exceptionalism in Modernizing Theory

When the WHO decided to adopt of California's media campaign and methods of community mobilization, it was acting on one of the truisms of globalization theory, as proposed by Anthony Giddens, Stuart Hall, and others, namely that the global flow of goods, services, capital, labor, military forces, tourists, information, images, and vectors of disease, far from simply erasing local identities and conditions—as in traditional modernization theory and its narrative of "convergence" along a one-way timeline of development—actually heightens differences (Giddens 1990; Featherstone 1990; Hall 1991; Buell 1994; Kahn 1995; Appadurai 1996; Sakai 1997).[3] According to the old theory, local culture was what got in the way of development.[4] By contrast, in the TFI's ambition to globalize California, "culture" emerges less as an obstacle than as an opportunity, as even the very condition of possibility of global tobacco control. For example, strategies designed to address the East Asian or Latino communities in California would in theory be readily adaptable to other locales within the diaspora. "Culture" is unmoored from its traditional association with a bounded geographical location and is redefined in terms of an international dispersion of communities and nationals (Appadurai 1996, 48; see also Gupta and Ferguson 1992, 1997). Still, it would seem that in this theory, cultural difference extends only so far: it stops at the newly acknowledged borders of diasporic communities themselves, for, internally, they are each understood to be relatively homogeneous. And it is within a fundamentally homogeneous field—dispersed as it is across national borders—that the WHO's Tobacco Free Initiative seeks to deploy its local/global strategies. The dream of homogeneity makes a return here within cultural difference and risks overlooking other nonmajority practices within these very same communities. Such omissions have already hampered many social marketing programs sponsored by the state or NGOs to reach out to underserved populations, especially in industrializing countries and the U.S. (Altman and Piotrow 1980; Luthra 1988; Resnicow et al. 1999). Moreover, the assumption of homogeneity also suggests the persistence of the model of the modernizing nation-state at the heart of the project of "sustainable globalization." This has worried some critics of globalization theory, according to whom within the global/local binary both the local and global operate as homogenizing frames. Inderpal Grewal and Caren Kaplan warn,

> Global-local as a monolithic formation may also erase the existence of multiple expressions of "local" identities and concerns and multiple globalities. In this particular way, global-local binaries dangerously correspond to the colonialism-

> nationalism model that often leaves out various subaltern groups as well as the interplay of power at various levels of sociopolitical agendas. (Grewal and Kaplan 1994, 11)

The finer point of Grewal and Kaplan's remarks is that processes of homogenization and differentiation are at work simultaneously at *both* global and local levels. Thus, their warning cuts both ways along the binary, at least as I read them: what takes place at the global level cannot to be understood as a single process of homogenization, nor can there be any romance of the local, either, as a privileged site of difference. So, for example, in the case of U.S. tobacco control, not only have non–Anglo-American activists disputed among themselves the validity of applying Anglo-American models of public health strategy in, say, France or Japan, but also anti-smoking advocates and state officials within the U.S. have contested California's statewide strategies as inapplicable to their regions or communities. Grewal and Kaplan's approach also suggests that the global/local binary may be inadequate to grasp the fluid nature of current conditions:

> What is lost in an uncritical acceptance of this binary division is precisely the fact that the parameters of the local and the global are often indefinable or indistinct—they are permeable constructs. How one separates the local from the global is difficult to decide when each permanently infiltrates the other. (Grewal and Kaplan 1994, 11)

Recent work on globalizing processes, especially in the Western Pacific and East Asia, seems to confirm their observations; it has highlighted how hybridization and indigenization of "global" products and transnational, regional exchanges work to decenter notions of globalization as merely a process of Westernization (Buell 1994; R. Foster 2002; Iwabuchi 2002, 15–16; Shome and Hegde 2002).

The question of the homogenizing tendencies of both the modern nation-state and modernization theory that legitimizes it leads to the related discourse of cultural particularism (or exceptionalism) and universalism according to which some cultural particularities are by nature able to transcend their parochialism to attain universal status, while others are not. Such a discourse surely underlies much of the media commentary cited previously on the introduction of anti-smoking measures that framed them as "American." According to Japanese studies scholar Naoki Sakai, the privilege of transcendence was what the so-called West has assigned itself since the beginning of its colonial expansion and, one could add, what Japan has attributed to itself in the Western Pacific and East Asia. Or as French social philosopher Etienne Balibar put it, the peculiar form of government called nation-state sovereignty (that of the U.S., France, Germany, the U.K., Italy, Japan) constituted one kind of particularism that expressed itself in one form of universalism called colonial empire. That is to say, the formation of nation-states and empires was one and the same process

(Sakai 1997, 153–158; Tanaka 1993; Balibar 2001, 110 n36).[5] Thus, the legacy of this common genealogy is that the nationalism of nation-states often leads them to view themselves as uniquely universalist, aggregative, even cosmopolitan by nature (the right to empire) and to dismiss both other national traditions and communities within the nation as irredeemably particularist and singular. This discourse can easily take an ethnic turn, as in the case of eighteenth-century philosopher Johann Herder, whose influential formulation of German ethnic nationalism was in response to the Enlightenment discourse of universalism (Balibar 2001, 49; see also Buell 1994, 64–68).

Globalization and the Liberal Arts of Government

Crucial to my project have been not only revisionist theories of globalization but also a vein of theoretical and empirical research that takes as its object the liberal arts of government. These arts seek to govern society at once through direct state intervention and at a distance by delegating activities to organizations and institutions of civil society as a way to manage populations. They overlap with public health's goal to mobilize citizens to act upon themselves and each other in order to improve collective well-being. Aggregating singularities and managing differences so as to render citizens governable or even self-governable lie at the heart of the liberal enterprise, especially in its most recent and radical form, the neoliberal revolution (in the classical sense of "liberal"—free markets, limited government—which is how it is still understood outside the U.S. today) that began in policy circles in industrialized nations in the 1970s.

Neoliberal policies' focus on tax reduction, deregulation of industries and markets, privatization of state services and publicly owned enterprises, the cutting of social and welfare benefits, and the introduction of market-based management principles in state agencies is well-known. Less acknowledged in public discussion is that these policies go hand in hand with an activist sense of governance that both positively shape the conditions of the free market and seek to forge new relations of governance with citizens in which the latter would become not simply objects of government but also its active partners (C. Gordon 1991; N. Rose 1996; George 1996). Citizens would learn to govern themselves, and in so doing treat themselves as an enterprise to be pursued and made to maximize their possibilities of existence. It is a question *not* of governing citizens too much but rather of the *state* doing too much governing (Dean 2002, 42). Contributing to these modes of neoliberal self-government are initiatives and movements that include everything from the retraining of workers for the eventualities of the labor market to neighborhood policing to consumers rights groups and the self-care and self-help movements of alternative medicine to the empowerment movement in social policy and public health (Castel 1991; Armstrong 1993; E. Martin 1994; O'Malley 1996; Cruikshank 1999). This is the form of governance favored in consumer societies in which the management of populations shifts from the older defined disciplinary spaces of paternalistic institutions (family, school, factory, hospital, army) to the fluid spaces of consumer-based routines of what French philosopher Gilles Deleuze called "control societies" (Deleuze 1995). However, in the past two decades the transfer of responsibility to individuals and communities has occurred contradictorily, for at the same time corporate mergers and the free movement of investment capital, factories, and goods in and out of countries have greatly diminished communities' and citizens' control over their economic fates. Such a situation courts the danger of opening the door to socially punitive policies directed at citizens and groups—who enjoy little effective power—to counter their "poor behavior." With respect to the transnational tobacco industry and its practices, the Tobacco Free Initiative and the Framework Convention on Tobacco Control proposed by the WHO were meant to address precisely some of these contradictions by setting up mechanisms of transborder governance (the Framework Convention) that would underwrite NGO and local activism and be underwritten by them in turn.

One example of an early point of conversion between new state interventions in health promotion and neoliberal modes of governance that stress citizens' agency is the reception given to the 1978 Alma Ata conference on public health held in the Soviet republic of Kazakhstan. The public health document issued at its closing has been hailed by adherents of activist state-sponsored social medicine as a "return to Virchow" (the nineteenth-century founder of European social medicine and public health) but also by social marketers, who viewed it as the conceptual opening of public health to the contributions of private-sector marketing methods of assessing and addressing public needs (Young and Whitehead 1993, 114; Manoff 1985, 3–4). The points of intersection would be in the state promotion of citizens as self-acting agents who in realizing their personal goals enhance community welfare. Another example was the very influential statements released by John H. Knowles, president of the Rockefeller Foundation, who preached the virtues of individual self-management in matters of health as a way to reduce the spiraling costs of health care in the U.S. (Knowles 1977a, 1977b). Of all the issues in tobacco control, secondhand smoke incorporates the most powerfully such a dynamic in which personal and collective, singular and aggregate practices are inextricably intertwined in a palpable way. It offers an invaluable window onto the tensions in tobacco control that I examine in this book.

Characteristic of liberal and neoliberal government alike in capitalist societies is, according to French philosopher Michel Foucault, the government of "all and each" that at once "totalizes" and "individualizes" populations or, in another language, that aggregates and singularizes, homogenizes and particularizes them at local, national, and international levels (Foucault 1981; C. Gordon 1991, 3). Foucault's provocative point is that there is no way to separate meaningfully the two processes; rather, they are mutually constitutive, if tensely so. The activities of governance and population management (statistical surveys; health and welfare policies; military conscription; juridical, medical, educational, and insurance practices; philanthropic activities; trade union and citizen activism; and so on) create aggregates through amalgamating singularities but also by the same token—and this is Foucault's original thesis—create singularities through aggregates. Thus, these practices can singularize population groups or subgroups (in U.S. population studies as Hispanic, black, white, Asian American, Native American, youth, immigrant, of color, and so forth) by aggregating citizens or population segments; conversely, they can particularize citizens by assigning them specific identities (for example, in U.S. public health there are the "hard to reach," "at risk," "vulnerable," and "underserved") that homogenize differences between individuals or between groups. Ultimately, there is no one or no group that can't be viewed simultaneously as a singularity and an aggregate of some kind (after all, even individuals can be aggregates of identities, risk factors, what have you).[6] Still, when it comes down to policy and scientific disputes, arguments generalizing from aggregate numerical data—"economies of scale"—win the day (Boltanski and Thévenot 1991, 18–23; see also T. Porter 1995). Again, California's tobacco control education program that at once addresses all state inhabitants, acknowledges differences between communities statewide, and encourages local activism would be a perfect example of this process of liberal governance at a distance through population singularities and aggregates, as would be the TFI's attempt to adopt this same program on a global scale.

Internationally, as in the case of California, other public health responses to tobacco use are often framed in the contradictory terms of both universal trends and local imperatives. With vastly fewer resources than their Californian colleagues, French and Japanese anti-tobacco advocates, researchers, and health officials negotiate the peculiar legacies of tobacco use (growth of heavy smoking in the 1960s and 1970s), state formation (centralized government, directed economies), public health policy and research, and citizen activism as they elaborate their responses to tobacco-related diseases, shifting patterns of cigarette consumption, and tobacco industry marketing practices. In all these sites, one way to gauge the nature of (neo)liberal practices of government will be to follow the collaboration between health officials and those from outside government, including anti-tobacco NGOs, epidemiologists, social marketing firms, and (especially in the U.S.) community representatives. Tracking these collaborations is one of the threads of this book's fabric.

Interdisciplinarity and Studying across a Public Health Controversy

The tensions underlying the population segmentation practices of tobacco control in the context of the neoliberal arts of government and globalization are the subject of this book. To grasp best the emergence of nonsmoking through antismoking campaigns in California, France, and Japan, I focus on public health knowing practices in relation to the social body with an eye to asymmetrical relations of power and privilege on the one hand and modes of collaboration between participants in tobacco health promotion initiatives on the other. My study is both interpretative and data-driven, and it stands at the intersection of cultural studies' focus on signifying practices in their historical specificity, communication's study of the technologies and infrastructure of media, and science studies' interest in the social and material nature of scientific and medical practices. As such, each chapter presents a different emphasis not only in approach, materials, and data but also in style of written argument. For example, chapters 1 and 2 on the conceptualization and implementation of the California campaigns will be more familiar to those working in the fields of history, communication, public health, and policy studies, whereas scholars in cultural studies, critical studies of mass media, and the humanities will perhaps feel more at home in chapters 3 and 4, which focus on California counteradvertising and U.S. visual media. Chapters 5 and 6 on France and Japan are perhaps the book's most evenly interdisciplinary in style and method. Finally, throughout the book, a certain "opacity" of objects and practices is respected: they do not yield fully to a hermeneutic of signification and cultural identity or to the transparency of scientific social worlds, strategic networks, and communication infrastructures (Grossberg 1993).

Originally trained in French studies, an interdisciplinary, text-based field, I had written a historical study of French demographic, sociological, philanthropic, psychiatric, public health, and literary discourses on family household practices before turning to research on contemporary anti-tobacco campaigns. At the start of this project, I focused on print and visual materials first in the U.S. (California) and then in France and Japan. My focus on the representational practices of public health stood close to that of several edited volumes in science studies and cultural studies that examine visual objects and apparatuses both in themselves and as a part of a large web of practices and knowledge (Crimp 1989; Lynch and Woolgar 1990; Gilman 1995; Treichler, Cartwright, and Penley 1999). However, at the urging of interdisciplinary-minded colleagues in anthropology and sociology, I decided to look at the processes that led to the products of the health promotion campaigns. In conjunction with archival research, I conducted more than seventy-five field interviews (mostly in English or French) of those involved in the design, production, and evaluation of these campaigns in order to understand the collaborative efforts of parties coming from what used to be called in U.S. sociology different "social worlds" of public health (epidemiologists, public health officials, advertising executives, anti-smoking advocates, and community representatives) who actually had a hand in creating the campaigns that first attracted my attention. As such, my study seeks to move beyond strictly text- and audience-based approaches to media such as film and television toward a plural method, as outlined by critical communication studies scholar Toby Miller: "When it comes to key questions of texts and audiences—what gets produced and circulated and how it is read—policy analysis, political economy, ethnography, movement activism, and use of the social science archive are crucial" (T. Miller et al. 1998, 15).[7]

In the end, my hope is that this experiment will also contribute to expanding the interdisciplinarity of public health by exploring questions that standard institutional and disciplinary commitments within public health and health communication do not always allow. In the late 1980s and early 1990s there was a lively debate within public health circles, primarily in Anglo-American countries, on problems in past and emerging methods in public health, or what Nurit Guttman would later call the invisible "ethical dilemmas" endemic to health campaigns (Guttman 1997). Some commentators denounced the legacy of colonial attitudes in public health (Young and Whitehead 1993; Tones 1993) and the persistence of expert-driven paternalism in state-sponsored health education and promotion programs, even in those that claim community empowerment as their goal (Tones 1986; Becker 1986; Minkler 1989; Braithwaite et al. 1989; Grace 1991; Beattie 1991). Others questioned the wisdom of adopting social marketing techniques that treat citizens as consumers who are unaware of their needs and discount conflicts among policy stakeholders (Wallack 1984, 1990a, 1990b; Grace 1991; Hastings and Haywood 1991; Got 1992; Lupton 1995). In short, there was a call for public health to examine the ethical side effects of its own programs of social persuasion (Pollay 1989). However, by the middle of the decade, with the exception of the issue of community empowerment, the debates had tailed off, and discussion seemed to settle on dealing with practical issues of audience segmentation, outreach, and so forth. The present study seeks to reprise some of these older debates but within a new context—that of globalization and liberal governance—and to contribute to scholarly debate in the field of critical communication studies (T. Miller 1993, 1998; Lupton 1995).

However, reviving such issues in contemporary tobacco control has proved to be no easy task. Complicating my research effort has been the remarkable politicization of the smoking issue, especially in the U.S. It would seem that contemporary tobacco control stands as a particularly intense example of what is called in science studies a public scientific controversy. Australian sociologist

Brian Martin, who has studied the dynamics of such controversies, reminds us that to conduct research on an emerging dominant discourse or scientific consensus—here, anti-smoking counteradvertising and those who produce it—automatically courts the accusation of "working for the other side." Even practicing scrupulous symmetrical analysis of both sides of a controversy recommended by conventional science studies does not provide safety from such accusations (B. Martin 1991, 163–165; Nelkin 1995). In tobacco control what is controversial is not the health hazards smoking poses to smokers. Annually, 420,000 Americans are reported to die of tobacco -related causes; in the language of social studies of science, the hazards of smoking to smokers have been "black-boxed"—that is to say, they are a relatively stabilized scientific fact that is no longer open to scientific debate. Rather, what has been debated is the threat of tobacco smoke to nonsmokers. The present study departs from the conventional study of the black-boxing process (say, of environmental tobacco smoke) to focus on the circuits through which scientific findings on the hazards of tobacco circulate, especially through public health campaigns. Just how involved matters can be for any prospective researcher is made clear not only by the tobacco industry's funding of scientific studies on tobacco but also by the industry's long-stranding practice of framing cumulative scientific findings confirming the health hazards of smoking as a "controversy."

With the increase in scholarly studies of middle class–dominated institutions and bureaucracies (research labs, museums, universities, corporate workplaces, state ministries, and so forth) have come reconsiderations of the traditional fieldwork narrative that have underlain research, especially in anthropology and sociology. Scholars have begun to recognize, in the words of the late U.S. anthropologist Diana Forsythe, who studied scientific labs, that "social worlds of fieldworker and informants may overlap considerably" and that fieldworkers study "those whose work skills are similar to their own" (Forsythe 1995, 10, 19), leading to the collapse of conventional narratives based on notions of radical otherness. It is striking that in the revisionist fieldwork literature (see Nader 1972 for an early example), the privileged metaphor for designating these new research circumstances has been "studying up" (replacing the older metaphor of "studying down"), as in the title of Forsythe's essay, although Forsythe's own remarks suggest that what is as much at issue is studying "across" or "over." In this discourse it would appear that if one is no longer studying "down" (and "out" from the metropolitan center) the old way from a position of guilty privilege, then one must studying "up" the new way from a position of unsettling powerlessness.[8] The question then becomes what happens when researchers study other researchers and knowledge producers, when one does interviews and fieldwork in the absence of radical differences in entitlement, privileges, and professional culture, particularly in one's "home" country.

One answer might be that the conventional model of power, in which asymmetries of marginalization and privilege are always clear-cut and fall along a single axis, is unworkable. In the absence of radical otherness, researchers are no more distant enough than they are other enough: intimacies, ambiguities, and ambivalences abound in the midst of persistent differences and power relations. Globalizing currents in public health can only intensify the hybrid dynamics of studying "across," "up," and "down," and they draw out the limits of these spatial metaphors and of the concepts of otherness and whiteness (in the context of many Euro-American researchers) that underlie them for grasping the ethical and political as well as epistemological dimensions of studies of scientific and medical practices.

A case in point is a panel at the annual meeting of the Society for the Social Studies of Science in which I presented an early paper on secondhand smoke TV spots broadcast by the California Department of Health Services. When I screened the ads, their high production values, punchy narratives, and visual humor, to my puzzlement, elicited bursts of laughter from the audience. Upon reflection, the laughter struck me as double-edged: appreciative of the humorous manipulation of media culture by public health professionals to educate citizens about the hazards of ETS but, by the same token, quite possibly condescending toward the intended audience among whom the listeners of my paper may not have included themselves (and it wasn't clear that the condescension didn't extend to the authors of the ads as well). I was led to wonder whether this complicated movement of identification and disidentification didn't stem in part from academics' sense of proprietorship over the education of citizens in public issues, a position that would put them (especially those of us in state-funded institutions) in rivalry with state officials, colleagues in public health, and anti-smoking advocates. Studying other knowing practices of fellow researchers and professionals entails questions of professional investments and identity that are perhaps not always so visible in other circumstances.

The Book's Structure

Part 1 is devoted to the California campaign that has served as a model in national and global tobacco control. In this section I focus on the multilingual media component, which was one of the signature features of the California program and arguably marked the ascendancy of social marketing techniques in U.S. health promotion. As a object of public controversy, the media campaign serves as an exceptional window onto many of the issues that shaped the entire program: the challenge of resisting transnational tobacco corporations' marketing practices targeting youth, minorities, and recent immigrants; expanding the anti-smoking movement and tobacco control community's narrow social base to diverse populations in an era of globalization; tensions between social marketing segmentation practices and the goals of community empowerment and between a policy bureaucracy's sense of prerogative and its commitment to community input; and the recasting of cigarettes by U.S. public health and mass media discourses as belonging to an earlier moment of industrial modernity.

Chapter 1 starts off our study of the California program as managed by the Department of Health Services in conjunction with the Department of Education. It analyzes the history and rhetoric of the 1988 ballot initiative, Proposition 99, and its media campaign that led to the establishment of the Tobacco Control Program in 1989 but over which the Proposition 99 campaign was to cast a long shadow in terms of conflicts between what U.S. public health researchers Constance Nathanson and Alisa Klaus call the inclusive, solidarist tendencies of public health and its sectarian, stigmatizing ones (Nathanson 1996, 618–619; Klaus 1993, 16, 29). I also examine the various conceptual and practical influences that went into the new health promotion program's design, especially the pooling of public health expertise such as social marketing strategies and community empowerment approaches from within the U.S. and abroad that gave the program its "Californian" identity. Finally, I detail the structures of collaboration and oversight that have served as the setting in which these issues and tensions have been played out.

Chapter 2 gives a critical account of the actual collaborative work during the early years of the California effort (1989–96) between state bureaucracies, public health researchers, private-sector ad agencies, and community advocates. This entails analyzing some of the first ads of the media campaign, which upon their release created a sensation in public health, advertising, political, and media circles, in terms of the social and ethical framing of the hazards of tobacco. I pay particular attention as well to points of crisis in collaboration around the management of evaluation, political pressures from Sacramento and the tobacco industry, the absence of meaningful epidemiological data, and the dominance of private-sector advertising agencies and their working assumptions in their dealings with state-sponsored programs involving community input. Indeed, it was the media campaign and its reliance on the mainstream public media sphere as defined by social marketers that became a major stumbling block to active participation by the ethnic networks and advisory committees, which the California Department of Health Services had set up to facilitate community participation. I close with a consideration of the possible limits of the California model of media campaigns in future health promotion.

Deemed by all parties to be one of the more successful components of the California program, the secondhand smoke media campaign, by virtue of its discourse of healthy practices, modern citizenship, family life, and community accountability, was one of the most volatile aspects of the media campaign. Chapter 3 deepens my inquiry into the social and ethical issues involved in translating scientific findings into lay terms for the purposes of community mobilization, by looking at the structure, form, and content of televised spots and billboards and the audiences, bodies, and subjectivities they tend to project. In these ads certain groups of citizens and residents (in terms of ethnicity and gender) emerge as amenable to the calls of public health authorities to act upon themselves and others to ensure the well-being of their communities, while others are framed as failed pedagogical subjects caught in modes of conduct associated with nonmodern cultures. At the end of chapter 3, I glance at anti–secondhand smoke spots of the Massachusetts campaign begun in 1992, to argue that some of the ethical and social issues raised by the California anti–secondhand smoke ads may help explain why relatively few of them were adopted by other anti-tobacco efforts in the U.S.

Chapter 4 expands the focus to link certain ads of the California campaign to the wider U.S. media culture and to the persistent question of smoking's relation to something called "late modernity," a globalized moment presumably dominated by "risk societies" (Beck 1992). I explore how these ads worked in conjuncture with U.S. television and film in the early and mid-1990s to reframe smoking as the tragic persistence in the postindustrial present of practices belonging to the recent but discredited past, namely that of industrial-based economies, the Cold War, and their illiberal modes of governance of populations. Television series include Fox's immensely popular *The X-Files* and NBC's critically acclaimed *Homicide: Life on the Street*, and films range from Hollywood mainstream cinema (*Waterworld, Demolition Man*) to independent cinema (*Kids, Clerks*, and *Smoke*) to new black, Latino/Chicano, and Asian American cinema (including *Boyz n the Hood, To Sleep with Anger, American Me, My Family, The Wedding Banquet*, and *The Joy Luck Club*). However, I argue that this revisionist narrative of industrial modernity symbolized by cigarettes plays out unevenly across these different areas of U.S. film and television production and that there may be no common media culture or narrative of either smoking or modernity. I suggest that cigarettes don't serve as privileged tools for all citizen-consumers for rethinking the past and present and projecting a better future.

The goal of part 2 is to sharpen the question of the underlying tensions of tobacco control and liberal government in a globalizing context by looking at two highly industrialized nations outside of the British-American sphere. France and Japan present quite different tobacco control landscapes marked, among other things, by long-standing tobacco monopolies, the devastation of the Second World War, and the massive entry into domestic markets of U.S.- and British-manufactured cigarettes in the 1980s. It is quite striking that in the eyes of many Anglo-American tobacco control advocates, France and Japan (as well as Germany) in the 1990s stood so far outside of what they deemed mainstream tobacco control that they could not serve as meaningful terms of comparison. That two of the world's oldest and most powerful industrial nations (three, if you count Germany) constitute radical exceptionalisms—or in our language, intractable singularities—in world tobacco control in and of itself gives pause and raises the question, what allows local sites of tobacco control to count or not in the global production and circulation of anti-smoking policies and campaigns?

Chapter 5 examines the history of French anti-smoking campaigns since 1976 in terms of particular practices of French public health and liberal government, citizen activism, medicine, and epidemiological inquiry. In particular, this discussion recasts the presumed contradiction between the ambitions of public health advocates rooted in statist, Republican ideology dating from the French revolution and the actual, more circumspect practices of the French state in terms of a tradition of governance and public health that has always borrowed policies and concepts from its European neighbors and the U.S. This provides a frame for reexamining French exceptionalism in public health and for understanding some intriguing paradoxes and surprises in policies, such as early attacks on secondhand smoke in 1976 and the banning of all forms of tobacco advertising in 1992, which complicate a simple narrative of global tobacco control. At the same time, French (as well as Japanese) anti-smoking advocates refrained from promising radical changes of self and society that characterize U.S. and Californian campaigns and from singling out groups and communities as "special populations" to the same degree. Chapter 6 considers the efforts of Japanese anti-smoking advocates from the late 1970s to the late 1990s. Japan offers an ideal complement to the French and U.S. cases, for it is a non-Western, highly industrialized country that, like France and the U.S., has been a major force in economic and cultural globalization and has a major tobacco industry (the state-controlled Japan Tobacco Inc.) that has important exports abroad, especially to Korea and East Asia; also, in the early twentieth century many Japanese smokers converted to cigarette smoking at the same time as consumers did in France and the U.S. However, in Japan there is a much larger gender gap in smoking rates and a different social configuration of smokers, which has had consequences for citizen activism and health promotion: the Japanese group Women's Action on Smoking is one of the rare feminist anti-smoking organizations that exist in highly industrialized countries, and anti-smoking rhetoric features women and children as the groups most at risk. But these anti-tobacco efforts take place in Japan in the shadow of the state's overt sponsorship of smoking, which effectively marginalizes anti-smoking groups in both the political and the public media spheres.

From the study there emerges a multiple narrative that complicates standard accounts of anti-smoking health promotion campaigns in the public health literature and in scholarly and journalistic writing. Although the study is comparative to some degree, I have tried not to situate the U.S., France, and Japan in terms of a simple baseline or standard of measurement (be it the progressive timeline of modernization; the expanding nimbus of the tobacco epidemic; or even a totalizing, utterly defining moment such as "global capitalism" or "neoliberalism"). Rather, I've opted for a narrative of overlapping but far-from-identical problems of public health population management and liberal governance that highlights local differences, if only because these differences, when evoked, are rarely dealt with at length in public health and health communication discussions, or in other arenas they become the crude instruments of facile polemics or cultural commentary. Each site stands as a "global singularity" that is neither merely the product of (or inexplicable departure from) universal trends produced by the circulation of personnel, expertise, studies, and policies, nor simply the result of local practices and histories that remain impervious by nature to outside processes.

Part I
California

Given that California is the destination for about 25% of immigrants to the United States and that 150,000 new students enroll in the state's schools each year, it should be apparent that California's campaign offers an unsurpassed opportunity to reduce tobacco use. It has the potential to infuse other state health agencies with sorely needed models for technology transfer and service delivery. Moreover, with the range of community strategies and the interventions being developed for groups at highest risk of tobacco-related disease, this landmark public health measure carries with it a wealth of information for future worldwide study and application.

—Dileep Bal et al. 1990

"California. We Know Better."

—Concluding title of California anti-smoking TV spot, 1997

Right now, Californians who still dare to smoke can still have a smoke under the stars or in wide open spaces. But for how long? That's not an unreasonable question according to smokers who are up in arms over the new legislative measure going into effect January 1st that forbids lighting up a single cigarette in bars, gambling houses, casinos, and night clubs in California. If smokers have grown accustomed to the ban on smoking in restaurants, in force since 1995, it's precisely because they've been able to withdraw to bars. . . .

—*Le Monde*, 2 January 1998

one
Global and Local Strategies

State and NGO Initiatives, Community Mobilization, and Social Marketing

Social marketing is a strategy for translating scientific findings about health and nutrition into education and action programs adopted from methodologies of commercial marketing. The opportunity is worldwide: only the urgency of its need may vary. . . . For example, infant mortality in Washington, D.C., at 26.9 per 1,000 births, is as high as Malaysia where the rate is 30.3. . . . The root cause is poverty but the poor diet and sanitation it breeds can be eliminated through health education.

—Richard K. Manoff, *Social Marketing: New Imperative for Public Health* (1985)

The way in which health promotion strategies are deployed in the modern state both reflects and helps to reproduce fundamental features of the distribution of social power; and that current policy debates need to be examined as a matter of cultural politics rather than (as health promotion debates are so often presented) as matters of technical rationality.

—Alan Beattie, "Knowledge and Control in Health Promotion" (1991)

The cultural audience is not so much a specifiable group *within* the social order as the principal site *of* that order. Audiences participate in the most global (but local) communal (yet individual) and time-consuming practice of making meaning in world history. The concept and the occasion of being an audience are textual links between society and person, for viewing involves solitary interpretation as well as collective behavior. Production executives invoke the audience to measure success and claim knowledge of what people want. But this focus on the audience is not theirs alone. Regulators do the same in order to administer, psychologists to produce proofs, and lobby groups to change content.

—Toby Miller et al., *Global Hollywood* (1998)

In November 1999 Derek Yach, project manager of the World Health Organization's (WHO's) Tobacco Free Initiative (TFI), announced the WHO's decision to "globalize" California's tobacco control program to 400 heads of local anti-smoking projects in California assembled at Lake Tahoe. What drew him was the program's ten-year track record of success in accelerating the decline of smoking in the state at a rate faster than that of the U.S. (from 24 percent in 1989 to 18 percent in 1999) and in greatly reducing Californians' exposure to environmental tobacco smoke through a multiple strategy of cigarette tax increases, local ordinances mandating clean indoor air, mobilization of ethnic communities, and the extensive use of social marketing techniques including paid advertising in an aggressive media campaign against secondhand smoke and tobacco industry marketing practices targeting communities of color, recent immigrants, women, and youth (Pierce et al. 1998, 2–13, 2–16). In essence Yach took a singular, highly visible campaign and exported some of its most successful features (the media campaign and community mobilization) in an attempt to articulate a public health project across vastly different international sites. It was part of a struggle that went toe to toe with the tobacco industry to replace the global smoker with the nonsmoker as the new international norm (Yach 1999b; Yach and Bettcher 2000).[1]

The week in California was only one of several important global tobacco control events organized by the WHO's TFI office in the fall of 1999. In September Asian youth leaders met in Singapore with U.S. teenagers involved in the design of the Florida 1998–99 "Truth" campaign meant to discourage teenage smoking through ads satirizing the tobacco industry; and in mid-November the WHO sponsored an international conference on tobacco and health, "Avoiding the Tobacco Epidemic in Women and Youth" in Kobe, Japan, with the express purpose of recruiting for the TFI women's nongovernmental organizations (NGOs), a constituency that sponsors claimed had little involvement in antismoking efforts to date.

The Identity of California in Public Health: Between Local Exceptionalism and Transregional Influences

Tracing the development of the California campaign—especially the media component—that eventually prompted the WHO to "globalize California" involves analyzing the history and rhetoric of the 1988 ballot initiative, Proposition 99, that led to establishment of the Tobacco Control Program in 1989 and examining the various conceptual, discursive, and practical influences that went into the health promotion program's design. It would be misleading to view the WHO's decision as simply the global articulation of local state practices in the U.S.; for in the eyes of Yach and California state officials, the power and interest of the Tobacco Control Program lay in part in the fact that it was operating in a social and cultural context that was already in some sense a local articulation of global conditions (the large presence of transnational, diverse populations and the dominance of media/entertainment industries).

What is less widely understood is that the Tobacco Control Program itself was from its inception in 1989 the product of local, national, and global collaborations with many groups: accumulated expertise from other states, the federal government, and other countries; public health researchers and activists (who often had extensive international experience); advertising and media professionals; and, later on, community representatives (American Indian, Hispanic/Latino, African American, and Asian/Pacific Islander). In their joint work over the years, these collaborators drew on the example of social marketing programs in nutrition and family planning implemented in industrializing nations during the 1970s and 1980s and on U.S. studies of public health interventions (COMMIT and ASSIST) commissioned by the National Cancer Institute (NCI). They also took inspiration from examples of community activism in the U.S. (notably Americans for Nonsmokers' Rights' success in passing local clean indoor air ordinances and the Philadelphia Uptown Coalition's victory over R. J. Reynolds' marketing of its Uptown brand targeting African Americans), from comprehensive media and community campaigns conducted in the 1980s in Australia, and from coalition building in Canada and advocacy in France (Kaufman 1999). In a sense, what Yach and the WHO decided to "globalize" was itself partly a product fashioned across global sites: the TFI was taking around the world—especially to industrializing nations—public health methods and approaches that had their beginnings partly in the so-called developing world twenty-five years before by U.S.-based social marketing firms. It is striking that the wide deployment of social marketing techniques within U.S. tobacco control took place during a decade of intense globalization in which if not American at least Anglo-American policies and approaches took the lead in international anti-smoking initiatives and U.S. hegemony over communication and media infrastructures in key arenas (cinema and television production and distribution) increased dramatically even as sites of production were increasingly relocated outside the U.S. These simultaneous developments raise the question of the force of U.S. and Anglo-American communication industries and technologies and how their new international division of labor has shaped global tobacco control (Miller et al. 1998, 1–16, 44–82).

Recurrent Strategies and Tensions in the Anti-smoking Campaign

The local and global development of key concepts, strategies, and goals of the state-sponsored anti-smoking effort emerged in the last quarter of the twentieth century, a period marked by the ascendancy of market-based neoliberal policy agendas. In this and the next two chapters I broach the question of the power and limits of the deployment of market-inflected approaches to populations in the arena of public health by exploring the tensions that arose in California's goal of ensuring a smoke-free environment for increasingly diverse citizens and residents in an era of globalization.

What interests me are the contradictions that surfaced in the formulation and realization of these democratic goals through the program's reliance on state and NGO policy bureaucracies, social marketing tools developed by private-sector firms, public health research data, and concepts derived from the community empowerment movement. In particular, I ask the following questions: what happens when a relatively homogeneous environmental and health social movement sponsors a state health promotion campaign that not only entails extensive collaboration between civil society organizations, experts, advertising agencies, and state policy bureaucracies but also involves attempts to expand the original citizen base of the movement to wider communities and populations? And what happens to notions of "health," "citizenship," and "community" when public health campaigns articulate them across different segments of society, especially through electronic and print media such as television, radio, and outdoor advertising?

Arriving at partial answers to these questions leads me to examine early strategies developed by anti-smoking and environmental groups, the rhetoric and tactics of the campaign to pass Proposition 99, and the design of the Tobacco Control Program. This involves tracking the inflection of health promotion strategies by the policy and rhetorical legacies of previous public health interventions with respect to particular populations and by dominant interpretations of epidemiological data. It also entails looking at various sites of tension including the intersection of community empowerment strategies, social marketing approaches, and the very structure of the public media sphere and how the latter segments and amalgamates citizens and community members into consumer markets. Finally, some answers to these questions lie in articulating the contradictory structures of evaluation and oversight.

Prop 99: Translating a Health Movement into State Policy

The Department of Health Services (DHS) launched its comprehensive anti-smoking campaign beginning with statewide media spots broadcast in April 1990 that attacked industry marketing and lobbying practices; decried the industry's exploitation of the African American community; and addressed problems concerning underage access, teenage peer pressure, the difficulties of cessation, and the hazards of secondhand smoke. The health promotion effort stemmed from the passage by California voters of Proposition 99 in November 1988, known as the Tobacco Tax and Health Promotion Act, which raised the cigarette tax from 10¢ to 35¢ per pack.[2] Of the resulting Cigarette and Tobacco Products Surtax Fund, 20 percent was formally allotted by Assembly Bill 75 (AB 75), passed in October 1989, to comprehensive tobacco health education. This amounted to around $130 million annually, of which approximately $14 million was earmarked for the media campaign component.

At that time California offered fertile ground for tobacco control initiatives: it had low cigarette taxes, no local tobacco industry involving the livelihood of thousands of workers and farmers, a burgeoning service economy, large numbers of white-collar workers and middle-class professionals for whom quality-of-life issues had become paramount, a strong environmental movement, and the second-lowest adult smoking rate in the nation after Utah (24 percent in California versus 28 percent nationwide in 1988; see CDC 1998; Pierce et al. 1998, 2–16). Moreover, the national culture of smoking was also undergoing profound changes. Like their counterparts in the U.K. and Canada who were spared the worst deprivations of the Second World War, U.S. smokers became heavy smokers in massive numbers during the 1940s (3,500 cigarettes per person per year), some twenty years earlier than among populations in countries (such as Japan, Germany, France, China, Italy, Belgium, and Russia) that would live with the economic consequences of the devastation of the war for decades. Consequently, the domestic U.S. cigarette market had already peaked in part because the health consequences of heavy smoking had already begun to make themselves felt in the way of the daily experience of tobacco-related illnesses and death among family, friends, and celebrities whose cause of death was widely publicized; around 1980 a general decline began in adult smoking rates nationwide. In terms of global tobacco control, this set the U.S., the U.K., and Canada apart from other leading industrialized nations and constitutes one of the underlying conditions of U.S. "exceptionalism" in the 1980s and 1990s. Also spurring changes in U.S. smoking culture were the creation of nonsmoking sections on U.S. domestic flights in the 1970s; the publication of the surgeon general's 1986 report on secondhand smoke or environmental tobacco smoke (ETS), which radicalized the anti-smoking movement; and the 1988 report on nicotine addiction.

Overall, however, the U.S. federal government had an uneven record on the regulation of the production, marketing, and consumption of tobacco products. As in many countries, tobacco industry influence had bought its product a unique status as neither a food product nor a drug, thereby placing it outside of government oversight (Kluger 1996). Moreover, unlike for France or Japan, there is no federal law or clause in the U.S. Constitution that makes the health of its citizens the responsibility of the state. Nor is there quite the equivalent to the British Public Health Act of 1848 that committed the national government to supporting public hygiene efforts at the local level (Hamlin 1994). Furthermore, tobacco control advocates in the U.S. were and are still working within the legacy left by the colonial experience and westward conquest in the form of the doctrine of states' sovereignty and the privilege granted to property and individual rights in an emerging capitalist economy, all of which have made public health matters in the U.S. by and large a local and urban affair and left them in the hands of local business and political elites and voluntary organizations (Fee 1994). The earliest serious attempts in the U.S. at banning outright the sale of tobacco between 1895 and 1915 never surpassed the state level. In the late twentieth century it fell to local anti-smoking groups, environmental organizations, and state and municipal governments to seize the initiative.

In California, prior to the campaign for Proposition 99, the forerunner of Americans for Nonsmokers' Rights (ANR) had made successful attempts to bypass national and state governments, where the transnational tobacco industry dominates the political process, by establishing local ordinances (starting in 1977 in Berkeley) modeled on Arizona and Minnesota laws restricting smoking in health facilities, workplaces, schools, theaters, hotel lobbies, and public transportation (Berkeley City Council 1977, Americans for Nonsmokers' Rights 1996).[3] However, it had failed to pass statewide clean indoor air referenda in 1978 and 1980 as a result of massive industry lobbying and well-funded campaigns, but the resulting publicity and debates laid the groundwork for Proposition 99 and forever framed industry representatives as outsiders intruding on local affairs.

Proposition 99 was launched by the Coalition for a Healthy California, a group comprising the Planning and Conservation League, the American Cancer Society, the American Lung Association, and the American Heart Association. The initiative formulated a new approach borrowed from Minnesota's smoking cessation program,[4] a tactic that was designed to finesse public hostility to new state expenditures in a political climate, following the passage of California Proposition 13 in 1978, that was dominated by neoliberal policy agendas stressing tax cuts and budget caps. The new referendum did not propose statewide clean indoor air regulations but instead focused on reducing smoking through "user" cigarette taxes, the funds from which would be devoted to environmental and health promotion programs and to indigent care. Through the user tax on cigarettes, the tobacco control program was to be completely self-financing and would presumably disappear once the market for tobacco products dried up (Traynor and Glantz 1996, 550). Proposition 99's fiscal measure is but one of several ways in which the anti-smoking forces negotiated a state-sponsored intervention in an unfavorable political environment; other ways included attacks on the tobacco industry and drawing attention to the hazards of secondhand smoke.

On 16 December 1987 the Coalition for a Healthy California launched its successful petition drive to collect 600,000 signatures to put Proposition 99 on the ballot in November 1988.[5] Throughout the fall of 1988, the coalition's television and radio spots, press conferences, leaflets, and brochures emphasized what passed for standard fare in tobacco control discourse at that time: the hazards of smoking and its annual costs to smokers, nonsmokers, and California residents in terms of illnesses and their treatment ($2 billion), deaths (30,000 in California), fires, lost productivity in the workplace, and higher taxes and insurance rates (another $4 billion).[6] More crucially, it also developed a highly effective anti-tobacco industry rhetoric that was to have a great future in the California anti-smoking campaigns in the 1990s: the campaign framed the tobacco industry as outsiders intruding on local affairs and decried its systematic deception of citizens, its sinister need to recruit new smokers every day in the U.S. to replace those who quit or died, and thus its practice of targeting "vulnerable groups" such as women, minorities, and teenagers (Troyer and Markle 1983; Coalition for a Healthy California 1988a, 1988b; American Cancer Society 1988; Advocacy Institute 1988; Salmon 1989b).

Much of the rhetoric of the campaign to pass Proposition 99 projected an image of Californians unified against a shared threat from the transnational tobacco industry. The official press release launching the campaign, "Initiative Will Take On the Tobacco Goliath in a Fight for a Healthier California," makes industry opposition a rallying cry: "In the face of tobacco industry threats to spend $16 million against the initiative, the Coalition for a Healthy California launched a drive to recruit nearly a million Californians to sign the initiative petitions and join the battle for a healthier California" (Coalition for a Healthy California 1987). What the press release underestimated was tobacco industry spending, which would exceed $21.4 million by the time the November election was over; the industry outspent the coalition nearly 15 to 1 (Traynor and Glantz 1996, 570). The industry contracted with various public relations agencies to set up organizations opposing the initiative, such as the Smokers' Rights Alliance and the Coalition against Unfair Tax Increases; one of its spokespersons, Kimberly Belshé, would later be appointed in 1993 by Gov. Pete Wilson as director of the DHS, which is responsible for the Proposition 99 tobacco control education effort (Ellis 1993).

The Proposition 99 spots (some of which were in Spanish as well as English) varied in their approach and pitch. Some used calm, authoritative voice-over presentations of Proposition 99 and rebuttals of tobacco industry objections (which claimed that the initiative was anti-smoker, would unleash a crime wave of smuggling, was a regressive tax that would enrich physicians and deprive the poor, and so on). Other spots capitalized on endorsements by the famous (actor Jack Klugman of *The Odd Couple* TV series), recounted melodramatic stories (a child visiting her mother in a hospital), or conveyed personal testimonies of the outraged (Patrick Reynolds, grandson of R. J. Reynolds and an anti-smoking activist), the bereaved (State Senator Alan Robbins about his late father), and even the dying (the James Almon spots that reprised the approach of the famous 1985 American Cancer Society spot featuring a terminally ill but still handsome Yul Brynner that was aired after he had passed away; in the Almon spots, the physical ravages of disease are visible for all to see).

The James Almon spot (which had several versions) ran as follows; it emphasized the need for educating the young:

> ***Almon***: [extreme close-up of Almon with oxygen apparatus] "I wanted to do this commercial because I am dying."
> ***Voice-over***: "James Almon, cigarette smoker for twenty-two years."
> ***Almon***: "When I was growing up, schools didn't teach the dangers of smoking. I wish they had. There's nothing you can do for me, but we can teach our kids not to smoke. A twenty-five-cent tax on every pack of cigarettes will pay for that."
> ***Tag***: "Vote Yes for the Tobacco Tax and Health Protection Act." (Coalition for a Healthy California. 1988c)

The spot was sometimes integrated into a long presentation of the initiative that appealed for campaign contributions. It was prefaced by the title, "This presentation is dedicated to James Almon who died March 30th, 1988." It is worthwhile quoting large segments of the appeal because it rearticulates the dangers the tobacco industry and smoking pose to youth in terms of the neoliberal emphasis on the costs of smoking to taxpayers and the business sector:

> ***Voice-over***: "Like Jim Almon, another 360,000 will die this year from tobacco-related diseases—lung cancer, emphysema, heart disease. And because another two million quit every year, the tobacco companies must hook 5,000 new smokers every day just to keep tobacco sales at the present level. [Shot of twelve-year-old white girl on couch eyeing cigarettes on a coffee table; close-up of a twenty-year-old white woman lighting up; photo of two eighteen-year-old white women with cigarettes between their lips.] Five thousand new tobacco users every day: 60 percent start before they are fourteen, and 90 percent start before they are twenty. Two million young people get hooked every year to keep profits at current levels. To get those smokers and to keep them smoking, tobacco companies spend seven million dollars every day, two and one half billion in advertising every year.
>
> "At the same time, American businesses and corporations spend millions every year to deal with the problem of smoking. . . . Smokers increase the cost of doing business because of costly sick leave, stress on workers, annoyance to non-smokers, cigarette breaks, and lost productivity. That's why a group of concerned citizenry, organizations, and business people who care [on screen the names of

> sponsors scroll by] are sponsoring a statewide initiative to make tobacco users pay their share of the costs. . . .
>
> "When you consider what smoking and other tobacco habits cost corporations and businessmen every day, not to mention the cost of illness, suffering, and human lives, we're sure you will agree with James Almon. Your contribution can help get the message to the voters. A small investment to make for a new generation of smoke-free Californians. A small price to pay for someone's life."
>
> ***Tag***: "Support the TOBACCO TAX INITIATIVE. PLEASE HELP NOW." (Coalition for a Healthy California 1988c)

Here the focus on costs designates as the pitch's intended audience the coalition's core constituency: largely white, upper-middle-class voters. Interestingly, representatives of this group figure less prominently in future California tobacco control ads that targeted youths, communities of color, and other "hard to reach" populations but nevertheless haunt the entire media campaign as the tacit model of "acting, thinking" citizens and community members who are more amenable than most to the call to accountable behavior.

Finally, one Proposition 99 radio spot went a step further and sharpened the threat the tobacco industry posed to smokers, children, coworkers, and business profits into a menace against the life of neighbor and family—in a word, a caring community:

> ***Female voice-over***: "There are children in our community who get hooked on cigarettes before the age of fourteen. There are people in our community who cannot pay for medical care and hospital emergencies. They are our neighbors, our friends, our family. We can help them. That's why a group of concerned citizens who care about the health and protection of all of us ask you to vote 'Yes' on the tobacco tax that could save lives. A 'Yes' vote on 99 will pay the medical bills and hospital bills for people in our community who cannot pay themselves. A 'Yes' vote on 99 will teach children the danger of smoking. The tobacco companies are spending millions to stop the tobacco tax. They're fighting for more profits. We're fighting to save lives. We're the American Cancer Society, the American Heart Association, the American Lung Association, the volunteers who are part of the Coalition for a Healthy California who provided this message. Please vote 'Yes' on 99. You could save someone's life." (Coalition for a Healthy California 1988d)

Tobacco Control Isolation and Community Outreach

Yet achieving community unity was no simple matter. To begin with, the coalition, like the environmental movement, smokers' rights groups, and the medical establishment, was at that time not known for its social diversity. In particular, it operated in relative isolation from many of the Californians it was publicly committed to serving. While the proposition enjoyed support from African American State Senator Diane Watson (D–Los Angeles) and also Latino politician Tom Soto (J. Miller 1988; Soto 1988), it received no endorsements from major organizations belonging to designated "vulnerable groups," including minority and women's organizations. Indeed, several organizations of color, such as the National Black Caucus of State Legislators and the West Coast Black Publishers Association, had opposed earlier attempts to pass tobacco control legislation in the State Assembly and now rejected the proposed new excise taxes as burdensome on their community (National Black Caucus of State Legislators n.d.; "Black Newspaper Publishers" 1988; Glantz and Balbach 2000, 44–45). In the eyes of anti-smoking advocates, the legislators' and publishers' opposition was primarily due to the industry's long-standing ties to African American business and education leaders, including its role as donor to the National Association for the Advancement of Colored People (NAACP) and the National Urban League. This had already been the subject of heated exchanges in the national press in which black leaders claimed that their organizations were being held to a double standard by white anti-smoking activists (Williams 1987; Jackson 1997).[7] Later, between 1990 and 1991, state health public health officials would attempt to improve relations by adopting a strategy of state-sponsored community empowerment in setting up ethnic networks and advisory committees to establish mechanisms of communication and collaboration (I return to this point later). However, no comparable committees were set up for women and youth.[8]

Compounding matters were yet other factors related to earlier public health interventions in California and the U.S. First of all, most of the community-intervention studies sponsored by the NCI and other public health agencies during the 1970s and 1980s on high blood pressure, chronic heart disease, and smoking cessation had mixed success in reaching blue-collar workers and communities of color and thus had left little in the way of a legacy of public health "capacity" (health education and literacy, extensive data, health organizations, and community leadership) among those groups compared with the more mainstream population, which was largely middle-class, male, and of European descent. This was particularly the case for tobacco control. (Moreover, other researchers have noted that in the U.S. field of epidemiology the underrepresentation of minorities is severe [Beaglehole and Bonita 1997, 103].) As a result, in the late 1980s U.S. underrepresented communities entered the new era of tobacco control with virtually no experience and little infrastructure in tobacco control, which contributed to the isolation of the tobacco control community from nonmajority groups in the U.S. and heightened anti-smoking advocates' perception of these groups' vulnerability to the new marketing strategies of the tobacco industry (Robinson and Headen 1999; Robinson 2000).

Moreover, the coalition's picture of smokers stemmed from available smoking studies that stressed that those at low income and education levels as well as African American men had among the highest prevalence rates (32.5 percent for black men, compared with 29.3 percent for white men in 1986; Novotny et al. 1988); the health burden of African American men who smoked was dramatically higher in the way of tobacco-related diseases.[9] This seemed to lead coalition leaders to worry that the tobacco industry would play smoking demographics to its advantage and decry Proposition 99 as not only fiscally regressive but also racist.[10] In an internal memo, Jack Nicholl, the coalition director, expressed his alarm this way:

> The tobacco companies will do everything possible to take advantage of the demographics of smoking. The profile of a smoker is minority, blue collar, high school education, low voter participation. *They are preparing to register 100,000 blacks and will build their 'No' campaign on the assertion that Prop 99 is discriminatory, racist and regressive.*
>
> They will be backed up by a list of minority organizational endorsements like Latino Peace Officers Association, Mexican American Political Association (MAPA), and League of United Latin American Citizens (LULAC). The tobacco companies contribute significantly to minority organizations and demand their support in these kinds of battles.
>
> However, we cannot concede these communities to the tobacco companies. Tobacco marketing to young, poor people is successful and causes life-long, significant health problems to that population. (Nicholl 1988a, emphasis in the original)

In this scenario, blacks are portrayed as both victims and pawns of the tobacco industry—in a word, something less than autonomous political actors. However, in terms of available prevalence data, the memo overdraws the picture somewhat, for studies already indicated that current rates for black men had begun to drop dramatically, as had those of whites. (Later studies would reveal that Asian, Hispanic/Latino, and African American *youth* rates were actually *lower than those for whites*; U.S. Dept. of Health and Human Services 1998). This would suggest that these groups were not simply the past or future passive victims of industry marketing strategies that the director's rhetoric implies. Moreover, I think it is fair to say that we witness here a recognizable form of nervousness in the discourse emanating from isolated expert policy and activist circles in which one group or community that is the ostensible object of policy can easily slip from the status as victim and object of concern to that as a potential threat or source of betrayal. Echoes of this particularizing perception would return in future anti-smoking campaigns in California. Indeed, this would contribute to one of the underlying tensions at work in the expansion of the anti-smoking movement in California through state-sponsored health campaigns to ever wider groups between the inclusive tendencies in public health policy and stigmatizing ones, which, in the U.S., arguably have roots going back to health reformers of the Progressive Era (Nathanson 1996, 618–619; Klaus 1993, 16, 29).[11]

Midway through the fall, the tobacco industry's counteroffensive lost steam as a result of several events: the state attorney general publicly refuted the allegations that a successful referendum would precipitate a crime wave of smuggling, and several law enforcement officers' associations subsequently dropped their opposition to Proposition 99. Moreover, organized labor took an "open" or neutral stance on the issue. In the end, the coalition prevailed, 58 percent "Yes" to 42 percent "No," on November 8. The next day, in the coalition's victory statement, Jack Nicholl announced that the industry had decided to go to court to overturn the results:

> The message of Proposition 99's victory is that the tobacco companies cannot come into California and buy the voters with a slick $20 million campaign. They cannot scare the voters with tough-looking cops warning of a crime wave; and they cannot trick voters with arguments that Prop 99 picks on smokers.
>
> You would think RJ Reynolds and Philip Morris would call it a day. But no. . . . They are so arrogant and have so much money they think they can do anything to get their way. So they have decided to take Prop 99 to court.
>
> This proves what we have been saying throughout the campaign: the tobacco companies don't care about the people of California, all they care about is their profits. (Nicholl 1988b)

The local struggle against transnational tobacco corporations was not over.

Assembly Bill 75: Mandating the California Campaign

To obtain funding for an anti-smoking health promotion program was one thing; to carry it out was a different thing altogether. The implementation of Proposition 99 required the drafting and passage of enabling legislation and thus returned the tobacco control effort to the halls of the California State Legislature, where the crafting of legislation would be subject to considerable political pressures. A final bill did not exit Sacramento with Republican Governor George Deukmejian's signature until nearly a year later, in October 1989. AB 75 reflected most of the provisions of Proposition 99. The Cigarette and Tobacco Products Surtax Fund ($650 million per year) allotted 20 percent to tobacco control education, 5 percent to tobacco-related research, 5 percent to environmental concerns, 10 percent for treatment by physicians of patients not covered by private insurance or federally funded programs, and 35 percent to hospital patients and indigent health care. The remaining 25 percent was to be distributed by the legislature among the preceding categories (TEOC 1991, 5; California Statutes 1989, 5432–5436).[12]

Of the tobacco control education funds, 10 percent was earmarked for the media campaign and 30 percent for school education; the remaining 60 percent was to be appropriated for local health departments (local lead agencies, or LLAs) and community efforts. Efforts were to be directed at the following populations: children 6–14 years old, young adults, pregnant women, low-income people, blacks, Hispanics, Native Americans, Asians/Pacific Islanders, current smokers, and school-age youth no longer attending school classes (TEOC 1991, 12). The selection of these groups matched those deemed "vulnerable populations" by the "Yes on 99" campaign and reflected the program's stress on prevention, cessation, and reducing the health and financial costs of smoking and the common interpretation of the prevailing public health literature, especially that published in the U.S. at that time (U.S. Dept. of Health, Education, and Welfare 1979; U.S. Dept. of Health and Human Services 1980, 1986; Davis 1987).

Mobilizing Vulnerable Groups: Globalization and Community Empowerment

In the eyes of state officials and anti-smoking activists, the explicit commitment to bringing the anti-smoking campaign to "vulnerable groups" was not only in response to the poor health picture of communities emerging out of epidemiological studies; it was also in reaction to the rapidly changing composition of California's population, the majority of which was expected to become nonwhite by the year 2000 as a result of new waves of immigration in an era of economic globalization. That was one of the features that made the anti-smoking effort unique; as such, the California campaign was destined to be a national laboratory for a new model of health education. Echoing conventional commonplaces about California, namely that it functions as the leading edge of an ever-evolving modernity for the rest of the U.S. and even the entire industrialized world, DHS director Kenneth Kizer, a Republican appointee, and Dileep Bal, the head of the DHS's Chronic Diseases Control Branch who was responsible for overseeing tobacco control, declared that the future of societies could already be found in the social diversity, transnational movements of population, and burgeoning younger generations of their state. Concluding an early assessment of their campaign published in the *Journal of the American Medical Association,* they claimed,

> Given that California is the destination for about 25% of immigrants to the United States and that 150,000 new students enroll in the state's schools each year, it should be apparent that California's campaign offers an unsurpassed opportunity to reduce tobacco use. It has the potential to infuse other state health agencies with sorely needed models for technology transfer [e.g., introduction and diffusion of new ideas or practices to audiences] and service delivery. Moreover, with the range of community strategies and the interventions being developed for groups at highest risk of tobacco-related disease, this land-

> mark public health measure carries with it a wealth of information for future worldwide study and application. (Bal et al. 1990, 1574)

California was in the throes of acquiring a new identity, that of not only a socially diverse but also a radically nonsmoking state. In the words of the DHS ad featuring a black man holding his baby daughter, "We are California and over 80% of us don't smoke. We know better."

The depth of the Tobacco Control Program's ambition to do outreach to underserved populations stems in part from the community empowerment or development movement, whose roots go back in the U.S. to the 1960s, to the New Left and civil rights movements' emphasis on self-government and community power and to the Community Action Programs of the U.S. government's War on Poverty, launched during the presidency of Lyndon Johnson (1963–68; see Cruikshank 1999, 69–80).[13] It is also related to the broader U.S. discourse on "minority health," which bears the legacy of the use of racial and ethnic categories by the federal government, most recently in its efforts to monitor and enforce the civil rights legislation of the 1960s and 1970s: Asian/Pacific Islander, black, Hispanic, Native American, and (non-Hispanic) white, with all of the attendant problems of racializing health issues and of lumping and splitting subgroups (Montes, Eng, and Braithwaite 1995, 247; Root 2001).[14] Thus,

> "minority" is a policy construct, not based on a group's genetics or heredity but on a lengthy history with society as formal objects of policy. . . . In the 1960s and 1970s, racial and ethnic minority groups challenged society's policies and attempted to define their own status and identity. Since then, the political process has become the forum for examining and changing the social and economic structures that have created minority status. The implication for minority health is that it is not being born into a minority group that is a risk factor, it is being a member of a minority group in this society. Because society has created these processes, society can also change them. (Montes, Eng, and Braithwaite 1995, 248)

In this view, "minority health" efforts must be community-based and must draw on community practices and values—which have served well historically as a means of survival—as a powerful tool for improving health.

In principle, in the realm of public health, community development was a counter to the traditional paternalist tendencies of state and voluntary health agencies. Ronald L. Braithwaite and other researchers articulated the intersection between community empowerment and public health this way:

> Health promotion efforts are likely to be more successful in these populations when the community at risk is empowered to identify its own problems, develop its own intervention strategies, and form a decision making coalition board to make policy decisions and manage resources around the interventions. Further because health and illness behaviors are culture-bound, primary efforts to ad-

> dress preventable disease and illness must emerge from a knowledge of and respect for the culture of the target community. (Braithwaite et al. 1989, 57)

The discourse of collective empowerment in public health drew on alternative medicine's self-help, self-care, and holistic movements of the 1970s and 1980s that emphasized individuals' responsibility to become literate and active in the management of their own health, independent of mainstream medicine (Crawford 1980; Armstrong 1993, 1995; E. Martin 1994). It also reflected thinking in international public health circles at that time as set forth in the WHO Ottawa Charter (1986), which advocated a conception of health promotion stressing both personal and community development and empowerment (Beaglehole and Bonita 1997, 215–216). Finally, the HIV/AIDS pandemic would force rethinking in U.S. public health circles and accelerate consideration of the social environment and the need for community input concerning—in the case of the pandemic—everything from the sexual and drug-using practices of particular groups (about which the Centers for Disease Control and Prevention [CDC] and the National Institutes of Health [NIH] knew virtually nothing) to the definition of AIDS and the design of experimental drug protocols (Patton 1990; Epstein 1996).[15] In the U.S., the most famous example of community empowerment in the service of tobacco control up to that time was the victory in 1989–90 of Philadelphia's African American activists over the launch of R. J. Reynolds' Uptown brand, the first ethnically marketed brand ever. The Uptown Coalition proposed an anti-industry approach that emphasized community self-determination over health matters per se. According to one of its leaders, the advantage of the community defense approach was that it avoided the tendency of an exclusive focus on "health" to divide members of the community against each other (C. Foster 1990; Robinson 2000).

The Rise of Social Marketing as Neoliberal Health Technology

The paradigm shift that took place during the late 1980s in U.S. public health and in tobacco control, away from a focus on individual cessation toward a more environmental approach, both complemented and complicated the task of community empowerment by giving priority to prevention, combating secondhand smoke, and excise taxes (Robinson and Headen 1999). The overall move to prevention reflected the evolution of thinking in international public health beginning with the influential 1974 Lalonde Report, which argued for integrating medical services with the broader activities of "health promotion" and focused on health consequences related to individuals' "lifestyles" (Beaglehole and Bonita 1997, 214–215; Breslow 1990; Hornik 1997). The stress on prevention meshed with contemporary calls for reining in skyrocketing health expenditures through individual self-management and squared with the neoliberal revolution taking place in policy and governmental circles and even at the grassroots level in industrialized nations, especially in English-speaking countries at that time, which promoted citizen initiative and self-governance (Beaglehole and Bonita 1997, 227–230; Barry et al. 1993; Lupton 1995, 48–76; Osborne 1993; O'Malley 1996). In the words of John H. Knowles, president of the Rockefeller Foundation, in a famous essay in *Science* magazine in 1977 titled "Responsibility for Health,"

> The individual must realize that perpetuating the present system of high-cost, after-the-fact medicine will only result in higher costs and greater frustration. The next major advances in health . . . will be determined by what the individual is willing to do for himself and for society at large. If he is willing to follow reasonable rules for healthy living, he can extend his life and enhance his own and the nation's productivity. . . . If he participates fully in private and public efforts to reduce the hazards of the environment he can reduce the causes of premature death and disability. He can either remain the problem or become the solution to it; beneficent government cannot. (Knowles 1977b)[16]

This shift of the burden to rationally maximizing individuals to solve collective problems was a narrow interpretation of "health promotion" favored in the U.S. that fitted with the nation's dominant model of liberal medicine (private care providers, private insurance, and later, after the Second World War, a massive network of publicly funded private hospitals), which generally pushed public health in the direction of individual prevention, an approach promoted by chambers of commerce and medical societies over the provision of health and nutritional services. At that time the U.S. version of "health promotion" elicited charges of "victim-blaming" from some quarters, both domestically and internationally. Outside of North America many practitioners understood "health promotion" to be a question of equal access to primary health care and of structural and environmental factors rather than a issue of individual lifestyles. The Alma Ata conference in 1978 and the WHO's document, "Health for All by the Year 2000," issued in 1981, set forth these perspectives on public health practice, which public health editors were later to deem "a return to Virchow"—that is to say, to the ideas of one of the founders of European social medicine and public health in the nineteenth century (Young and Whitehead 1993, 114).[17] In terms of tobacco control, particularly in Australia and the U.S., the shift to prevention in international public health included setting anti-smoking public agendas through extensive use of social marketing–inspired health promotion that would stimulate public debate with the goal of shifting community norms away from the practice and tolerance of smoking (Carlyon 1984; Becker 1986; Minkler 1989, 19; Lupton 1995; Beaglehole and Bonita 1997).

From a historical perspective, it seems only fitting that social marketing should play such an influential role in anti-smoking efforts, for historians have long pointed out not only that machine-made cigarettes were the first truly mass consumer product but also that their appearance in the late nineteenth century in the U.S. marked the beginning of modern marketing campaigns (Schudson 1984; Brandt 1990; Kluger 1996). As an interdisciplinary field based in the methods and tools of the behavioral sciences and the entertainment industry, social marketing has roots that go back to the 1960s (Manoff 1985, 31, 124; Ling et al. 1992). Definitions of *social marketing* abound. An influential one from the early 1980s by Philip Kotler reads:

> Social marketing is the design, implementation, and control of programs seeking to increase the acceptability of a social idea, cause, or practice in a target group(s). It utilizes market segmentation, consumer research, concept development, communication, facilitation incentives, and exchange theory to maximize target group response. (Kotler 1982, as quoted by Hastings and Haywood 1991, 136)[18]

In terms of its direct interest to public health promoters, proponents of social marketing claim that it focuses on the "perceptions and perceived needs" of consumers rather than on the traditional public health insistence on what the audience "should do."[19] As such, some claimed that social marketing should therefore be seen—not unlike community empowerment—as a powerful tool of democratic education and social welfare: an instrument of health and social literacy that by virtue of its attention to consumers' specific needs and community feedback (through survey research, focus groups, monitoring, and other mechanisms) overcomes the presumptuousness and ignorance of top-down paternalisms of traditional government and public health (Manoff 1985; Becker 1986; Minkler 1989; Atkin and Arkin 1990; Tones 1993; Lupton 1995; Beaglehole and Bonita 1997). For social marketers, "community" largely means an aggregate of individual consumers modeling choices in a free-market economy. Richard Manoff, an influential proponent of this new approach to communicating with the public, saw social marketing as a crucial tool of community development, democratic governance, and the creation of an open society:

> To breed freedom and strength in a people requires policies of openness, of which most important may be the right to the power of knowledge and the right to put that knowledge constructively to work for one's self, which is to say for one's role in society, and the right to the tools to make it happen. Social marketing supports enterprises to strengthen the capacity of people to be the custodians of their own health, to be armed with self-confidence and self-reliance that are, finally, the only true defense against the wrong, the misleading, and the deceptive. (Manoff 1985, 89)

Here, Manoff positions social marketing as a successor to public health education if not to public education altogether as a mode of citizen empowerment.[20]

But some skeptics at that time remained unconvinced. One student of marketing discounted the populist-sounding rhetoric often heard from marketers when asked about their profession:

> The marketing literature is not populist but pseudo-populist; it is not the voice of the people and it cares nothing for enhancing the autonomous, rational-choice of the people, but it insists that decision-makers should anticipate what people will want. Of course, "want" is narrowly conceived. (Schudson 1984, 31)

Still, although within the Anglo-American-dominated world of marketing there had been considerable debate about the wisdom of expanding the definition of *marketing* to include the selling of noncommercial, intangible products such as ideas, social causes, and behavior,[21] by the late 1980s major working concepts of social marketing had become central to U.S. health promotion, as evidenced by the immensely influential NCI's practical manual (the so-called Pink Book), *Making Health Communication Programs Work: A Planner's Guide*, published in 1989.[22] Moreover, the role for social marketing in combating poor health becomes all the more crucial insofar as lifestyle illnesses (chronic disorders linked to consumption and daily routines) are closely tied by public health practitioners to knowledge and norms embedded in language and culture:

> With the advent of lifestyle illnesses, social marketing, which depends heavily on the media, is likely to play a bigger role in public health. Lifestyle illnesses, such as cancer, heart diseases, psychological disorders, malnutrition and overnutrition, accidents, and sexually transmitted diseases, are, in fact, transmittable by the impact of words and images on lifestyle. Similarly, words and images are needed to combat them. (Ling et al. 1992, 358)[23]

It is precisely in these terms that many public health officials described the interventions made by the California anti-smoking media campaigns in reaching poor and immigrant populations: helping set agendas, shaping public speech and discussion, and, in the long term, shifting norms by stepping in at a point just ahead of the curve of social change already under way in California in the form of local ordinances (Wallack and Sciandra 1991; California health official A 1996; Pierce 1997; Burns 1997; Glantz 1997).[24]

The Global Ambition of Social Marketing

Moreover, from its inception, social marketing itself was a health technology elaborated to address the needs of diverse underserved populations. Its pedigree is a transnational one going back to public health programs in international development. Before being adopted in the U.S., Europe, and Australia, many of the contemporary commercial marketing methods and techniques used in health promotion were first developed during the 1970s and 1980s for family planning and breastfeeding campaigns in South and East Asia, Latin America, and Africa by private firms (such as the Academy for Educational Development, Manoff International, Porter Novelli Associates, and others) and adopted by U.S.-based foundations (Ford Foundation, Rockefeller Foundation) and international NGOs (International Planned Parenthood Federation, UNICEF, CARE, Population Council, Population Services International, Westinghouse Health Systems), often in conjunction with the World Bank and U.S. government agencies, such as the U.S. Agency for International Development (see Altman and Piotrow 1980; Manoff 1985; Ling et al. 1992).[25] Indeed, according to Manoff, the challenge that the urban poor and minorities in the U.S. posed to government and public health agencies was identical to that of the poor in so-called developing countries (Manoff 1985, 36). Underlying this comparison was the practical assumption—widely shared, according to U.S. communication scholar Brian Goldfarb, by educational experts in North America and Europe throughout the twentieth century—that visual and aural electronic media (television, radio, cinema) could cheaply circumvent obstacles posed by certain populations' low literacy rates and could exploit these groups' presumably more "primitive" relation to symbols (Goldfarb 2002, 1–8). In this regard Manoff writes:

> In Third World countries, the most vulnerable populations cannot be reached through newspapers, magazines, and other print communications. Even in a country like the United States, millions of people are functionally illiterate—and almost all are among the most disadvantaged—unable to understand or carry out simple printed instructions. Millions more who can read have undoubtedly been weaned away from the reading habit by the more convenient, entertaining appeal of radio, TV, and the cinema. (Manoff 1985, 213)

Moreover, proponents of social marketing claimed that their methods of studying local sites in all their singularity (economic and social infrastructure, customs, patterns of consumption, health concepts and practices, channels of communication, linguistic use, etc.) enjoy a potentially limitless arena of activity. Social marketing's expansive role in democratic literacy thus encircles the globe from industrializing nations to disadvantaged subpopulations within the highly industrialized world. Its methods, when used sensitively, are applicable anywhere (Manoff 1985, 7; Romano 2000).[26] This global scope of social marketing is reflected in the very international careers of social marketers and public health professionals in health promotion and communication at the end of the century, including those I've interviewed for this book (Elder 1997; Connolly 1997; Hornik 1997; Pierce 1997; Romano 2000; Slama 1997; Health promotion researcher A 1997; Meurisse 1999; Tominaga 2000).

Community Health Advocates' Critique of Market-Based Approaches

The translation of expertise into community practices through social marketing was not without dangers in the eyes of some community health advocates. At that time within public health circles internationally there was debate over the wisdom of shifting to a social marketing–based approach to health issues. Advocates formerly identified with women's and community health movements dating from the 1960s and 1970s pointed to the danger of transforming inherently conflictual political questions into marketing issues based on consensus and on expert medical and public health knowledge that tended to dissolve questions of power and erase community voices (Grace 1991, 329).

Moreover, some public health advocates closely aligned with the community empowerment model criticized public health social marketing for adopting the terms and concepts of the advertising industry itself (notably, viewing citizens as "consumers") and the consequent temptation to resort to mechanisms of audience manipulation and risk of smuggling back in a form of state-sponsored paternalism that social marketers' claimed their approach has eschewed (Salmon 1989b; Montes et al. 1995; Braithwaite, Bianchi, and Taylor 1994).[27] Regarding these dangers, a much later study claimed that culture-based messages, such as those stressing for an African American public "how smoking and drugs are modern forms of slavery . . . while on the one hand potentially salient, must be pre-tested, as some segments of the population may find them irrelevant (if they don't place high priority on community issues), inflammatory, or offensive" (Resnicow et al. 1999, 14). Finally, advocates and marketing practitioners alike cite troubling but persistent findings in the social marketing literature that media health campaigns presumably designed to communicate with the working poor, the unemployed, the poorly educated and illiterate, and recent immigrants—in 1980s and 1990s U.S. parlance, the "hard to reach"—all too often managed to reach only a small portion of the overall population, many of whom turned out to be largely middle-class or urban elites. It would appear that the very phrase "hard to reach" constituted in the U.S. an inadvertent admission of some of the problems dogging social marketing's assumptions and methods. These findings cropped up in early nutrition and contraception programs in Bangladesh, India, and elsewhere as well as in community interventions in the U.S. (Altman and Piotrow 1980, 424; Luthra 1988; Ling et al. 1992; National Cancer Institute 1995, 199; Peterson and Stunkard 1989, 823).[28] According to Lawrence Wallack, the rational-choice model of the individual underlying social marketing that focused on lifestyles seems to privilege well-educated citizens who are free from pressing material constraints:

> Just as lifestyle theory tends to reduce complex health issues to individual behaviors, television takes social and health issues and reduces them to personal emotional dramas. Problems are ultimately defined as a lack of information or as people with the right information making the wrong decisions. . . . The logic of the remedy suggests that if we could just get the right information to the right people, each one of them would change and eventually the larger social problem would be diminished. Reason would win out! For a select part of the population this may apply. Yet for the vast majority, this is an inadequate and partial strategy. (Wallack 1990, 45; see also Guttman 1997, 170)[29]

Or, as one U.S. sociologist has suggested, public health analysis can unwittingly produce a very middle-class social geography in which, in our language, one cultural or social particularity can threaten to erase all others (Goldstein 1992, 127).[30]

These skeptical assessments stand at odds with government officials who were early proponents of community empowerment and saw little conflict between their goals and marketing-inspired approaches. They actually invoked the marketer/consumer relation to describe the problem of overcoming the obstacles blocking communication between state agencies and citizens. For example, Sargent Shriver, director of the Office of Economic Opportunity under President Lyndon Johnson (1964–68), put the government's difficulty in reaching the poor in the War on Poverty this way: "It isn't reaching the consumer—the poor themselves. And so, we have to engage in a new kind of market research. We have to find out why the old product didn't appeal to the consumer—the one-fifth of the market. And only the poor—the consumer—can tell us" (cited in Cruikshank 1999, 77). In the end, the California campaign gave equal weight to the themes of community and health, producing mixed results, for in many of California's anti-smoking ads targeting specific communities the theme of community self-defense against an outside threat gave way to narratives of destruction and betrayal by smokers within the community, especially in ads on secondhand smoke, which provoked dismay among community anti-smoking activists.

The Prominence and Design of the Media Campaign

The major role played by the media campaign in the California Tobacco Control Program was a matter of both circumstance and choice. As a product of circumstance, the prominence of paid advertising in broadcast, print, and outdoor media in a comprehensive health promotion campaign comprising many other elements (local ordinances, school programs, community mobilization, excise taxes) stemmed in part from tobacco industry attempts to eliminate any media component in the enabling legislation (AB 75) through aggressive lobbying efforts, and from its success throughout the 1990s to get state officials such as Gov.

Pete Wilson to cut funding for the media campaign (Balbach and Glantz 1998). As a result of the ensuing public controversy, in the eyes of the public and the press, the media effort became something of a symbol of the entire campaign and often enjoyed greater visibility than many tobacco control activities by local health departments, NGOs, and citizens groups. Moreover, in the wake of the DHS's scandalous failure to respond to the onset of the HIV/AIDS pandemic with effective health promotion campaigns, some state health officials felt that the Tobacco Control Program offered the chance to "do something" (California health official A 1996).

In many respects, the advent of the HIV/AIDS pandemic cast a long shadow over tobacco control efforts in the 1980s and 1990s, not only in the U.S. but in many other countries, including France and Japan. The public health community's poor response, marked by government scandals related to uncontrolled blood supplies, left it discredited in the eyes of the public and those affected directly by the pandemic (Setbon 1993; Epstein 1996). While campaigns in Europe showed some success in encouraging condom use, in the U.S.—one of the early major sites of the pandemic—health education was much more uneven; often little was done in the way of major efforts by the federal government or states, and what was done came under strong criticism for its homophobia and skittishness around sexuality on the one hand and was attacked by the Christian Right for "condoning homosexuality" on the other (Grover 1989; Crimp 1989; Patton 1990).

As a matter of strategic choice, the prominence of the media campaign in California was based most obviously on the crucial role played by political advertising in the successful campaign to pass Proposition 99 but also on the experience and goals of social marketing programs in nutrition and family planning, which had been initiated in the 1960s and 1970s in South Asia, Indonesia, Africa, and Latin America, and the recent and current community intervention studies and anti-smoking campaigns in Europe, Australia, and North America.[31] Some of the personnel of and many consultants to the California program had worked on these health education projects. For example, the lessons of the successful Australian use in the 1980s of high-impact paid media with weekly monitoring in conjunction with local efforts were made directly available in the person of John Pierce, the former director of evaluation for the Victoria campaign in Sydney (Pierce et al. 1987; Pierce, Macaskill, and Hill 1990). When he left his position as chief epidemiologist at the Office on Smoking and Health at the CDC in Rockville, Maryland, to join the faculty at the University of California, San Diego, from the campaign's inception he served as one of its main evaluators. The DHS brought in two officials from the Minnesota Department of Health to share their department's experience as the first in the U.S. to do a statewide media campaign with purchased spots (based on pre- and post-testing but including no formal evaluation) in a comprehensive campaign. One would later become a member of the Tobacco Control Section's (TCS's) Media Advisory Committee. Finally, there was the presence of Terry Pechacek of the CDC, who was a senior investigator of the NCI's Community Intervention Trial for Smoking Cessation (COMMIT) intervention study on smoking and served as chair of the Tobacco Control Program's Evaluation Advisory Committee (Pierce, Macaskill, and Hill 1990; Pierce 1997; Minnesota health official 1997).

The NCI's ongoing COMMIT community-based intervention (1988–92) was an important model for California tobacco control, for it was a comprehensive campaign that was focused on local communities. However, it had its drawbacks, of which researchers were well aware early on, concerning its inability to reach non-middle-class and minority audiences and the low exposure of citizens to various interventions, especially media-based ones (National Cancer Institute 1995, 199; Cooper and Simmons 1985, 344–345). As this intervention study was at the local level, media campaigns were disallowed by the original research design (there is little in the way of "local" television and radio in terms of broadcast areas).[32] Many researchers involved in the California campaign were eager to overcome these faults in design and implementation in an attempt to address the great diversity of the state's population. They favored extensive use of media in conjunction with media advocacy (the staging of newsworthy events designed to catch the eye of the press and to exert pressure on specific state officials, politicians, firms, or agencies) and actions at other health promotion venues (schools, workplaces, community centers, places of worship, physicians' offices, etc.). They argued that the media campaign served as "air cover" for the "ground troops" of local and state activism (Erickson, McKenna, and Romano 1990; California health official A 1996; Pierce 1997; Burns 1997; Glantz 1997; Minnesota health official 1997). This reflected closely the understanding of the role of media as put forth in the NCI's "Pink Book," which synthesized current wisdom in social marketing and health communication and made it available to practicing health advocates interested in launching health education programs (National Cancer Institute 1989).[33]

Finally, there was the historical example of anti-smoking public service announcements aired over U.S. broadcast television networks between 1967 and 1970, when a Federal Communications Commission (FCC) ruling, based on its "fairness doctrine" requiring that opposing viewpoints be broadcast on issues of public interest, forced broadcasters to allot anti-smoking messages free airtime to counterbalance the tobacco industry's massive advertising. The American Cancer Society and other health voluntaries produced hard-hitting ads with the help of pro bono work from Madison Avenue, some of whose approaches would have long careers in future anti-smoking campaigns: warnings addressed to smoking parents as poor role models; satires of oversold macho images of smoking; thoughtful accounts by smokers of the difficulties of quitting; elegiac testimonials by survivors about loved ones who quit too late or by terminally ill smokers; and ads deglamorizing smoking as a smelly, unglamorous habit and citing tobacco smoke as an indoor pollutant. When the industry decided to preempt a proposed ban on broadcast advertising by voluntarily withdrawing its spots, anti-smoking ads lost their free airtime and were relegated to off-peak broadcast hours on commercial networks, where they languished until Minnesota and then California began to pay for media buys of airtime at full commercial rates fifteen to twenty years later. In the years immediately following, cigarette sales went up in the absence of both TV ads and anti-smoking messages, and this helped convince anti-smoking advocates of the power of anti-smoking media campaigns (Warner 1980).

Vulnerable Populations, I: Market Segmentation versus Community Media

The recourse to the techniques and tools of marketers had consequences not only in terms of constructing citizens and residents of California as consumers and risk groups but also in terms of "audience" segmentation and overall media campaign design. Not only was it one of the first campaigns to use paid advertising at going commercial rates, it also aired some spots in Spanish, Mandarin, Cantonese, Korean, and Vietnamese over non-English stations. Still, the media component of the Tobacco Control Program followed the standard commercial marketing practice of segmenting the state into eight major media markets. This approach tended to ignore regions north of San Francisco, the Central Valley, and the border area with Mexico, where much of the nonmajority population resided. Moreover, the DHS decided to run most of its TV and radio spots and newspaper ads in mainstream media venues rather than in community-based radio and print media.

The state's reluctance to make extensive use of community and non-English media led it to replicate some of the problems associated with then longstanding marketing practices that routinely favored mainstream, middle-class, and upscale audiences (Schudson 1984, 31, 239), for market segmentation was traditionally driven primarily by the bottom lines of mainstream TV, radio, and print media, which did not reach many communities, let alone address specific populations' needs. Moreover, marketers at that time rarely surveyed nonmajority populations or multilingual communities using questionnaires in languages other than English.[34] As a consequence, mass media buys tend to be biased against ethnic media (Health promotion researcher B 2000). While major surveys were conducted for the DHS in English and Spanish, others, such as the DHS in-house California Adult Tobacco Survey, were English-only (California health official A 1999; Ethnic network director A 2000; Ethnic network director B 2000; Ethnic network director D 2000; Pierce 2002a, 2002b).[35] As a result, the tobacco control effort found itself confining its ads to a "statewide" public media space that did not readily address and accommodate audiences out of the mainstream. Thus, although it ran anti-smoking ads in seven languages, it relied heavily on mainstream television and radio stations located along the coast between San Francisco and Los Angeles and in the Sacramento area to get its message across.

Vulnerable Populations, II: Tobacco Education Ethnic Networks

To promote citizen mobilization and build an infrastructure for tobacco control in underrepresented communities in 1990–91 the TCS set up ethnic networks. In order of creation, they are the Asian/Pacific Islander Tobacco Education Network (APITEN), the Hispanic/Latino Tobacco Education Network (H/LaTEN), the American Indian Tobacco Education Network (AITEN), and the African American Tobacco Education Network (AATEN). The networks encouraged local community initiatives through grants, and with the TCS sponsored local and statewide conferences and workshops to provide technical assistance and training in health education, assessing community needs, media advocacy, and countering tobacco industry influence. Ethnic advisory committees to the TCS were also created to improve community input into both the networks and the TCS itself.

By all accounts the H/LaTEN was the largest and most powerful. The size and linguistic unity of the communities it represented lent it great power, but it faced considerable challenges stemming from their geographical and cultural diversity. Much of its efforts focused on processes of acculturation of recent immigrants and refugees to the U.S., with its relatively higher smoking rates, especially for women and youth. Formed in 1996 at the University of San Francisco, it had more than 700 members statewide by 1999 (Baezconde-Garbanati et al. 1999). Similar problems confronted the smaller APITEN (more than 60 organizations and 300 members as of 1999), problems that were compounded by members' great linguistic diversity (Hong and Yu 1999).

In principle, the AATEN was poised to exercise great input into tobacco control, given the African American community's strong ties to mainstream media and the statehouse and its high level of community organization and visibility. Viewed as a community of color at high risk (high smoking-related cancer rates among adult men), it was nonetheless the last network to be set up and received comparably little funding. This was cause for comment among members of other networks. Although since the mid-1980s the African American community had been the only community to be the object of straightforward targeting by cigarette manufacturers and although the TCS's commitment to the black community was actually higher than in most states, some speculated that the African American community's smaller size (7 percent of the population) and rate of growth in California relative to other communities of color gave it lower priority in the eyes of public health officials.[36]

The AITEN represented one of the smallest underserved populations in California but one whose values stemming from its historical experience with tobacco as a means to achieve economic well-being and as a sacred herb with healing properties and an important element in religious activities required novel approaches at variance with those of mainstream tobacco control (Shorty 1999, 77–78).

Contradictory Structures of Collaboration, Oversight, and Evaluation

Like many state-sponsored programs in the U.S., the tobacco control effort in California set up structures of consultation, accountability, and assessment of its performance. Together with the ethnic networks, these structures would constitute the primary settings in which many of the underlying tensions around the governance of populations through tobacco control between state bureaucratic prerogatives, social marketers' assumptions, anti-smoking ideology, public health research and data, and democratic goals of community empowerment were played out. They are worth reviewing briefly here. The enabling AB 75, passed in 1989, mandated an elaborate, contradictory framework that combined outside consultation and oversight with top-down management from Sacramento. This structure functioned in part as a mechanism through which various "stakeholders" and observers of California tobacco control could monitor the implementation of the goals of Proposition 99 as they understood them: California affiliates of voluntary health organizations, tobacco control advocates and researchers, health care professionals, and local "target communities." As we will see, tensions built into this contradictory structure of input, oversight, and management would reach a high pitch in the statewide media campaign.

The DHS appointed Lester Breslow, former director of the DHS and former dean of the University of California, Los Angeles (UCLA) School of Public Health, as chair of the Ad-Hoc Tobacco Prevention Interim Advisory Committee to oversee the establishment of the tobacco control effort and the first year of the program. The committee began to meet immediately after the passage of

AB 75 in November 1989 and was dissolved one year later to make way for the permanent Tobacco Education Oversight Committee (TEOC, later renamed the Tobacco Education and Research Oversight Committee, or TEROC, in 1994), which started its activities in September 1990. AB 75 charged TEOC with holding public meetings at least four times a year and with making a formal report to the legislature and issuing periodically a Master Plan outlining the goals and achievements of the entire education program. Seven members were appointed by the governor, five by the Assembly and Senate, and one by the state superintendent of public instruction. Carolyn B. Martin of the California affiliate of the American Lung Association was appointed chair, and Breslow was made vice chair.[37] It took a while for the committee to reach its full complement, and as of September 1990 it still had unfilled appointments for official members from "target communities" (Bal 1990). AB 75 allotted no independent funding to TEOC itself but rather mandated that the DHS provide the committee with staff and material services. This arrangement, together with the presence of DHS director Kenneth Kizer on TEOC, granted the DHS considerable control over the committee's agenda and schedule.[38]

Two other committees were also constituted. Their functions were advisory: the DHS appointed a large Evaluation Advisory Committee (EAC) of tobacco control experts mostly from outside of California. It contained no members from the DHS. Technically speaking, it was a voluntary scientific "task force" that advised the DHS on health promotion methods and procedures used in the campaign as well as the approaches adopted in the contracted evaluation studies. It met twice a year and reported directly to the DHS. Over time its membership expanded and was composed essentially of tobacco control researchers and advocates from other state health departments, the U.S. federal government, and Canada. As previously noted, some members were significant actors in the NCI's COMMIT study, including Terry Pechacek, who served as the advisory committee's first chair. Meetings were called not by the chair but by either Dileep Bal or Michael Johnson (chief of evaluation at the DHS), who established the agenda (Tobacco control researcher A 1997).[39] Most of the committee's work involved giving advice to the DHS on issues such as weighting data in studies, evaluating data, and establishing future campaign goals in response to presentations made by contractors to the DHS.

The third committee was the Media Advisory Committee (MAC). It reviewed materials, sometimes before production but most often just prior to their release for broadcast or printing. Its membership was smaller than TEOC's; it included roughly the same representation of voluntary health organizations, anti-tobacco advocates, researchers, and target communities; and like the EAC, it contained representatives from health departments outside California.[40]

In addition to these monitoring and advisory mechanisms, AB 75 established a multiple evaluation process with the mission of providing an assessment of the campaign's "results" before the sunset date of 30 June 1991, at which point the legislative authorization would expire unless renewed by the Assembly. As those involved in the evaluation process would later point out, this was a very political decision by the Assembly and rendered properly scientific evaluation of the program's efficiency quite difficult within such a short time period. The requirement of addressing a large number of specific groups of citizens ("target populations") through the campaign rendered the evaluation task even more daunting. The one immediate effect of the eighteen-month deadline was to put pressure on the DHS to get something up and running as quickly as possible. Given the challenge posed by the introduction of large sums of money into already existing organizations with their embedded priorities and established routines, as well as the need to recruit and train new staff and to collect data on various groups and communities in order to get such a campaign off the ground, it should be no surprise that the media component of the campaign was out the door first by many months (California health official A 1996). Contracted out to a commercial advertising agency, the media campaign had the fewest organizational hoops to go through, and it was executed by a private-sector firm accustomed to fast turnarounds of clients' orders. The campaign's quick start was also favored by the fact that it enjoyed little monitoring from the DHS apart from that of Dileep Bal, the head of the DHS Chronic Diseases Control Branch, who struck up a strong relationship with Paul Keye, the ad agency's project director and one of its principals, who shared his goal of denormalizing tobacco use.

Quoting the original enabling legislation (AB 75), the DHS issued its request for proposals (RFP) to prospective agencies on 1 December 1989. The RFP declared "social marketing and affecting change in core values" as key components of the campaign and that "the campaign shall focus on health promotion, disease prevention, and risk reduction, utilizing a 'wellness' perspective that encourages self-esteem and positive decision techniques." It specified the target populations, giving priority to children (50 percent of ads), followed by minorities (25 percent), and the general population (25 percent). The goals were to dissuade initiation by youth; encourage cessation by pregnant women, minorities, and school dropouts; warn of the health hazards of passive smoking; and contribute to a 75 percent reduction in tobacco consumption by 1999 (to an adult prevalence rate of 6.5 percent). The selected agency would be in charge of a comprehensive media campaign (market research, advertising, public relations, and media evaluation), including research on targeted groups, and would use an advisory group comprising DHS staff and minority and other target population representatives (Tobacco Education Media Campaign 1989, 2–8).

These structures of oversight and evaluation were shot through with potential conflict between state agencies, advertising firms, researchers, and communities that were to collaborate together on one of the largest public health campaigns in recent U.S. memory. They heightened some of the underlying tensions that characterized late-twentieth-century tobacco control as it negotiated both local constraints and global goals, engaged in the dynamics of aggregating and particularizing established and emerging populations and subgroups, and struggled to address the needs of citizens in a neoliberal political environment that favored cutting taxes and welfare services, implementing government initiatives that would nourish free-market forces in civil society, and encouraging citizen and community autonomy and self-government. As tensions arose, these structures complicated the Tobacco Control Program's mission to reduce rapidly the smoking prevalence rate among adults and adolescents. This was to be especially the case of the statewide media campaign that, for better or for worse, became the political symbol of the overall anti-tobacco efforts by the health department and its allies.

two

The Dynamics of Collaboration and Community Input in the Media Campaign

The prerequisites and prospects for health cannot be assured by the health sector alone. More importantly, health promotion demands coordinated action by all concerned: by governments, by health and other social and economic sectors, by non-governmental and voluntary organizations, by local authorities, by industry and by the media. People in all walks of life are involved as individuals, families and communities.

—WHO Ottawa Charter for Health Promotion (1986)

But what about the really dangerous stuff—all those carefully polished, fatal illusions the tobacco industry has crafted to mess with our minds so they can mess with our lives? . . . This is going to be a media campaign about a media campaign—as much about hype as hygiene. It's going to talk about a shared community opportunity and a shared community menace.

—Newspaper ad announcing California's anti-smoking campaign (April 1990)

From the outset, the media campaign was an important component of the California Tobacco Control Program, for several reasons: the major role television and radio spots enjoyed in the "Yes on 99" campaign to pass Proposition 99; the tobacco industry's repeated efforts to remove any media-based health promotion from the Tobacco Control Program's budget; and a consensus among health promotion researchers that electronic, print, and outdoor media could—in conjunction with local ordinances, excise tax increases, and community mobilization—help stir up public debate and set new health agendas. By virtue of its exceptional visibility, the media campaign affords us invaluable access to the modes of collaboration between state bureaucracies, private-sector ad agencies, public health researchers, and community advocates and to the ethical and social issues that arise in any health campaign (Pollay 1989; Guttman 1997).

Moreover, the institutional and organizational structures of collaboration, evaluation, and input provide some of the most important settings in which to observe on the ground a series of related tensions in the campaign, some of which I have already explored in the preceding chapter: between the social composition of public health officials and anti-smoking advocates and the nonmajority populations to which they were publicly committed to serve; expert-based and community-based knowledges; the stated democratic goal of making health available to all and the traditional sense of state prerogative in matters of public health; community empowerment strategies and social marketing techniques; and the strategy of fomenting community unity against industry marketing practices and the effort to combat secondhand smoke emitted by smoking citizens. Studying these collaborative practices and the tensions that emerged through them involves looking in detail at the launch of the media campaign in 1990 and the dynamics created by the recourse—unusual at that time—by a state health department to the services of a powerful advertising agency. I then explore conflicts over evaluation and oversight of the media campaign that, moreover, were worsened by the relentless political pressure of industry lobbyists and hostile politicians in Sacramento. Finally, I turn my attention to the input of the tobacco control ethnic networks into the media campaign and their effectiveness in shaping it. In California the making available of populations for public health interventions in tobacco control through media-based health promotion and outreach programs threw into relief the contradictions that underlay the aggregating and particularizing tendencies in liberal government of "all and each," which seeks to promote citizens and their communities as autonomous and self-regulating actor-subjects. "Building responsible communities" (N. Rose 1996, 332–333) to fight the smoking epidemic would be no simple task.

The Media Campaign Takes on the Tobacco Industry

In February 1990 the California Department of Health Services (DHS) awarded the first contract for the statewide media campaign to the Keye/Donna/Pearlstein advertising agency of Los Angeles. The selection process had been simplified by the stipulation in the enabling legislation (Assembly Bill 75) that disqualified agencies with tobacco companies as clients, which reduced the number of prospective contractors. Moreover, Keye/Donna/Pearlstein already had an account with the California Tourism Bureau and had a proven track record in public health promotion media in the U.S. War on Drugs. The agency's anti-heroin TV spot for the Partnership for a Drug-Free America in 1987, titled, "This Is Your Brain," made its reputation by featuring the striking image of an egg simmering in a frying pan.

The expectation had been that the campaign would initially broadcast ads from other media campaigns during the first three months of the official effort. Instead, Keye/Donna/Pearlstein came out on 10 April 1990 with its own, including the "Industry Spokesman" spot that was to make the reputation not only of the media campaign but of the entire DHS tobacco education effort. In press conferences around the state, the DHS released a first wave of general market TV and radio spots that dealt with industry marketing ("Industry Spokesman" and "Subliminal"), industry misinformation dispensed by lobbyists ("A Couple of More Good Years" and "Testifiers"), and industry exploitation of the African American community ("Rappers"). Other spots focused on underage access ("Vending Machine"), teenage peer pressure ("Smart Kids"), cessation ("Quitting Takes Practice," an animated cartoon), and secondhand smoke ("Kitchen," "Bedroom," "Car," and "Livingroom"). The following June non-English spots were aired that targeted Chicano/Latino and Asian American communities.

To drive the originality and aggressive nature of the new campaign home to public health colleagues and supporters of Proposition 99, the next day the DHS published full-page ads in thirty major California dailies (but none in community-based newspapers) explaining the approach, targets, and messages of the fifteen-month media effort then under way. In this fashion did the campaign kickoff literally perform what some public health observers have termed the double-faceted nature of most health promotion campaigns: namely, the mobilization of citizens around a particular problem and the publicizing of the sponsoring state or voluntary agency's own activities (Lupton 1995, 125–126; Robinson 2000). It embodied all of the aggressiveness advocated by Australian evaluator John Pierce and U.S. social marketer William Novelli (Pierce, Macaskill, and Hill 1990; Novelli 1990).

The newspaper ad featured in large print across the top of the page "First the Smoke. Now the Mirrors," below which a large reproduction of a U.S. quarter surrounded by stills of the "Board Room" and "Rappers," secondhand smoke spots and billboards, Steve McQueen with a cigarette, a Mexican postage stamp alluding to the necessity of increasing American tobacco exports abroad as U.S. sales drop, and, finally, in small print, the details of the campaign.

FIRST, THE SMOKE. NOW, THE MIRRORS.

In less than a generation, the bad news about cigarettes has become no news. Most Americans—even the very young—know the unavoidable connection between smoking and cancer, smoking and heart disease, smoking and emphysema and strokes.

Somebody taught you to smoke. Someone powerful and attractive and important to you. Someone you wanted to be.

Today a surprising number of us can tell you that cigarettes are our #1 preventable cause of death and disability.

So, we seem to know about the smoke. But what about the really dangerous stuff—all those carefully polished, fatal illusions the tobacco industry has crafted to mess with our minds so they can mess with our lives?

"Smoking is important. It makes you beautiful and fun and sexy. (Okay, it's dangerous. But lots of exciting things are dangerous.) Smoking makes you powerful. It says you're sensitive and grown up."

That's one hell of a message.

How can you fight it?

Today, the California Department of Health Services begins a fifteen month advertising campaign that goes right at the tobacco companies' predatory marketing—the selective exploitation of minorities, the seduction of the young, the selling of suicide.

Well, won't the tobacco industry fight this campaign?

Sure. The smokescreen has already begun.

One of our new commercials is a tobacco industry spokesman telling you business is terrible. He needs your help.

"This effort pits smokers against non-smokers."

Wrong. This program would have never happened without the active support of California's smokers. Despite their habit, or maybe because of it, they wanted people to know the truth about addiction and discomfort and disease and death. (Ask smokers if they want their children to smoke. Or their grandchildren. Ask them if they'd start smoking if they could have the decision back.)

Do you know anyone in Latin America? Or India or Africa or Korea? Tell them that—as U.S. consumption of cigarettes drops—the tobacco companies must have new customers, new markets, new revenues. Tell them what that means.

If you think that tobacco companies' exploitation of Blacks began with Uptown last year in Philadelphia, this rapper needs to talk to you.

"This is a threat to our First Amendment right to advertise a legal product."

On the contrary, we intend to make you *more* aware of the tobacco industry's advertising. And, if we pinch the right nerve, we expect them to make you more aware of ours.

One of our new commercials shows what happens to innocent bystanders, that smokers aren't the only ones who smoke.

This is going to be a media campaign about a media campaign—as much about hype as hygiene. It's going to talk about a shared community opportunity and a shared community menace.

There's never been anything quite like it. But this is California. We don't need to do it the way it was done before.

In 1988, California voters asked for a 25¢ per pack tax on cigarettes. About 2% of the $1.47 billion in revenues will go to this media campaign. 81% goes to medical care for the indigent and 17% for tobacco education programs.

WARNING: THE TOBACCO INDUSTRY IS NOT YOUR FRIEND

California Department of Health Services

Print advertisement announcing the California anti-smoking media campaign.

Courtesy of California Department of Health Services.

The text reads as follows:

In less than a generation, the bad news about cigarettes has become no news. Most Americans—even the very young—know the unavoidable connection between smoking and cancer, smoking and heart disease, smoking and emphysema and strokes.

[Small image, left margin, of TV set with Steve McQueen smoking. Caption beneath set: "Somebody taught you to smoke. Someone powerful and attractive and important to you. Someone you wanted to be."]

Today a surprising number of us can tell you that cigarettes are our #1 preventable cause of death and disability.

So, we seem to know about the smoke. But what about the really dangerous stuff—all those carefully polished, fatal illusions the tobacco industry has crafted to mess with our minds so they can mess with our lives?

"Smoking is important. It makes you beautiful and fun and sexy. (Okay, it's dangerous. But lots of exciting things are dangerous.) Smoking makes you powerful. It says you're sensitive and grown up."

That's one hell of a message. How can you fight it?

Today, the California Department of Health Services begins a fifteen-month advertising campaign that goes right at the heart of the tobacco companies' predatory marketing—the selective exploitation of minorities, the seduction of the young, the selling of suicide.

[Small image, left margin, of a perforated filmstrip with a close-up of a tobacco executive. Caption: "One of our new commercials is a tobacco industry spokesman telling you business is terrible. He needs your help."]

Well, won't the tobacco industry fight this campaign?

Sure. The smokescreen has already begun. "This effort pits smokers against non-smokers."

Wrong. This program would never have happened without the active support of California's smokers. Despite their habit, or maybe because of it, they wanted people to know the truth about addiction and discomfort and disease and death. (Ask smokers if they want their children to smoke. Or their grandchildren. Ask them if they'd start smoking if could have their decision back.)

[Small image, center, of Mexican postage stamp. Caption: "Do you know anyone in Latin America? Or India or Africa or Korea? Tell them that—as U.S. consumption drops—the tobacco companies must have new customers, new markets, new revenues. Tell them what that means."]

[Small image, right margin, of African American rapper. Caption: "If you think tobacco companies' exploitation of Blacks began with Uptown last year in Philadelphia, this rapper needs to talk to you."]

"This is a threat to our First Amendment right to advertise a legal product."

On the contrary, we intend to make you *more* aware of the tobacco industry's advertising. And, if we pinch the right nerve, we expect them to make you more aware of ours.

[Small image, right margin, of a perforated filmstrip of a seated man smoking and watching TV next to a child. Caption: "One of our new commercials shows what happens to innocent bystanders, that smokers aren't the only ones who smoke."]

This is going to be a media campaign about a media campaign—as much about hype as hygiene. It's going to talk about a shared community opportunity and a shared community menace.

[Large image, center, of U.S. quarter (with profile of George Washington) along the top rim of which is "LIBERTY" in capitals, on the left "IN GOD WE TRUST," and along the bottom rim "1988," the date of the passage of Proposition 99. Caption: "In 1988, California voters asked for a 25¢ per pack tax on cigarettes. About 2 percent of the $1.47 billion in revenues will go to this media campaign. 81 percent goes to medical care for the indigent and 17 percent for tobacco education programs."]

There's never been anything quite like it. But this is California. We don't need to do it the way it was done before.

[Right margin, in capitals: "WARNING: THE TOBACCO INDUSTRY IS NOT YOUR FRIEND."]

California Department of Health Services.

The ad serves notice that the media campaign will not be business as usual. Not so much an information campaign as an attack on the culture and glamour of smoking, tobacco education programs will take on the tobacco industry's promotion of lifestyle illnesses "that mess[es] with our minds so [it] can mess with our lives." In so doing, the campaign would avoid frequently cited pitfalls of public health campaigns that ignore the symbolic value of cigarettes in the wider culture and the cultural and political power of the tobacco industry (Beaglehole and Bonita 1997, 65; Lupton 1995, 149). In the words of social marketing proponents, it will be a combat of "words and images" over the fates of the bodies of California's inhabitants (Ling et al. 1992, 358).

Under the banner of the icon of the founding father of the nation (George Washington) and slogan ("Liberty") and a new revolutionary date on the coin (1988), the ad enjoins nonsmokers and smokers alike to use their freedom of speech and become allies in a community defense against the tobacco companies' "predatory marketing practices, the selective exploitation of minorities, the seduction of the young, the selling of suicide." Here, the ad stole a page from the tobacco industry's own book of strategies: frame anti-smoking in terms of core American values and freedoms while rearticulating them (Burns 1997).[1] The individual right to smoke is no longer an innocent one but is counterbalanced by the right of nonsmokers to breathe clean air.[2] What is more, the ad picks up where Jack Nicholl and the Coalition for a Healthy California campaign left off seventeen months before, in November 1988: it's about Californians defending themselves against predatory outsiders. And in true Californian fashion, it is ahead of its time:

> This is going to be a media campaign about a media campaign—as much about hype as hygiene. It's going to talk about a shared community opportunity and a shared community menace.
>
> There's never been anything quite like it. But this is California. We don't need to do it the way it was done before.
>
> California Department of Health Services.

The "Industry Spokesman" Spot: Cigarette Manufacturers as Abject Citizens

As for just who these shadowy outsiders are, the "Industry Spokesman" English-language TV ad fills in the physical and ethical features. Of all the ads released by the DHS in the 1990s, it enjoyed the highest recall among the viewing public. The ad pictures tobacco industry executives (all white men) plotting the recruitment and death of future customers in a smoke-filled corporate board room.

In a marked non-Californian accent, the chair presents the main item on the agenda to the assembled board as the camera cuts back and forth between the speaker and his smoking colleagues:

> Gentlemen, gentlemen, the tobacco industry has a very serious multi-billion-dollar problem. We need more tobacco smokers. pure and simple. Every day, 2,000 Americans stop smoking and another 1,100 also quit. Actually, technically, they die. That means that this business needs 3,000 fresh new volunteers every day. So, forget all about that cancer, heart disease, emphysema, stroke stuff. Gentlemen, we are not in this business for our health.

A peal of eerie laughter erupts around the table, and the Mephistophelean chair joins in as the camera shifts to slow motion and the image fades out to a black screen against which appears the white title, "Paid for by the California Department of Health Services."

The discursive framing of the industry as beyond the pale of legitimate citizenship and community membership is powerful and complex. Speaking in distinctly white ethnic, non-middle-class inflections, the assembled executives are a strange hybrid of "Mafia" and aristocratic stereotypes who body forth the dread felt in the presence of a death-dealing class of citizens who themselves do not share our terror of disease, dying, and death. For them the defiance of death in the act of killing others (they, too, are smokers) defines their bodies, desires, and lust for power and gain. To our horror, they stand at a complete remove from the democratic, communal pursuit of health and happiness. Here, the tobacco executives are so many corporate murderer-suicides to whom no reasonable appeal may be made.[3] In a sense these executives appear to belong to a bygone era more associated in the U.S. with the shamelessly exploitative practices of industrial society, with its lingering feudalisms, secrecy, and white ethnic identities (from earlier European immigration), that are now clearly out of place in contemporary California and its service-based economy. (The discourse that articulates smoking in terms of a flawed modernity linked to the recent historical past is explored in detail in chapter 4.)

The visual drama of a rogue industry plotting the demise of fellow citizens to meet the bottom line is delivered with all the hallmarks of professional pro-

"Industry Spokesman."
Courtesy of California Department of Health Services.

duction afforded by advertising purchased at going commercial rates: high production values; clever, fluid visuals; smart sound tracks; punchy narratives and dialogue; and polished final product. In the minds of its creators, these qualities set the California media campaign apart from its predecessors, not only in the U.S. but from around the world:

> There had never been anything quite like it. First, the campaign was professionally produced. This is important. When my associates and I first began this effort we looked at anti-smoking campaigns from all over the world. Very quickly we lost count of the powerful concepts and critical messages buried in low quality technique and amateurish production. . . . The second reason the campaign hit hard was that we paid cash for our media placements. . . . Our spots didn't run in the middle of the night. They weren't placed against marginal, irrelevant audiences. We weren't public service petitioners. We were marketers. (Keye 1993, 6–7)

What is more, as the project director would state many years later, the agency was faced with a very media-savvy population in California that could name the director of a particular commercial ad (Keye 1999).

The media effort's signature anti-industry approach had its origins less in the formal guidelines established by AB 75 or the DHS than in the anti-tobacco discourse of the "Yes on 99" campaign examined in chapter 1 and the desire to draw citizens together against the tobacco industry in California. It also stemmed from the internal evolution of the DHS campaign design and development:

> The cigarette companies were never in any of our original thoughts or conversations with one another. You can't find the topic in our first work. We were dutifully engaged in responding to the client's epidemiological grid: chronic smokers, pregnant women, little kids, adolescents, newly arrived Americans and ethnic groups at special risk.
>
> What happened was that—as we dug into each topic—there, right in the middle of everything were the Smokefolk, making their quaint, nonsensical, fraudulent arguments and—by sheer weight of wealth and momentum, power, and privilege—getting away with it. . . . Frankly, the tobacco industry pissed us off. They insulted our intelligence. (Keye 1993, 9)

The wisdom of doing anti-industry spots received confirmation, acknowledged in the newspaper ad, from the experience of the African American coalition in Philadelphia that had just won its fight in January 1990 against R. J. Reynolds' plans to test market among the city's black community Uptown, the first cigarette specifically designed for an ethnic or racial group. One of the coalition leaders explained to the DHS and Keye/Donna/Pearlstein that their principal strategy had been an anti-industry one that emphasized community defense against outside forces and the right to self-determination over health issues, properly speaking (Robinson 2000). Later, public health advocates would admit that anti-industry ads initially didn't work so well for other groups (notably Asian Americans and Hispanics/Latinos) and that such an approach perhaps would not work in other cultural contexts outside of the U.S. (California health official B 1998; Romano 2000).

Since the DHS ads were intentionally "rude" (in the project director's words), it would appear that one of their desired effects was to overcome some of the operative divisions that often arose in health promotion that pitted teachers against students, physicians against patients, parents against children, and public health educators against citizens by aligning all of these groups of Californians against the tobacco industry (Keye 1999). However, this projected unity, which dispensed with distinctions between citizens faced with a community health menace, came into tension with the campaign's very mandate to address the needs of particular "vulnerable populations" (women, minorities, and youth). It was also undercut by the anti–secondhand smoke ads that inadvertently tended to divide California citizens between smokers and nonsmokers (as one might expect) but also between men and women and along social and ethnic fault lines that were exacerbated by the steadily deteriorating social climate in the state in the 1990s.

The effects of the first ads were dramatic and rippled in all directions. A spokesperson for the cigarette industry's Tobacco Institute termed the ads "nasty, bitter, hateful, and in one instance, racist" (Hinsberg 1990a). An internal R. J. Reynolds memo interpreted reactions to the ads among various communities this way:

> It is our understanding that many leaders within the minority community are outraged by the patronizing and denigrating tones of many of the proposed anti-tobacco "health" advertisements.
>
> The black and Hispanic communities are also upset because very few Prop 99 dollars have found their way to the poor in the form of new health services. A substantial amount of Prop 99 revenues are being use [*sic*] to "replace" state funding from other sources.
>
> In addition, black professionals, who are represented by our consultant, have been quietly urging legislators to make changes in Prop 99 funding, including the use of media money to support health care programs. We are factoring all these circumstances into our plans to combat the current media attack.
>
> Further, a backlash may be building within the Asian community. Our consultant with this group suggests that many of this heritage are concerned about the use of Orientals in the advertisements. We are continuing to work with groups within the Asian community. (Chilcote 1990)

Several TV stations refused to air the "Industry Spokesman" spot because it "trashed" the tobacco industry. Even two years later, some stations remained hesitant about broadcasting the spot. Meanwhile, in the advertising industry trade publications *Advertising Age* and *Adweek*, the campaign met with mixed and even hostile reviews ("Anti-Smoke Torch Flickers" 1990; Garfield 1990; Hinsberg 1990a, 1990b). The minority press wasn't particularly pleased, either. Black newspaper publishers voiced anger over the small portion of the media budget allotted to African American newspapers and radio ($1 million out of $29 million), a point repeated by Philip Morris Communications' opinion piece published in the *Sun Reporter* titled, "Is It Health, or Is It Racism?" (Philip Morris 1990). The publishers claimed that it was a campaign that perhaps addressed minorities but not through their own media, whereas fully 37 percent of tobacco advertising targeted blacks through these outlets. Moreover, some community leaders found that the use of rap music in spots was discriminatory and that the campaign should have had more black input earlier on ("Anti-Smoking Campaign" 1990). Finally, in public health circles, advocates were surprised by the ads' aggressiveness. The first spots even startled a former president of Americans for Nonsmokers' Rights (ANR), who thought they were almost too strong even for him at that time (Braxton 1990; California health official A 1996; Glantz 1997).

The Dynamics of Private Sector–State Collaboration

The 1990 anti-industry ads, together with those on the hazards of secondhand smoke, generated a great deal of media attention and public commentary. They were considered by DHS officials, anti-smoking advocates, and ad executives alike to be among the most effective spots in helping set an agenda for the overall campaign and stimulating public discussion. They viewed the ads as contributing to shifting social norms on secondhand smoke, accelerating the adoption of smoke-free local ordinances, getting smokers to quit in record numbers, and shifting public attitudes toward the tobacco industry (California health official A 1996; Pierce 1997). In the eyes of the larger public, these ads *were* the media campaign and the media campaign *was* the tobacco control effort (California health official A 1996; Burns 1997; Glantz 1997; California health official B 1997; Minnesota health official 1997; Tobacco control advocate A 1997; Advertising executive A 1997; Bal et al. 1990; Popham et al. 1993; Popham 1997; TEOC 1991; TEROC 1995).

The new policy of contracting for paid advertising at commercial rates introduced an unknown variable to the public health media campaign that created an element of instability in the management of the anti-smoking effort. One inadvertent but perhaps predictable effect of the national and international attention garnered by the ads was to identify the DHS with the media campaign and the very advertising agency that created it. Nearly from the outset it would appear that Keye/Donna/Pearlstein enjoyed great authority in its dealings with the DHS, an outcome that may have been made all but inevitable by the unequal relationship between a high-powered Los Angeles commercial firm and a state agency with little or no permanent staff professionally trained and experienced in health communication and media until after the first contract had been issued (Stuyck 1990). Moreover, there was a sense of shared strategy between Paul Keye and Dileep Bal concerning tobacco industry marketing practices as well as among DHS staff. Bal gave Keye great freedom in designing and getting out the first wave of ads quickly, in accordance with the deadlines set by the enabling legislation (AB 75), with a minimum of bureaucratic interference.

In a sense, the very conditions that gave Keye/Donna/Pearlstein a relatively free hand to develop a hard-hitting campaign may also have made the Tobacco Control Section (TCS) and the DHS something of a prisoner of their own offspring and of the midwife they had hired. In other words, the DHS acquired an *identity*—the very "Californian" media campaign—and a very powerful collaborator, whose predominance would be inherited by successor ad agencies. As one member of the Evaluation Advisory Committee (EAC) put it, "The company that gets the contract dictates what's going to be done" (Tobacco control researcher A 1997). In the eyes of some observers and participants, the identification of the TCS and the DHS with the agency was only to deepen as their collaboration continued in the following years. In turn, this outcome may have rendered more difficult the other broader collaboration called for by the enabling legislation between state agencies, media professionals, health communication experts, anti-smoking advocates, voluntary health organizations, and community representatives, if only because those parties wishing to have effective input or commentary on the conduct of the media campaign by Keye/Donna/Pearlstein may have found in the DHS at times less a client passing orders to a contractor than a co-contractor, as it were, of the ads themselves.[4]

To be sure, the collaboration between Keye/Donna/Pearlstein and its state client was far from a simple one. It was marked by tensions stemming from private-sector impatience with state bureaucratic requirements and legislative mandates and by public officials' distrust of a commercial firm's handling of public-interest issues. Indeed, in the project director's own mind, his involvement in the campaign stood as a shift from the "ethically trivial" routine goal of efficiently marketing consumer products to the "ethically dense" issue of helping set public agendas in tobacco control (Keye 1999). In this fashion, the media campaign represented not only a relatively new type of collaborative work between public and private sectors in the U.S. (contracted use of paid advertising by a health department) but also the blurring of the usual division of labor and of cultural and political interests: the private sector formally contracted to carry out a public mission. Although in many other countries in Asia and Latin America, media involved in early social marketing programs were state-owned and not so market-oriented, some retrospective evaluations of social marketing programs highlighted similar problems (Altman and Piotrow 1980, 398–399; Manoff 1985, 7).

Moreover, it is worth noting that this blurring of boundaries was perhaps rendered not only necessary by the task at hand (countering the glamour and culture of cigarettes promoted by the tobacco industry's aggressive marketing toward women, youth, recent immigrants, and communities of color) but also possible by the fact that many of the collaborating actors regularly crossed boundaries as part of their work and thus had several "identities." (Keye, for example, was a major donor to the Americans for Nonsmokers' Rights Foundation; many tobacco control researchers were anti-smoking advocates; and many public health officials were members of voluntary health organizations—Dileep Bal, the DHS official who oversaw the Tobacco Control Program, is a former president of the California affiliate of the American Cancer Society.) In other words, within the public health community, people had long been in the habit of wearing many hats. The lines of division that *did* persist were those cited previously in the campaign to pass the Proposition 99 referendum: between organizations committed to local activism such as ANR versus those who preferred to operate at the state and federal levels as tobacco control lobbies (voluntary health organizations), and then between both of these groups together and the medical and hospital associations.

Tensions around Oversight and Evaluation

Whatever ambivalence may have existed among the contracting parties at the beginning of the campaign concerning the nature of the collaboration between state officials and a commercial ad agency, it was further fueled by AB 75 and the first request for proposals (RFP), which assigned responsibility to the contractor for the evaluation of the cumulative media campaign and for managing the minority agencies that were handling 60 percent of the media contracts.

It would appear that Keye/Donna/Pearlstein was not happy with either assignment. Moreover, the DHS didn't relinquish control of the media evaluation process, perhaps in order to correct the utterly contradictory mission of assigning evaluation of the campaign to the agency charged with designing and placing all the ads, which led to recurrent tensions. As for the evaluators themselves, they chafed under what they perceived to be the ad agency's and DHS's interference in their work (Keye 1999; California health official A 1996; Popham 1997).[5]

Moreover, the multiple oversight and advisory mechanisms established by AB 75 complicated the broad collaboration mandated by the same bill. To begin with, as it was, productive collaboration between so many stakeholders and experts of different backgrounds would have been no mean feat, and in the eyes of one outside participant, it might have been a mistake to have interpreted this mandate as inclusively and generously as the DHS had done, for the consultative process slowed things down (Minnesota health official 1997). As noted before, these oversight and consultative mechanisms were themselves a contradictory mix of control by and autonomy from the DHS. The state agency appointed the members, called the meetings, set the agendas of the two advisory bodies—the EAC and the Media Advisory Committee (MAC)—and in the case of the independent Tobacco Education Oversight Committee (TEOC), because it lacked its own funding and staff, the TCS and the Chronic Diseases Control Branch of the DHS exerted control over its agenda, wrote up the minutes, and literally had a member of its upper administration sitting as a voting committee member.

In terms of the advisory committees' meetings on the media component, although there was some discussion of strategies and even storyboards, matters were not helped perhaps by the format that was often adopted, which consisted of the agency's making presentations of largely finished products to the committees for their approval, if not necessarily for their feedback. It would appear that both public health advocates and ad executives felt that discussions were not very satisfactory, for they entailed unacceptable violations of the division of labor between them. On the one hand, the ad agency didn't deem it appropriate for committee members to constitute themselves as "media critics" and begin to make suggestions, say, as to the broad framing of messages, let alone which actor to select for a spot; on the other hand, some members (especially health researchers) felt that the agency's use of data stemming from focus groups it conducted foreclosed true discussion and that, from their point of view, the media professionals had begun to constitute themselves as so many "epidemiologists."

Finally, participation in the deliberations of the three committees by members of the targeted communities got off to a slow start. And communication between these particular members and the agency left both parties dissatisfied. In fact, communication between community advocates and agency executives apparently deteriorated to the point that the TCS had to meet with each of the two groups separately. The MAC had perhaps the rockiest history, and, according to one account, it stopped meeting altogether by 1992–93, when the non-DHS members, frustrated by lack of serious input, seemed unable or unwilling to schedule a meeting. As this was happening, the DHS had initiated the ethnic networks and advisory committees that would reopen the process of input into both the media component and the overall program (see below).

Outside Political Pressures

Vastly complicating matters was a changed political climate in Sacramento. When Republican Pete Wilson assumed the governor's office 1 January 1991, he abandoned the neutral stance toward the anti-tobacco education program adopted by his Republican predecessor George Deukmejian. Throughout Wilson's eight-year tenure as governor, his office and the state legislature made repeated attempts to interrupt or shut down entire components of the program, thus politicizing it along party lines. In 1991–92 they began by cutting the media component's entire budget, thus effectively stopping its activities and diverting funds to indigent health care, which at that time was facing a crisis because of revenue shortfalls due to the California recession ("State Set to Scrap" 1992; Glantz and Balbach 2000, 150–152). Meanwhile, preliminary evaluation data on the overall program released at that time gave the first indications that the combined effect of increased cigarette taxes, the media campaign, and local ordinances had reduced the number of adults who smoked by 17 percent, nearly double the national rate of decline at that time (Balbach and Glantz 1998; Pierce 1997; see also Pierce et al. 1994).

After that tactic failed, two years later Wilson, in a brazen move, replaced the DHS director with the former spokesperson of the anti–Proposition 99 coalition ("No on 99 Committee") sponsored by the tobacco industry, Kim Belshé. At that time, Wilson had begun his first unsuccessful presidential campaign bid, and his close ties to the tobacco industry were confirmed by the fact that he appointed a vice president of Philip Morris and chairman of the 1992 Republican national convention to run his campaign (Pierce 1997; Glantz 1997). In 1995–97, under its new director, the DHS never spent the full amount of restored funding, produced and aired few new ads, and from those few withdrew several anti-industry ads, including "Nicotine Soundbites," based on the April 1994 Waxman hearings in the House of Representatives in which industry chief executive officers (CEOs) stated under oath that nicotine wasn't addictive. Between September 1996 and March 1997 no ads were produced at all, forcing the TCS to air old ads, which goes against the common wisdom of marketing practices—namely, that old ads wear out and are rarely effective. Finally, from about mid-1995 to mid-1997 the EAC held no meetings (Balbach and Glantz 1998; Connolly 1997).

Subsequent studies confirmed what public health advocates feared: as a result of the defunding of the Tobacco Control Program and a cut in cigarette prices introduced by manufacturers in 1993, the decline in adult prevalence rates flattened out and even reversed, while adolescent rates went up. Adult prevalence even increased briefly, from 16.7 percent to 18.6 percent between 1995 and 1996. As one EAC member commented with regard to industry pressure, "They performed a natural experiment for us all: they showed that when you stop spending money, you get what you pay for" (Tobacco control researcher A 1997; Pierce et al. 1994; "Blue Haze" 1997). Meanwhile, the governor's office began exerting direct control through the DHS Office of Public Affairs to which all storyboards had to be submitted for prior approval, and the Tobacco Education and Research Oversight Committee (TEROC) was effectively shut out of the process (by then the MAC was moribund). In fact, faced with TEROC protests and its call for an emergency meeting, Governor Wilson and Republican Assembly Speaker Curt Pringle fired the three physicians on TEROC who were leading anti-smoking advocates—Lester Breslow, Paul Torrens, and Reed Tuckson—and replaced them with members with closer ties to their offices (Morain 1997). However, despite changes in its membership, TEROC maintained pressure on the Wilson administration to allow the media campaign to go forward with strong ads, and the voluntary health organizations and ANR took out a full-page ad in the *New York Times* denouncing Wilson's interference in the Tobacco Control Program. However, the governor's office persisted in its close scrutiny of the TCS even after 1 January 1999, when Gray Davis, a Democrat, took office (Balbach and Glantz 1998).

Eight years of political pressure from the governor's office and the State Assembly forced tobacco control advocates in the TCS, the Chronic Diseases Control Branch of the DHS, and the California Department of Education to work together under very difficult conditions: on the one hand, they had to respond to the inquiries and urgings of TEROC and the EAC as well as voluntary health organizations and ANR, which supported a hard-hitting tobacco education program; on the other, they had to contend with the interference and hostile scrutiny of the governor's office and the state legislature, whose representatives worked as their own superiors in the Preventive Medical Sciences Division and the DHS director's office. By 1995 it would appear that the TCS and the Chronic Diseases Control Branch lost control over the content of the media component. What little internal leverage they were able to maintain in their dealings with the upper administration of the DHS, the legislature, and Pete Wilson was attributable to the active interventions of TEROC, the officially constituted oversight committee.

The Challenge of Evaluation as an Administrative Technology

It was over evaluation and input by targeted groups that the ambitions and weaknesses of the collaborative process and the liabilities of prior practices of tobacco control and social marketing became most apparent. With respect to evaluation, as political pressures mounted after the first wave of spots, the TCS seemed to resist input from outsiders, even those with whom it had established formal contracts. Regarding community input, poor collaboration with specific communities was partially remedied by the establishment of the ethnic networks and advisory committees, which actually fostered communication.

In the public health literature, researchers and practitioners repeatedly stress that "evaluation"—the assessing of the effectiveness of a campaign—is utterly crucial to the success of health promotion efforts and, it should be added, to justifying renewal of public funding to skeptical state authorities (National Cancer Institute 1989, 1995; Wallack and Sciandra 1991; Mattson et al. 1991; Corbett et al. 1991). Evaluation can be understood as one of those administrative technologies that work in the interface between political rationalities (for example, government of populations, reduction of health expenditures, improving citizens' health, community empowerment, and so forth) and concrete operations (such as media health campaigns) and that involve a flexible, pragmatic process of "experiment, invention, and adjustment," which characterizes liberal governance (Miller and Rose 1990, 172). It intersects with the demand for transparent "good government" that is accountable both to its own legislative mandates and internal procedures and to the general public. However, reliable evaluation often requires time and a commitment of significant resources. What's more, it is not well liked by ad agencies, state bureaucracies, politicians, and concerned industries, for, on the one hand, evaluation inserts an unknown quantity or variable (the research process and resulting data) into the work and political processes, and, on the other, it often restricts the freedom of action of concerned parties by binding them to the published results.

Moreover, "academic"-based evaluation is performed by researchers who are presumably independent, if not of their own research institutions and professional culture, at least of the worlds of business, advertising, state agencies, and statehouse politics. Thus, by its very nature, independent evaluation is not made to put at ease parties used to exercising a great deal of control in their dealings with clients, contractors, and colleagues, let alone evaluation performed in a highly politicized arena involving theretofore unheard-of sums of money for anti-tobacco education. The first sign of reluctance to integrate independent evaluation fully into the campaign was the decision to reject a proposal to hand over control of evaluation to an independent academic body (Pierce 1997). Ten years later, some anti-smoking advocates would return to the substance of this proposal and issue a call for outside assessment (Kleinschmidt 2000).

Most media and health communication experts agree that it is difficult to evaluate the media component alone of a comprehensive campaign that also involves passing ordinances, raising taxes, school health education, and community activism. Perhaps in response to political pressures and the reluctance of some actors to be bound by the outcome of a well-executed evaluation, the enabling legislation and resulting RFP ended up complicating evaluators' task on several scores: by putting no one person in charge of the studies from outside the DHS; by breaking up the evaluation into three areas with, eventually, three contractors (an ongoing baseline survey of attitudes and behavior to establish California prevalence rates [University of California, San Diego], a statewide media evaluation [IOX Associates, Los Angeles], and the independent comprehensive "process" evaluation [San Diego State University]); and by requiring results within the sunset time of eighteen months (California Statutes 1989; Tobacco Education Media Campaign 1989). One evaluator even felt that the timeline was deliberately set up at the behest of the tobacco industry in order to frustrate the generation of any demonstrable results, thereby giving the legislature leave to cancel the program (Pierce 1997). Complicating the evaluation picture further were numerous in-house surveys conducted by the DHS.[6]

The imposition of a tight deadline was seen by some researchers as an aid to mounting an effective campaign and by others as an impediment. Some researchers, impatient with what they saw as the academic tendency to rely on control studies and to emphasize the purity of data over the more practical concerns of tobacco control, welcomed the legislative pressure that, in their view, forced the DHS to rely on qualitative research and evaluation methods more commonly used in the private sector, which routinely operated under strict deadlines (Burns 1997). Even so, this was no easy task, for the DHS mandated not only the tracking of many different populations and groups but also that the tracking had to be *statewide*. Both aspects of tracking went against the practical experience of marketers, who usually would use limited test markets for evaluation and target just a few groups in order to develop their spots. Here we witness a clash between marketing's own presuppositions and practices (strategic but narrow sampling) and the hybrid political obligation that combined a traditional commitment of the state to "the public" (all Californians) and a new commitment to California's diverse populations.

In the end, two standard forms of formal evaluation favored by researchers and health communication experts were absent from tobacco education programs. First, AB 75 ruled out an ongoing form of "formative" evaluation of media (as the ads were being broadcast or put up). Moreover, the first RFP explicitly stipulated a "summative" evaluation "following [the campaign's] culmination," thereby eliminating any formal mechanism for corrective feedback and fine-tuning of spots (regarding venue, exposure, etc., through focus groups) as would be done later in the Australian national campaigns in the 1990s (Tobacco Education Media Campaign 1989, 5). And second, the Tobacco Control Program funded virtually no longitudinal studies to investigate smoking practices over time in relation to other factors (the comprehensive campaign, tobacco industry marketing, and so forth). The campaign's experimental partnership with the private sector and its methods of self-correction and timely execution did not yield a campaign accountable to established parameters and responsive to evolving data concerning the needs of targeted populations. The flow of information and paperwork repeatedly broke down in the face of long-standing work styles, bureaucratic imperatives, and political pressures. TEOC attempted to correct the omission of formative evaluation and longitudinal studies but was rebuffed by the DHS upper administration (TEOC 1991, 1993).

Clashes between the Practices of Social Science, State Policy Bureaucracies, and Marketing

Tensions over evaluation remained high with virtually every study undertaken for the Tobacco Control Program, thereby exposing the fault lines among collaborators. For example, the results of the first evaluation by IOX Associates, published in 1993, left no party satisfied, including the evaluators themselves. Recruited by Keye/Donna/Pearlstein advertising, which needed an evaluator to meet the terms of the RFP guidelines, IOX got the ad agency to accept a "mixed" evaluation that would involve some ongoing evaluation as well as an assessment of the outcome. They agreed upon a set of items (the choice of issues, how to reach audiences, etc.) as guideposts in the development of the ads. In the end, of the twenty or so items identified, the agency admitted to have covered less than five. According to the evaluation project director, it was very hard to get the agency to agree to track exposure of particular groups to the ads (through polls), for it would appear that the agency was more committed to a notion of "subliminal" campaign focused on a hard-hitting theme and less on audiences' exposure per se and how they received the ads (Popham 1997).

Feedback in the form of written reports was also hampered by the fact that the TCS was apparently fearful of tobacco companies' gaining access to the campaign's working documents through the Freedom of Information Act (FOIA). As a result, most of the IOX reports were left in draft form (and thus exempted from public access by FOIA), and, according to the project director, none of them reached TEOC. One consequence was that ad campaigns that were apparently testing poorly with targeted groups were allowed to go on (Popham 1997). One such campaign was the group of "Clifford" ads released in 1990–91 targeting youth. They starred a well-known redheaded teenage TV actor playing "Clifford," presumably the embodiment of adolescent cool style. Clifford staged in locker rooms short, ironical vignettes deglamorizing smoking by youth. In the ads he mockingly compared his smoking peers with a genetically deficient and slovenly dressed non-middle-class kid or likened their future fates to those of postmenopausal women and postclimacteric men speaking in rural or uneducated accents. Much later, both DHS officials and ad executives admitted that the ads did not work, without offering specific reasons (California health official A 1996; Advertising executive A 1997).

Between IOX, the TCS, and Keye/Donna/Pearlstein, the various parties seem to agree that there was something of a clash of assumptions, approaches, and work styles. TCS officials felt that ad executives weren't always bound by the health communication research literature (California health official B 1998). As if in corroboration, the ad agency's project director, in an interview, emphasized the value of creative, intuitive work done with the information supplied to the agency by its clients over research and data in order to grab audiences' attention and to avoid the tendency to lecture the public about hygiene (Keye 1999). One student of the advertising world's practices had this to say about the way of making ads in agencies' creative departments:

> Ideas for copy or art do not derive from a philosophy. Creative workers tend to say that a good, intuitive understanding of human nature is what matters most. . . . If ideas for campaigns come from any identifiable place, it is from the culture of advertising. (Schudson 1984, 85)

In that world, the "eclecticism of common sense" even prevails over categories from advertising agencies' own research departments like "market segmentation" and "positioning" (Schudson 1984, 66). For the project director at IOX, this emphasis of "creative" over "data" was a startlingly new approach. He summed up his retrospective account of IOX's collaboration with the ad agency this way:

> One of the things that I didn't understand is that the ad agency personnel are guided by their creative intuitions. What has been reinforced for them over the years is the "good idea." The "idea" withers when inspected in a data-oriented fashion. So the creative folks at Keye/Donna/Pearlstein had no interest at all in whether the damn things worked as judged by data, or judged by evidence, as judged by any kind of criterion. They had their feelings and that was what governed them.
>
> . . . It's a different mind-set. You have to rely on the imagination, the perceived creativity of the effort, irrespective of external, data-based repudiations of that effort. That doesn't make any difference. What's important is "I've grasped the keenness of the idea" and that for me was a very different way of thinking about the world. (Popham 1997)

Commenting on one of the flaws of the study—the lack of connection between the content of commercials and the surveys conducted—TCS officials did not see the strategic necessity of ongoing formative evaluation and interpreted IOX's approach as being more committed to its own methods than to getting the job done. As one member of the TCS's Media Unit remarked:

> This is one of the big differences of this campaign compared to most of the other kinds—and by campaign I mean the whole Tobacco Control Section—this has never been a control-group kind of study or intervention. We did an intervention across the whole state of California and used the United States as the control group. We didn't sit around to prove something worked. We took the best research, the best minds, the best information available and we said, "Based on the best knowledge and science we have right now, what should we do?" And we tried updating that, as opposed to most things that happen in the academic world, where you have a very specific control group, a very specific intervention. So, the IOX study tried to do that after the fact, tried to back that into a control-group thing and it didn't work.[7] (California health official A 1996)

Here, it is striking how, in all three instances, a way of working is discredited by collaborators as impractical and more devoted to maintaining its own routines and professional identity than to carrying out the task at hand.

The worst crisis over evaluation would occur in 1998, as administrative interference from both the DHS upper administration and the TCS in the Tobacco Control Program reached its height. It would appear that politics exacerbated the fault lines between collaborators in California tobacco control. When the director of the TCS-funded California Tobacco Survey refused to soften his study's conclusion that confirmed a worry expressed by many—namely, that the program's effectiveness had dropped, particularly among youth, during 1993–96 after the diversion of funds from the program beginning in 1992 and that teen rates had risen during that same period—the TCS reversed its decision to renew his contract.

Apparently, the TCS felt that the report would make it vulnerable to further political pressures even as it validated advocates' contention that the results stemmed from defunding of the program in the first place. The report's publication was complicated by the fact that the contract contained no funding for printing. Finally, more than a year later after the report was submitted, the DHS agreed with the University of California's suggestion that it be published electronically on the Internet (Pierce 2002a, 2002b; Glantz and Balbach 2000, 362–366; Glantz 2000). At this juncture, it remains impossible to say whether the TCS was simply giving in to pressures from the governor's office or attempting to exert bureaucratic control over the unpredictable outcome of the evaluation process, or both.[8]

Expanding Community Governance: Ethnic Networks and the Tobacco Control Section

If collaboration across disciplinary, organizational, and public/private sector divides presented difficulties, these same difficulties only compounded the task of addressing the public health needs of California's diverse population. In the 1980s and through the 1990s, populations of non-European descent grew. By 2001 the population broke down as follows: 49 percent white, 31 percent Hispanic/Latino, 12 percent Asian American/Pacific Islander, 7 percent African American, and 1 percent Native American. Geographic and linguistic diversity were equally great: as of 2000, 26 percent of the California population had been born outside the U.S., and more than one third of school-age children spoke a language other than English at home (FAIR 2001).

The California Tobacco Control Program's mission proved to be a formidable one, especially with respect to communities of color, whom the public health literature and legislation constantly invoked but who in the 1980s were relatively absent from the tobacco control community in the way of reliable data, researchers, leaders, advocates, and programs—a fact acknowledged to this researcher by advocates and state health officials. They were also absent in the form of detailed epidemiological data (Cooper and Simmons 1985; Braithwaite, Bianchi, and Taylor 1994; Robinson and Headen 1999).[9]

As recounted previously, to promote citizen mobilization and build an infrastructure for tobacco control in underrepresented communities, the TCS set up ethnic networks and advisory committees to the TCS: the Asian/Pacific Islander Tobacco Education Network (APITEN), the Hispanic/Latino Tobacco Education Network (H/LaTEN), the American Indian Tobacco Education Network (AITEN), and the African American Tobacco Education Network (AATEN). During the first meetings between the TCS and the networks, an atmosphere of mutual mistrust apparently prevailed. It could hardly have been otherwise, given the very newness of such a collaboration with so many stakeholders and community representatives. (Beginning in 1994 the Centers for Disease Control and Prevention [CDC] Office of Smoking and Health enlisted California's experiment in community mobilization by funding national organizations, but as of 2000 California still was the only state to have such a formal collaborative structure with communities in tobacco control.) The clash of assumptions—and work styles—between state officials, community advocates, researchers, and ad executives was virtually inevitable. Community representatives seem to have been of two minds regarding these meetings. Although they expressed delight in the TCS's gestures of inclusion, nonetheless they faced a situation that looked all too familiar: in the language of one participant, it seemed that various disenfranchised groups were there more by invitation than by right and were told by the TCS, in so many words, "you work for us." It was to be a long road to the Lake Tahoe meeting in November 1999, when Derek Yach, project manager of the World Health Organization's (WHO's) Tobacco Free Initiative (TFI), would praise the joint efforts of the TCS and the local community project directors as a model of community mobilization and collaboration in global tobacco control (Yach 1999a).

At the beginning, available tobacco control funding for local and statewide projects actually outstripped the existing infrastructure of both state agencies and community health organizations. As one participant commented, "We were resource rich but people poor." Moreover, existing nongovernmental organizations (NGOs) weren't always able to take on yet another pressing issue, and in local health departments "on the ground" training in ethnic-specific issues was slow to start (eliciting a scathing memo from the chair of TEOC in October 1992; C. Martin 1992). The networks had the power to make local grants, and they cosponsored conferences and workshops in assessing community needs, conducting health education, and resisting the influence of cigarette manufacturers.

In the course of time the networks adopted various strategies. Like the other networks, the H/LaTEN did not adopt the modernizing view that community practices were in and of themselves obstacles to getting its members to refrain from smoking in front of others or to quit. Rather, given their internal cultural and regional diversity together with the high rate of two-parent households relative to other communities, the Hispanic/Latino network stressed common Hispanic/Latino "cultural values," such as the importance of "family," to deliver messages across all of its internal communities concerning the influence of smoking parents as role models and the dangers of secondhand smoke. The APITEN deployed similar family-based strategies but, facing communities with a multitude of languages and national backgrounds, it also devoted its energies to creating the Asian-Language Smoker's Help Line in Mandarin, Cantonese, Vietnamese, and Korean and organized several Southeast Asian Summits and a

Gathering of Pacific Islanders, as well as organizing press conferences to press for tobacco control research and education in their communities (Hong and Yu 1999).

The African American community faced it own particular issues, including the deliberate targeting of its members by tobacco companies from the R. J. Reynolds Uptown brand in 1990 to the X cigarette (packaged in black liberation colors) in 1995 and the introduction of mentholated versions of Marlboro by Philip Morris and Camel by R. J. Reynolds in 1996, which angered leaders and mobilized constituents over issues of community control. Moreover, the tobacco control education campaign's youth focus didn't square well with community experience where youth rates were extremely low (in 1997, 7.2 percent were daily smokers, compared with 19.9 percent of white students). With respect to cessation, one challenge was to convince the community that tobacco was not better than other drugs (Ethnic network director C 2000). In the early years of the program, the AATEN funneled funding through well-known organizations such as the National Association for the Advancement of Colored People (NAACP) and local organizations such as the Watts Foundation (which served both African American and Latino populations).

Finally, in response to the special place that tobacco has historically enjoyed in Native American cultures as a sacred herb and means of livelihood, the AITEN pursued a general approach that attempted to separate commercially produced tobacco from ceremonial tobacco raised within the community by encouraging community tobacco production for ceremonial needs (Shorty 1999, 77–78).

Beginning in 1992 the ethnic advisory committees began to hold joint meetings with the TCS as the Tobacco Control Program matured, to argue for funding, improve health education competence, share tactics, and develop future strategies. These meetings were by no means simple affairs, for group dynamics were complicated by the fact that in any one room there would be present vastly different styles of speech, self-presentation, agreement, and dissent. For example, some groups tended to be outspoken with state officials and each other during meetings, whereas others preferred expressing their positions privately only after the meetings had been adjourned. Some participants admitted that they feared that once again they would find themselves "competing for crumbs" from the table of a white-dominated funding source. As it turned out, divisive competition between communities abated over time.[10]

Epidemiological Data as Currency of NGO-State Collaboration

In their dealings with the TCS, community representatives discovered that what served as common currency in policy discussions was epidemiological data.

Since there was precious little of these data concerning their respective groups at either the national or the state level, they found themselves in a quandary. Some prevalence studies did exist, such as the one made available to the "Yes on 99" campaign that put smoking rates nationwide at 32.5 percent for black men and 29.3 percent for white men (Novotny et al. 1988); others were under way (the California Tobacco Survey and the California Adult Tobacco Survey). However, ten years into the program, the lack of available data on the practices and meanings of tobacco use, industry marketing, and effects of tobacco control efforts in their communities often came up as a concern in interviews with directors of the ethnic networks and with TCS officials as a problem still unresolved. Without detailed aggregate data, it was difficult for network directors to make arguments to the DHS for funding new programs, and measuring current programs' effectiveness was quite hard. Moreover, the paucity of relevant information made it easier for public health officials to slip back into their customary role of making health decisions for communities—that is to say, to shift away from promoting "community empowerment" to imposing a "top-down" program (Breslow and Tai-Seale 1996). A glance through TEROC reports and evaluation studies indicates the depth of the problem: for the purposes of generating meaningful data, Native Americans, Pacific Islanders, and Asian Americans are often lumped together under various rubrics or simply absent all together. While major surveys were conducted for the DHS in English and Spanish, others, like the DHS in-house California Adult Tobacco Survey, were English-only (California health official A 1999; Ethnic network director A 2000; Ethnic network director B 2000; Pierce 2002b).[11]

Diversity within Diversity: The Structure of Data

Moreover, for community advocates, the issue of meaningful information on specific practices cut two ways: it was a crucial tool for them not only in their negotiations with state agencies but also in their dealings with their own populations. If they could not help but look upon the DHS as quite remote from their communities because of tobacco control's lack of cultural competence concerning their communities, in turn, because of their own growing collaboration with the DHS, some advocates stated that they worried that they would appear almost "mainstream" or "white" in the eyes of their own constituents. And as they began to tackle the task of tobacco control in California, they felt overwhelmed by the *internal* diversity of their groups—the "diversity within diversity" as one put it—and were confronted immediately with the limits of their own cultural competence.

For example, in California the rubric "Asian/Pacific Islander" actually groups together populations who hail from many different regions of the U.S., South Asia, Southeast Asia, Northeast Asia, Oceania, and Polynesia and who speak more than thirty languages. While "Hispanics/Latinos" presumably share a common language (with the exception of Brazilian Americans), they come from various regions in the U.S. and from more than twenty Latin American and Caribbean countries with distinct linguistic and cultural inflections. In both sets of communities, these differences are enhanced by histories of immigration, exile, and assimilation in the U.S., which in turn comprise different histories of smoking and tobacco use, especially from one generation to another. These conditions render health communication approaches that assume—as one health promotion researcher put it—"one size fits all" quite useless (Robinson 2000).

The uses and limits of data remind us of the paradoxical nature of aggregative and singularizing tendencies of the process of constituting populations as objects of public health intervention through data collecting: only through treating citizens as aggregates can the specific needs of groups and communities be addressed; by the same token, the singularization of groups and their needs through aggregating citizens exposes the somewhat arbitrary nature of these classifications and risks erasing internal differences within each group (Bowker and Star 1999).

Media Campaign as Stumbling Block

From the outset, the media campaign itself posed a major obstacle to successful collaboration between the ethnic networks and their advisory committees and the TCS, for several reasons. First, the fact that the campaign unfolded in a climate of political controversy caused the Chronic Diseases Control Branch and the TCS to circle their wagons and maintain a posture of vigilant wariness. Moreover, even as the media campaign was conceived to target youth, women, and communities of color, by virtue of its recourse to commercial ad agencies and marketers' segmentation practices, it tended to marginalize underserved populations, which rendered input and joint work difficult. And, if there were few extant epidemiological studies of the tobacco practices of California's diverse population, matters were scarcely better in the field of (formative) market research in the private sector (Cerone 1994; Weinstein 1995). The agencies' methods contained little to encourage fine-grained studies of nonmajority audiences, and according to one former tobacco control official, the agencies' bottom line could push them to do things quickly and efficiently:

> I think what [we] and some others really learned is how little . . . depth there was in corporate marketing, because these firms reflect what they are asked to

> do, and their skills and talents and capacities are a reflection of what they are generally asked to work on. I think what they are asked to do, every corporation sort of knows, they come to California and they know it's kind of close to chaos in terms of diversity. So what they want is a simple, straightforward cheap marketing campaign. To do that you have to say, oh there is really only one key message that fits all groups. They don't really want to get into knowing about each of the niches, it's expensive to do all that niche marketing. So we had firms who basically lumped Asians together. (California health official B 1998)[12]

Then there were problems created by the statewide nature of the campaign, which imposed the difficult task of tracking outcomes for many groups in the major electronic markets of large population centers and spread resources quite thin.[13] (However, in the eyes of one evaluator, it wasn't clear that finer-grain studies would have produced more effective media [Pierce 2002a].) In the end, the ad agencies contracted to do the media campaign—even those ethnic agencies with presumably strong ties to communities of color—were quite isolated from Californians. As one official stated, "They don't appreciate the subtleties in the communities."

Across all networks, members struggled throughout the 1990s to convince the TCS of the strategic necessity of going beyond the limits of the media campaign's statewide orientation and of making innovative use of print media and local radio broadcasts that are community-based and relatively cheap.[14] An early crisis erupted around the allotment of media contracts and the placement of ads, especially in community newspapers. Substantial funding was at issue. The crisis threatened to cripple any further collaboration with the TCS. On the one hand, community representatives perceived strongly that the TCS was reluctant to commit any resources to community print media.[15] On the other, TCS members expressed distrust of the newspapers, especially African American and Latino, because of their financial ties with the tobacco industry (through advertising revenues and donations) and their vulnerability to industry pressure. Moreover, state officials weren't convinced by claims made by publishers for their circulation figures and community influence. They thus tended to view publishers' demands for ad placements in their magazines and newspapers as a straightforward expression of commercial interest.

In turn the TCS's reluctance provoked charges of racism. At one point, the black and Latino press actually boycotted the press conferences staged by the TCS to kick off the second wave of the media campaign in spring 1993 (TEOC 1993). For TCS officials it became impossible to separate community issues from the influence of publishers' self-interest, all the more so because community tobacco control advocates sided with the newspaper owners. Moreover, at times public health officials came to view some of the advocates or "gatekeepers" as willfully creating obstacles by using their privileged knowledge of community needs and proper approaches to exclude consideration of all others. The debate over the media campaign seemed to have reached a standstill from 1997 till fall 1999, for the TCS simply produced no new ads for communities of color (California health official A 1999), a decision apparently made by the DHS upper administration.

Community Input into the Media Campaign

The oldest network was the Asian/Pacific Islander network. It was not authorized to contract directly with an ad agency responsible for anti-tobacco counteradvertising, and throughout the 1990s it enjoyed little effective influence over the design and treatment of particular spots. The difference in perception of community needs between the ad agency and the APITEN has been apparently quite pronounced. As one member put it, referring to the APITEN's weak position, "If [the ad] serves us, it serves us. If it doesn't, too bad." One possible result of poor collaboration, cited by network members and health officials alike, was the release of a billboard in English and Cambodian featuring a blindfolded man standing before a firing squad with a cigarette in his mouth. The caption read: "Save the bullets!" The evocation of Khmer Rouge massacres in Cambodia as an analogy of what smokers do to themselves hit too close for many community members, who still had vivid memories of the killing fields. The clever, aggressive appropriation of residents' experience for the purposes of their own good (tobacco cessation and prevention) by a state agency backfired, and the ad was pulled immediately. It crossed the line that researchers had drawn against the inappropriate exercise of public health professionals' prerogatives or expertise in community settings (Breslow and Tai-Seale 1996; Neighbors, Braithwaite, and Thompson 1995).

It would appear that the African American network's input into TCS and the media campaign was fairly limited. The placement of counteradvertising in community newspapers remained a point of contention throughout the 1990s, and the perception persisted that white tobacco control advocates held African American publications to a different standard with regard to tobacco money when compared to the white-owned press (Williams 1987; Ethnic network director C 2000). Moreover, similar issues over the TCS's and ad agencies' appropriation of a community's experience resurfaced in the media campaign. In 1994 the TCS released a billboard that provoked protests by black advocates. Produced by Carol H. Williams Advertising and very widely placed, it featured the half-lit close-up of young African American man facing the viewer with a lit cigarette between his lips; the caption read, "Eric Jones put a contract on his family for $2.65." The ad's spare visuals and terse commentary are accusatory and resemble a police profile. Blunt, it is meant to shock and compel assent. Community representatives complained that the ad "badgered" smokers and indulged in labeling other people (Ethnic network director C 2000). For their part, it was clear to TCS officials that community representatives felt such ads to be intrusive and arrogant. Part of the power of these ads stems from the fact that they address the issue of the hazards posed by secondhand smoke. The anti–secondhand smoke ads targeted smokers directly as threats not only to themselves but also to loved ones, colleagues, and strangers. That was the radical potential but also volatile nature of secondhand smoke as an issue: it immediately entailed the health of others and by extension that of members of the community. Ads warned citizens of the hazards of tobacco smoke in which viewers are called upon to govern themselves and each other in terms of new scientific knowledge concerning the dangers of smoke not only to smokers but also to nonsmokers. (The volatility and power of this issue are explored in depth in chapter 3.)

Another example was an earlier TV spot released in 1990 featured a working-class Latino household as its setting; there, a young man in a sleeveless undershirt smokes continuously in front of the TV set, oblivious to the coughing of the little girl sitting next to him. It was one of the most-recalled spots of all the media campaigns. We are watching a much older brother with his young sister.

In the context of dominant constructions of Latinos as family oriented (unlike African Americans), the smoker's self-absorption recasts him as a nonfamilial intruder who threatens not just the physical but also the moral well-being of the home. He constitutes a warning addressed to the parents who tolerate his habit, and his presumably male working-class indifference to matters of health and care of children places him on the margins of family life and outside the middle class. Focus groups of community members expressed strong reservations about the spot concerning its negative portrayal of Latinos (CDC 1995–98, vol. II, 102). Had the H/LaTEN existed at the time, it might have prevented the ad's release, for, reflecting the power and size of its constituency, the network became the most successful one in achieving direct input into TCS policies and the media campaign. It actually contracted directly with ad agencies of its own choosing and during the 1990s was involved in virtually every aspect of production, from conceptualization and design to final product. According to its director, the advisory committee to the TCS was effective enough to block poorly conceived or targeted ads (Ethnic network director D 2000).

More could be said about the kinds of challenges facing any state-sponsored health promotion initiative meant to mobilize particular communities or pop-

"Livingroom."

Courtesy of California Department of Health Services.

ulations, but it is important to note that regarding the California campaign, the stronger and more autonomous was the ethnic network working with the health department, the fewer the number of problematic anti-smoking ads, but, by the same token, from the point of view of state officials and anti-smoking advocates, the weaker were the resulting ads: they offended no one and thus didn't break "the surface tension" of public and private discourse especially concerning secondhand smoke (California health official A 1999; California health official B 1999).

Throughout the 1990s, relations between communities of color and the TCS did improve gradually, and collaboration and input increased (with the notable exception of 1995–96, during which there was virtually no input at all into TCS because of political pressures coming from the governor's office). Advocates and health officials alike seem to agree that not only did communities' "capacity" for tobacco control (education coalition building, media advocacy, and so on) grow, but so did the TCS's "capacity" for collaboration with them and dealing with community-specific issues. Between the inception of the program and the end of the 1990s, important changes took place, and toward the end of the decade TCS members deemed the subsequent mobilization of communities of color locally and statewide as one of the great successes and defining features of the California Tobacco Control Program, along with the passage of municipal ordinances against smoking in enclosed public spaces, cigarette tax increases, restricting youth access, and the media campaign.

Local Narrowcast Media: Overcoming the Limits of Social Marketing

In the late 1990s the TCS began to change priorities and started to shift funding toward community media and local coalitions. According to some network members, it took the tobacco Master Settlement Agreement (MSA), signed in November 1998, that retired tobacco industry billboard and transit advertising and cultural sponsorships and the industry's subsequent large increase in promotional allowances and sponsorship of public entertainment (*Federal Trade Commission Cigarette Report* 2002) to convince the TCS to begin to invest more in local media campaigns. Moreover, something of a paradigm shift took place in the world of market research that entailed jettisoning any notion of homogeneous consumer markets in favor of ever-finer segmentation of a fractured populace.

This trend most likely began in marketing departments of multinationals such as Nike, catering to youthful consumers, whose resources far outstrip those of ad agencies, which, moreover, remain largely wedded to traditional notions of the market (Health promotion researcher A 1998). Yet the fracturing is not simply the result of a belated discovery of contemporary U.S. and Californian demographics by the private sector but also the result of the proliferation of the media themselves (broadcast/cable/satellite TV; broadcast/cable/satellite/Internet radio; the Internet; CD-ROM and DVD formats, and so forth) and the popularity of paid cable TV channels such as HBO. In the view of one media consultant and former member of the MAC, this double "fragmentation" may indeed be a signal that the "California" media campaign model needs to be revamped and that media campaigns may work in the future only on a local basis using narrowcast media (CD-ROM, DVD, videotape, and so on) in close collaboration with community coalitions (Cunningham et al. 2001; Health promotion researcher B 2000).

This latter position is close to one long favored by some health advocates who are less convinced of the ability of general media campaigns to accelerate shifts in social norms around tobacco, foment public discussion, and so on. However, this new moment in U.S. tobacco control would have local coalitions generating and using media themselves and not simply adopting and tweaking to local conditions ads made elsewhere in Sacramento and Los Angeles for a "statewide" campaign.[16] Meanwhile, as the networks struggled in the late 1990s to "localize" media interventions, some of their members actually argued that "mainstream" counteradvertising needs to be changed as well to include ads that, in so many words, dislocate underserved groups from their status as "special populations" and redefine "mainstream" culture in general market ads by picturing single images that include the full diversity of Californians (Ethnic network director A 2000; Ethnic network director B 2000). The point would appear to be to avoid tokenism and inadvertent stigmatization and to shift the unstated norm away from a "general population" of European descent. The general population would thereafter include everyone.

Denouncing Philip Morris' "Find Your Own Voice" Multicultural Campaign

The question of diversity and tobacco control came to a head in November 1999, shortly after the Lake Tahoe conference, when the ethnic networks took it upon themselves to denounce the new Philip Morris Virginia Slims $40 million "Find Your Own Voice" print media campaign. As noted in my introduction, the ad campaign powerfully articulated smoking in terms of narratives of community diversity, immigration, and assimilation. This media advocacy project turned out to be one of the most successful collaborations between the TCS and the networks. They held a press conference to decry the ads' targeting of their communities. The ads consisted of up to five pages placed in major women's magazines (such as *Vogue*, *Glamour*, and *Ebony*) and other publications (such as *TV Guide*). In the lengthy ads, for example, the first and last two pages each featured a glamorous-looking model of color who "performed" her identity (through her dress and a pleasurable activity like a dance or a song) and "found her own voice." In the shots, no cigarettes could be seen. However, the middle third page featured a striking Caucasian blonde provocatively holding a cigarette who had also "found her own voice."

The ethnic networks claimed not only that the ads had deployed some of the worst stereotypes of people of color but also that in the longer ads the movement from the framing pages to the center and back again inscribed a familiar narrative of "inclusive" difference—in the manner of Benetton clothing advertising—but one that articulated a trajectory of assimilation and acculturation toward a desirable norm and identity centered on whiteness and smoking. According to the director of the APITEN,

> The ads promote smoking as a way of gaining cultural acceptance. The Asian American ad features a young Asian woman in what appears to be Chinese opera makeup and dress. The caption reads, "In silence I see. With wisdom I speak." "This ad is abhorrent on several levels," said Betty Hong, Director of the Asian & Pacific Islander Tobacco Education Network (APITEN). Hong notes that the ad plays into cultural stereotypes by portraying Asian women as mysterious and exotic creatures. The ads exploit minority women who may feel disenfranchised and

> seldom see images of themselves in the media. Additionally, the "find your own voice" theme is clearly aimed at girls and young women seeking to create their identities in a mainstream America that often does not see them as American. (APITEN 1999b)

In these ads there is presumably something for "everyone": for readers of European descent, the allure of "exotic" cultures that doesn't threaten but rather confirms their identity; for readers of color, the promise of both assimilation and protection of their cultural identities.

In a sense Philip Morris came close to resolving the challenge bedeviling general media campaigns meant for a very diverse population: its ads address each and every citizen *through* their community and ethnic differences. Moreover, to square this circle, the ads have taken the well-exploited theme of American individualism (presumably shared by all) and redefined it in relation to community, racial, and national affiliation. In a sense, the press conference can be seen as a response by California's diverse communities to the threat of global tobacco promotion that capitalizes on community differences and the dynamics of population movements and acculturation to promote its product. The ethnic networks sought to establish a strategy that would unite their communities against a common menace, not all of which—as we have seen—shared the strong anti–tobacco industry sentiments of the African American and white populations (California health official B 1999). From another perspective, the press conference can be viewed as the expression of a struggle between the tobacco industry and public health and community advocates over the bodies of entire populations in terms questions of health, good citizenship, and identity.

three
The Campaign against Secondhand Smoke

Family, Ethical Subjects, and the Social Body

. . . The way in which health promotion strategies are deployed in the modern state both reflects and helps to reproduce fundamental features of the distribution of social power; and that current policy debates need to be examined as a matter of cultural politics rather than (as health promotion debates are so often presented) as matters of technical rationality.

—Alan Beattie, "Knowledge and Control in Health Promotion" (1991)

Information campaigns and social marketing efforts represent attempts at planned social change, and it is insufficient to examine an inherently social phenomenon in a social vacuum. It is equally insufficient to evaluate campaigns only in terms of the criteria specified by campaign organizers.

—Charles T. Salmon, *Information Campaigns: Balancing Social Values and Social Change* (1989)

What has heretofore been avoided is the idea of multiple, simultaneous meanings, the very ambiguity that is inherent to visual images, no matter what their venue. What has been lacking is a comprehensive sense of the function that images can have in the contemporary study of the history of health and illness as it is practiced today.

—Sander L. Gilman, *Health and Illness: Images of Identity and Difference* (1995)

Throughout part 1, I have made a fairly complex argument involving a host of factors and issues that shaped the California Tobacco Control Program. The overall context informing much of the campaign has been the major population shifts that took place in the wake of a new wave of economic and cultural globalization in the 1980s, the spread of neoliberal policies in governmental circles in the 1970s and 1980s, the introduction of corporate marketing techniques into the public and nonprofit sectors, and the rise of new definitions of health among citizens and public health practitioners alike. In particular, within this overriding context I have focused on the media campaign and the tensions that shaped the general anti-smoking initiative. Permeating the program's efforts since their inception were tensions between California tobacco control and the underrepresented groups it was publicly committed to serving (communities of color, recent immigrants) and between the program's democratic vocation of ensuring health for all; the community empowerment movement; and the persistent paternalisms of traditional state agencies, epidemiological studies, and social marketing practices.

Together with the structures of collaboration, evaluation, and oversight established by the Tobacco Control Program, the television, radio, and outdoor advertising warning citizens of the dangers of secondhand smoke not only served as various settings for the unfolding drama of those tensions but also heightened them. One of the signature features of the tobacco control effort,[1] the large number of anti–secondhand smoke ads arguably played a role in dramatically lowering Californians' exposure to environmental tobacco smoke (ETS) in the workplace, schools, and households as well as in restaurants and bars, and in accelerating smokers' decisions to quit, which together led to a decline in the incidence of lung cancer rates in California during the 1990s (J. Coleman 2000; Cimons 2000).[2] The new experience of relatively smoke-free environments quickly became one of the tangible aspects of daily life in California in the 1990s that set it off from most other states in the U.S. and other countries. Resident and nonresident travelers remarked upon this frequently. Thus, the practice and result of the anti–secondhand smoke campaign's struggle against transnational cigarette manufacturers in California highlighted local and regional identity as virtually no other aspect of the Tobacco Control Program did.

Highly social in its implications, secondhand smoke entails the fate of all citizens. When the threat that smokers pose to themselves is understood to be posed to the general population—when their lungs threaten to become the lungs of every citizen—then a new health policy becomes an obligation enjoined upon all. This logic of risk was one of the factors that revolutionized and expanded the anti-smoking movement in the U.S. and elsewhere in the 1980s. Thus, the intensely social nature of secondhand smoke inevitably sharpened questions about how public health campaigns translate scientific findings into programs capable of addressing linguistic, ethnic, and regional communities and intensified the conflict between inclusive and stigmatizing tendencies in public health policy.

Projecting Ethical Subjects and Accountable Communities

Delving deeply into the structure, form, and content of the Department of Health Services' (DHS's) anti-smoking ads, this chapter and the next further refine the character of these tensions and show how they inadvertently ended up defining California's populations in terms of their ethical and social features. In particular I focus on the narrative and visual logic of these ads—what cultural historian Sander Gilman has called the multiple, ambiguous meanings inherent in visual imagery (Gilman 1995, 10; Williamson 1978)—and the epistemological and affective relations they establish with possible audiences. The trajectory of this inquiry takes me to the heart of the neoliberal project of community mobilization for projects of self-management and governance, for such projects require the recruitment of citizens in terms of who they presumably are and what they can do. According to U.K. political theorists Peter Miller and Nikolas Rose, "Governing involves not just the ordering of activities and processes. Governing is a 'personal' process: it operates by means of human beings. Political rationalities become effective to the extent that they can enroll individuals, in their professional and personal capacities" (Miller and Rose 1990, 175).

Thus, like the "Industry Spokesman" ad that provocatively filled in the ethical features of tobacco industry policies and the executives who formulated them, the anti-ETS ads paint a portrait of California's populations but in this case in terms of their amenability to education and reform in matters of public health. Here, ethical qualities are paired with intellectual ones: an accountable subject is also a knowing subject, able to comprehend public health information and convert it into action. Through an analysis of these ads, an entire ethical landscape or "imagined community" (B. Anderson 1991; Appadurai 1996) emerges in which different groups are made to occupy different places in terms of their differential ability to act upon themselves and others as worthy citizens in need of assistance, failing pedagogical subjects, or ethically bankrupt community members.

I argue that through medical and cultural markers of smoking, the media campaign tended to construct an internally differentiated community of Californian and U.S. residents (and nonresidents) divided roughly into two groups. In this projection, on the one hand there is the "general population" of knowledgeable, self-governing subjects who are already healthy or, if not, wish to be so; and on the other there are unhealthy subjects deemed threats to their households and communities and largely beyond rational appeal. The first group is largely marked as of European descent or middle-class, or both; the second is deemed mostly men of color and those who submit to their authority. In a further development of the tensions underwriting much of the anti–secondhand smoke campaign, the first group is cast as embodying the possibility of achieving a reformed "modernity" (understood in this discourse as the self-revolutionizing process of capitalist economic and cultural development and democratic governance) that will be tobacco-free; and the second group as embodying, in so many words, an incomplete or belated modernity still stuck in the customs of an older or more traditional culture that tolerated backward behavior, including smoking. And what is the visible measure of modernity and good citizenship in these ads? As we will see, it is healthy household and family life, which serves as the stage for playing out the hazards of smoking.

I conclude the chapter by exploring how the tense play of social differences embedded in the ads may well have increased the difficulties of transferring social marketing–based health promotion strategies targeting ETS from California to other regions in the U.S. and abroad. Indeed, as it turns out, few of the anti-ETS ads were adopted by campaigns elsewhere, especially in the years preceding the Master Settlement Agreement (MSA) in November 1998.[3] Some of the reasons for this unintended result may be found in the legacy of social marketing programs first implemented in industrializing and then later in industrialized countries. By some marketers' and health promotion advocates' own admission, although presumably nonprint mass media used in social marketing programs are able to bypass the obstacle posed by low literacy rates among poor populations, these programs are not always well equipped to address non-middle-class and nonurban groups (Altman and Piotrow 1980; Luthra 1988; Schudson 1984; Manoff 1985; Grace 1991; California health official B 1999). Other reasons may stem from the legacy of U.S. public health policies that have traditionally attributed the origins of health problems to unhealthy "others" on the margins of society (the poor, indigenous populations, descendants of former slaves, recent immigrants) and the unique practices of U.S. epidemiology that have segmented the population into various ethnic and racial groups—something quite rare, when not expressly banned (as in France), in other highly industrialized countries. In this case, an intractable U.S. and Californian "exceptionalism" in matters of public health may have rendered the anti-smoking ads untranslatable to other contexts.

Communities of Vulnerability, Ethical Responses, and Neoliberal Governance

By the mid-1990s, chances were relatively good that a regular afternoon or prime-time TV viewer in California's major media markets had come across a ten-second spot like the following: titled "Smokescreen," it features a vertically split screen with a gray background, divided into two compartments labeled "Smoking Section" and "Non-Smoking Section." As curling smoke drifts from left to

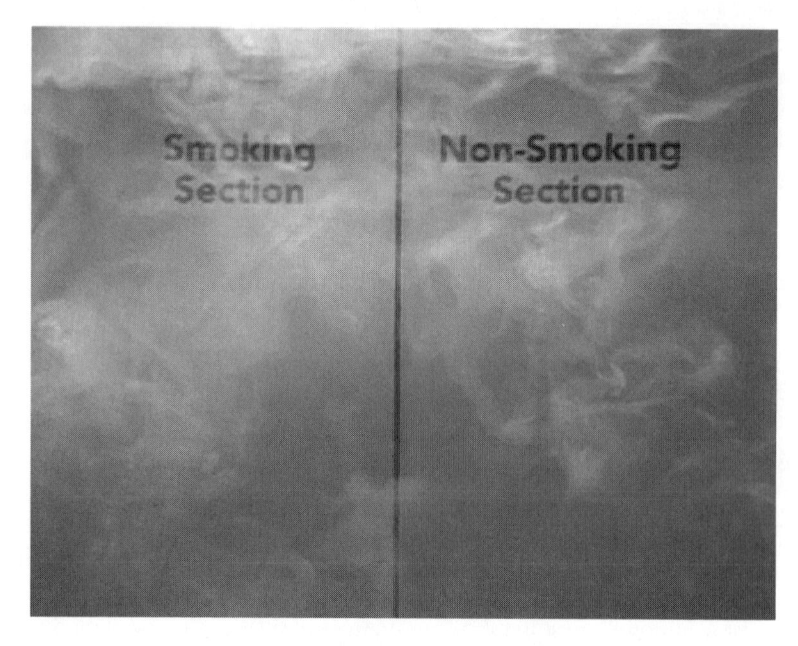

"Smokescreen."
Courtesy of California Department of Health Services.

right across the vertical line (filling both sections with thick smoke) against an ominous-sounding synthesizer, a male voice-over comments, "No matter what they tell you, smoke doesn't stay in the smoking section." A title flashes in black letters across the bottom of the screen: "Paid for by the California Department of Health Services."

I have chosen this ad because it serves as a good introduction to the style and tone of the anti–secondhand smoke media campaign. It also encapsulates neatly the preoccupation of U.S. environmentalists, the public health community, and alternative medicine with built spaces and the bodies—from humans and animals to pathogens—that inhabit or circulate through those spaces, which has driven the effort to combat smoking and the hazards of tobacco smoke.[4]

In this ad, secondhand smoke constitutes the perfect health threat; like all too many pathogens, from viruses and bacteria to environmental toxins, it freely violates the old boundaries and barriers meant to contain it and therefore, by implication, necessitates new policies to combat and eradicate it. Here, before our very eyes, secondhand smoke is transformed into a new entity: it is potentially infinitely extendible, is capable of going anywhere (presumably, even etiolated doses remain noxious at a distance), and therefore requires complete eradication through equally extensive vigilance and scrutiny. Not only does the smoke refuse to stay in its assigned place, it also fills the nonsmoking section with almost as many toxins: by the end of the spot, the smoke is virtually as thick on one side of the split screen as on the other. The ad suggests that there is no getting away from tobacco fumes. As the U.S. Environmental Protective Agency (EPA) states in its 1992 report, "Virtually all Americans are likely to be exposed to some amount of ETS in the home, at work, or in public places" (EPA 1992, 2–2). The ad inscribes at once a new perception and, by implication, a new policy. By transgressing the vertical bar, the drifting smoke establishes a social boundary that in a sense never existed before: an absolute barrier that *should* offer complete protection from secondhand smoke but one never afforded by the simple segregation of smokers from nonsmokers. The old built social space of the smoking policies of the 1970s and 1980s that tolerated secondhand smoke offers no safety and, so we are now to understand, never did.

This health narrative promotes new policies (eradication of smoking) and reconfigures built space (the futility of spatial barriers) in terms of a perception of zero rates of flow of smoke and zero tolerance of risk. And what is true of space is also true of bodies: the human bodies presumed by this ad are what U.S. anthropologist Emily Martin has termed "flexible bodies"—a concept of the physical and psychical self that emerged in U.S. biomedical research, business, and self-help discourses and practices in the 1980s and 1990s. This concept arose during the heyday of a neoliberal thinking in policy circles that preached deregulation of financial, labor, and commercial markets, leaving workers and managers alike to fend for themselves in the free play of market forces in the latest stage of capitalist development, what economist David Harvey has called "flexible accumulation" (Harvey 1990). In the realm of biomedicine, the "flexible body" models the body as a porous membrane whose boundaries are defined less by barriers than by rates of flow and—as is the case of nonsmokers—are open to an endless traffic of toxins and pathogens. Much like the enterprising individual in an uncertain and competitive environment, one must anticipate potential physical threats and engage in ceaseless vigilance of the environment and constant physical and mental retraining (E. Martin 1994).

Interestingly, the built space configured by the ad is defined by the most unbounded, invisible, and omnipresent vector of all: air. And, as ecologists have long insisted, air belongs to and is shared by "everyone." (Thus, tobacco smoke, in a sort of classroom scientific demonstration performed by the ad, is just exhaled air, which, like body tissue or cells, has been "stained" by a dye [tobacco smoke] to reveal its fundamental features and properties.) Here, air *is* democratic space par excellence and occupies the field of vision established by the spot.

In the case of carriers of pathogens such as smokers, like the smoke they exhale and the space they inhabit, the ad frames their bodies as uncontained and in need of discipline. The spot elicits and confirms the realization that under the long-standing smoking policy, if the nonsmoking section is ultimately identical to the smoking section, then, in terms of health, any meaningful difference between the bodies of nonsmokers and smokers vanishes. As a consequence, the ad contributes to the production of a very modern sense of danger as "risk" as an object of worry, calculation, and policy (Douglas 1992).[5] This sense of vulnerability is characterized, according to U.K. medical sociologist David Armstrong, as the "discovery of danger everywhere." Since the late nineteenth century, he writes, "modern dangers, as constructed by public health, are both less tangible and more pervasive" (Armstrong 1993, 407, 400). And in the late-twentieth-century modernity, the invisible and incalculable consequences of the thoroughgoing modernization processes of global capitalist development that have little respect for political borders or geographic divides have gone so far as to give rise to what a leading theorist of globalization, German sociologist Ulrich Beck, has termed "risk societies," whose communities are defined by a deep-seated sense of physical vulnerability (Beck 1992). In terms of health pedagogy, it could be said that through its visual strategies, the TV spot reprises and extends this discourse.

By calling on citizens to make the choice to regulate themselves, each other, and the tobacco industry through personal acts and local and state laws, the spot takes two strands of U.S. public health policy that have long stood in tension with one another—individualizing incitements to personal hygiene and communal policing of collective behavior (advocated particularly by Progressive reformers in the early twentieth century and their descendants today)—and weaves them together in a fabric that reconciles the political projects of traditional libertarianism (no state interference) and communitarianism (the involvement of the state and institutions of civil society in both public and private life). This is the scope and power of the new discourse on secondhand smoke: it finesses many of the old obstacles to U.S. public health initiatives in a political climate that favored cutting welfare entitlements, reduced state activism, and encouraged citizens to see themselves as active partners of government.

Scientific Discourse and Populist Epistemology

In many ways, the "Smokescreen" spot can be understood as a particularly effective democratic pedagogical device about public health matters. By means of its visual narrative, the ad—in creating an apparatus of perception and a field of vision, affect, and knowledge—constructs the place and identity of the knower and the known within that field in terms of ethical and intellectual features. In the age of HIV/AIDS and invisible pollutants, the ad seizes upon the persistent uncertainties and social paranoia of dominant health discourses in order to at once elicit and satisfy in viewers the very public health need to identify and name pathogens (here, tobacco smoke), infectious agents (smokers and the tobacco industry), and risk groups (nonsmokers and the "general population") and to extend to viewers reassuring matters of scientific fact (the discovery of ETS). In this spot, viewers get to watch a remarkably clear demonstration of scientific fact. New, expert knowledge of environmental tobacco smoke is transformed into a matter of everyday sight. *To see* is literally *to know*, and viewers are enlisted as virtual witnesses of scientific fact and are recruited as knowing and knowledgeable subjects in an effort to create a "tobacco-free" state (DHS 1995).[6] Their ability to learn actually validates what they already know and enfolds them into a world of democratic expertise and neoliberal governance in which citizens learn to take care of themselves and one another.

Thus, in this pedagogical narrative, citizens can enjoy the knowledge of public health experts and apply it in their daily lives so long as they have eyes to see. Each and all participate in a public consensus that mirrors the new consensus of scientific researchers. It fulfills perfectly the mission assigned by Richard Manoff to social marketing—"the translation of scientific findings . . . into education and action programs" (Manoff 1985, 36). At the same time, the ad supplies something of the social identity of its receptive audience, for the "common sense" of perception, confirmed by the all-knowing ironical tone of the voice-over ("No matter what they tell you, smoke doesn't stay in the smoking section"), belongs to "you," which marks us viewers as public health's "general population" of nonsmokers or smokers who desire to quit. They are distinguished from "them," those special interests singled out by the ad at that time: recalcitrant restaurant owners and airline companies, the tobacco industry, defensive or thoughtless smokers, and adherents to the old smoking policies.[7] In this fashion does the ad build on the perceived weaknesses of the first California ordinances and the strengths of the recent ones totally banning smoking from the workplace and non-bar areas of restaurants. In so doing, for some viewers the ad introduces a perception of ETS that was perhaps quite new, while for many others the spot articulates and lends the authority of the DHS to a perception that was already emerging in California. In a sense, an ecological sensibility and its accompanying politics have come indoors and now offer a revamped vision of modern democratic life with its configurations of bodies, spaces, and social relations.

The Spread of Anti–Secondhand Smoke Policies in California and the U.S.

Created by the Keye/Donna/Pearlstein advertising agency for the California DHS, this ad is but one among many that filled the California airwaves and print media throughout the 1990s with slogans in English, Spanish, Mandarin, Cantonese, Vietnamese, Korean, Tagalog, and Cambodian. The ads proclaimed that cigarettes are "one weapon that kills from both ends," that "secondhand smoke kills 53,000 people each year," that "*si tu fumas, ella fuma*" ("If you smoke, she smokes"), that "smokers aren't the only ones who smoke," that "Warning: when you smoke, your family smokes," and that "*en esta casa està un asesino*" ("There's a killer in this house"). Other slogans, not focused on ETS, aggressively emphasized the denials and machinations of the tobacco industry and its targeting of youth ("Do they think we're stupid?") and smoking as a legal, addictive drug ("Smokers are addicts, tobacco companies are pushers, and smoking stinks"). Some of these were picked up and amplified by local media as when, for example, the *Los Angeles Times* ran an editorial at the height of the U.S. War on Drugs that did not hesitate to call the consumption of alcohol, cigarettes, and drugs by pregnant women as "doing drugs in the womb" ("Doing Drugs" 1993).

The ads built upon and fed the momentum created by the new regulation of smoking in public areas and state institutions, which expanded at a rapid pace at that time. Beginning in the late 1980s, California restricted smoking in most indoor public areas and banned smoking in schools, universities, courthouses, and state-licensed child day care centers. Restaurants were covered in 1991 and all bar areas and bars (except owner-run) in 1998. Some cities, such as Davis and Palo Alto (both with large university communities), even banned smoking within twenty feet of public buildings, effectively eliminating outdoor smoking downtown. Across the U.S. ETS became a major political issue both locally and nationally. New York City was not far behind California in adopting new measures, and the major fast-food chains soon followed suit. In prisons many wardens banned smoking by inmates, and not only in California; the U.S. Supreme Court recognized the presence of secondhand smoke in prison cells as a form of "cruel and unusual punishment." And the nonsmokers' rights nongovernmental organization (NGO) Action on Smoking and Health (ASH) began efforts to expand these restrictions to include public beaches and parks; by the mid-1990s a movement was well under way nationwide to do so and had met with some success.[8]

These smoking restrictions were also spurred by a wave of new scientific studies of ETS that extended the previous focus on lung cancer to coronary heart disease in older citizens and ailments affecting children especially, such as asthma, bronchitis, allergies, and middle-ear infections.[9] ETS accounted for ten times as many cases of heart disease as lung cancer, substantially boosting the number of U.S. deaths attributable to secondhand smoke to 53,000 (Wells 1988; Glantz and Parmley 1991). These findings sparked a great deal of commentary in the U.S. press and in public health circles ("Passive Smoke" 1991; "Innocent Victims" 1991). During the 1990s, the press gave a lot of play to other studies that followed on the dangers posed to fetuses by ETS and on ETS as a cause of sudden infant death syndrome (SIDS) (Stolberg 1994a; Maugh 1995; "Smoking and Breast" 1996).[10]

Health, Common Sense, and the Rise of Smoking Populations

With the publication of successive U.S. surgeon generals' reports starting in 1964, over time a portrait of smokers and nonsmokers emerged in public health discourse. When the focus was still on men alone, early studies noted a correlation between smoking and the more manual trades and occupations and, conversely, an association between nonsmoking and white-collar jobs but little in the way of a correlation based on race or education. Later studies in the 1970s and 1980s began to draw strong connections based on education and race (with black men smoking more than whites) and reported higher rates among women in white-collar occupations (U.S. Dept. of Health, Education, and Welfare 1964, 363–364; U.S. Dept. of Health, Education, and Welfare 1979, 18–16; U.S. Dept. of Health and Human Services 1989, i–vii).

This picture suggests that that contemporary health discourses have associated particular social identities with health-related behavior, at least as public health studies have analyzed it. In the U.S. the impression began to emerge that the knowledge producers tend also to be the "healthy" ones. Speaking philosophically, it was as if to *know* (about health) was to *be* (healthy, middle-class, etc.) and conversely, by implication, to be (healthy, middle-class) was to already know (about health, etc.). As one health promotion researcher has argued, here knowledge of scientific matters of "fact" would somehow translate into "healthy" behavior, and both would seem to be the special purview of a particular class of citizens (Wallack 1990, 45; see also Goldstein 1992, 127). This would seem to imply that the "Smokescreen" spot's commonsense pedagogy might well be that of a narrow population segment and not the audience conventionally presumed by proponents of visual pedagogy who they claim are best reached by its nonverbal methods, namely the poor and illiterate (Goldfarb 2002, 1–8). Rather, the ad addresses preferentially an audience almost identical to that of the traditional public health community. Roughly speaking, this would be the "upper middle class" that, by virtue of its education, cultural capital, and capacity to influence and make decisions, enjoys the power, in U.S.-based sociologist Michèle Lamont's words, to "frame other people's lives in countless ways as they conceive, advise, hire, promote, judge, select, and allocate." This class of people engages heavily in the social boundary work that creates symbolic groups often through the exclusion of others (Lamont 1992, 11, 13).

Here, in present public health discourse, statistical truth, social identity, and, by implication, public policy court the danger of being forever bound tacitly together. In such a process, U.S. health promotion campaigns can become sites where expert and popular notions of "health," the social body, public and private space, racial and ethnic difference, and middle-classness mutually construct and authorize each other in powerful ways.[11] Thus, by the same token, an inability to "see" and act on the dangers of secondhand smoke would threaten to place viewers outside the community of knowledgeable citizens as unsuccessful pedagogical subjects. The issue of how to appeal to citizens who apparently do not match neoliberalism's image of rationally maximizing subjects—in U.S. parlance, the "hard to reach," who are presumably those addressed by TV and radio spots—emerges forcefully here. As we will see, the full range of ideal and failed subjects of this social pedagogy will inform all other ads in the anti–secondhand smoke campaign.

Some public health officials were quite aware of these problems concerning the tension between a populist rhetoric of common sense and the construction of the general population of viewers as (white) middle class. One state health official from outside California noted that they crop up even in anti-industry ads with their declared goal the uniting of all Californians against the menace presented by predatory cigarette makers. For example, one of the most famous anti-industry spots was "Nicotine Soundbites," released in 1995. This ad featured a famous news clip of tobacco industry CEOs denying under oath before Congress during the Waxman hearings in April 1994 that nicotine is an addictive drug and ended with another ironic voice-over, "Do they think we're stupid?" The all-inclusive "we" (or "you" in the "Smokescreen" ad) may not embrace everyone—or, at least, may embrace more one group than another. According to this official, spots like this—even though, or especially because, they base their appeal on common sense and viewers' native intelligence—risk putting those viewers who are not already confirmed anti-smokers in the uncomfortable position of being made to feel like foolish pawns of the tobacco industry by the California DHS at the very moment the DHS claims to speak on their behalf. In this case, the danger of provoking a backlash, especially among non-middle-class viewers, is quite real; they could turn on the health department and declare that the tobacco industry executives had a right to earn a living like anyone else and could question the DHS's entitlement to pass judgment on entire industries and groups of people alike. Things are even more complex in the anti-ETS ads, in which the hazards of secondhand smoke are played out in a social field marked not only by class differences but also by those of ethnicity, gender, race, and national origin.

Media Outreach: Using Family Narratives

Perhaps the most powerful figure for registering and dramatizing collective danger posed by secondhand smoke is the family. "Family" has long been the privileged site for playing out the cultural antinomies between self/other, public/

private, inside/outside, health/disease, normal/pathological, and so on, which seem to organize collective and individual existence in liberal, industrial, and postindustrial societies (Coontz 1988, 1992; Reid 1993).[12] Family has served as not only a target but also an instrument of policy in the modern configuration of the "social," that form of governance dominated by the welfare state and its experts. It has also been favored by policymakers wishing to reform the welfare state, for anchoring new initiatives in family household life has the advantage of simultaneously individualizing and socializing possible solutions of social problems by engaging citizens in their personal capacities and making them accountable to a group or community (Donzelot 1979; Donzelot and Roman 1992). Moreover, family has long been a favored narrative in public health campaigns, including earlier anti-tobacco public service announcements in the 1960s and 1970s, and will prove to be of great strategic value to the California campaign. Already many early studies of secondhand smoke had focused on family households and the risks to wives and children—the enabling legislation's "vulnerable populations"—created by the smoke of husbands and household members (Hirayama 1981). Finally, the focus on family intersects powerfully with advertising's own practices. Family melodramas are one of the advertising hooks long favored in the U.S. by Madison Avenue.[13]

That said, in the eyes of some tobacco control advocates, family narratives as vehicles of health promotion don't work equally well in all communities, and one of the dangers of domestic dramas is providing comforting stories of harmonious community and home life at the expense of aggressive denormalization of smoking behavior. In this view, the "surface tension" of the smoking issue would remain unbroken (California health official B 1999). This would appear to be the case with some DHS spots. Take, for example, the ad titled "Mary." It features a very young Asian American girl in a party dress peering quietly out a window at other children (off screen) playing outside during a birthday party. Against a background of sad piano music, a concerned male voice-over informs viewers that she can't go out because of chronic bronchitis due to secondhand smoke. In a similar vein is a spot titled "Baby in Playpen" that tracks tobacco smoke from one end of the home to the other, where it envelopes a crib, and the baby, eyes welling with tears, stands up and with a look of distress gazes outside the crib as the voice-over intones the dangers of secondhand smoke. Yet another ad titled "Poisons" humorously depicts the reproachful looks of members of a clean-cut Latino family stopping the husband and father from lighting a cigarette proffered by an unknown (off-screen) male acquaintance. The ad ends with smiles all around.[14] Or, finally, take the very melodramatic Spanish-language ad titled "Hidden Killer," which takes viewers through a half-empty house in search of a serial killer (secondhand smoke) and ends very simply with an anonymous, ungendered hand stubbing out a cigarette in an ashtray. In these spots and others like them, the source of secondhand smoke is rarely identified, and if it is named, as in the "Poisons" ad, the health threat is forestalled and family harmony is restored.

The Smoking Fetus: Families in Danger

Within the family context, other ads break the "surface tension" between smokers and nonsmokers completely with the goal of stirring up debate and galvanizing citizens. They are among the most recalled spots of the media campaign and would seem to confirm health officials' claims for the effectiveness of aggressive ads. For example, there is a thirty-second TV spot that directly addresses health issues in terms of California's diverse citizenry and the global migration of populations to the U.S. from sending countries with different histories of smoking. Produced by Imada Wong Advertising, it brilliantly deploys the new public health logic within the family household. Shot in a cool, tense style, it features an Asian American husband in street clothing who intently smokes while reading the paper at the kitchen table as his wife, in a bathrobe, begins to set the table. Through what could be called a series of shot/reverse shots (although their eyes never meet), which move progressively closer, viewers witness the drama of the husband reading and smoking while his wife coughs more and more violently, smoke pouring out of her nose and mouth as a threatening soundtrack builds (over the sound of her coughs and the faint clatter of dishes).

As the camera cuts back and forth between the husband and wife, her exhaled smoke matches his; they exchange not words but streams of tobacco smoke. At the crescendo, the camera tilts down her bent body, stopping at her visibly pregnant stomach, which she clutches with one hand. The final image is a title on a black background that reads (in English, Chinese, Vietnamese, or Korean): "Smokers aren't the only ones who smoke." The ad transforms the old domestic narrative of the working husband's absorption in reading the paper into the criminal obliviousness of a smoker poisoning his wife and future child. What was perhaps an all-too-familiar and tolerated though unsatisfactory state of affairs is made strange and thus sinister: the body of the nonsmoker and her fetus emit as much smoke as that of the smoker himself.

This ad is one of a group of four anti–secondhand smoke TV spots released together in 1990. They are among the most aggressive ads targeting ETS. The other three feature parents or children in various settings in which one family member is forced to breath another's cigarette smoke. One ad depicts a middle-class white couple sitting up in bed and the discomfort of the coughing wife, as the husband, disheveled and unshaven, smokes. Another features a middle-class black couple in a car, the husband puffing away as the wife struggles to escape

"Kitchen."
Courtesy of California Department of Health Services.

the fumes filling the vehicle. The third ad we have already met: it has a Latino household as its setting; there, a young man in a sleeveless undershirt smokes continuously in front of the TV, set oblivious to his little sister coughing next to him ("Livingroom"). This last spot and the "Kitchen" ad were among the most successful anti–secondhand smoke spots, enjoying among the highest recall of all of the TV spots in the DHS campaign's early years; they also garnered the Community Arts Magazine Award (CDC 1995–98, vol. II, 102).

With respect to the "Kitchen" ad, it is important to understand that the surreal visuals (the stark kitchen decor; the fact that the mother and the fetus "smoke") are not simply a heavy-handed rhetorical turn. Rather, in my view the style and tone of this ad and others like it mark, once again, a new public health vision in which porous bodies and space are in intimate, vulnerable relations. The husband's polluting body merges dangerously with that of his pregnant wife and fetus; his smoke is theirs. In this narrative, the ad positions the (male) smoker as unwitting murderer-suicide who does away with not only himself (the concern of the old anti-smoking policy) but also members of his family. In this kind of familial melodrama, characteristic of many DHS anti–secondhand smoke ads, viewers are returned to the most normative of family households, favored by Madison Avenue: the nuclear family of male breadwinner, housewife, and child, which characterizes a shrinking proportion of U.S. households today (Coontz

1992). And, as in U.S. advertising generally, in these secondhand smoke dramatizations, women are always wives, mothers, or mothers-to-be (and men are husbands, fathers, or fathers-to-be), and they, along with children and fetuses, are the risk groups of choice. They seem to constitute ideal "passive smokers" (nonsmokers who inhale tobacco smoke against their will).

The family hooks did not operate in an ideological vacuum. They intersected with and amplified discourses in the wider culture. During the 1980s and 1990s, "families in danger" was a cry in the U.S. across much of the political spectrum from neoconservatives and neoliberals to Democratic liberals, communitarians, and even some progressives. They deplored the decline of the traditional nuclear family household and the "lack" of "family values," especially among single mothers, the working poor, and people of color (with perhaps the exception of Chicanos and Latinos; see "Scapegoating" 1989; "World without Fathers" 1993; Stacey 1994; Reid 1995). This ad and those I examine below seem to draw on the same rhetoric for their reach and effect. Not only is the traditional nuclear family household no longer dominant (if it ever was [Coontz 1992]); to compound the horror, where and when such a family form actually does exist, it is being done in by its smoking members.

Powerfully, the "Kitchen" spot articulates the contemporary, expanded sense of the body's vulnerability to tobacco smoke in terms of individual behavior within a family context. In so doing, it extends the site of health beyond the indoor public space of the "Smokescreen" ad to the domestic sphere, which escapes the regulatory power of local ordinances and state legislation. The spot both replicates (social) marketing's tendency to define social problems in terms of individual behavior and finesses this narrow perspective by linking personal behavior to family and, by implication, community and even citizen responsibilities.

Gendering Secondhand Smoke: Male Threat, Female Rescue?

Through its short, dramatic vignette, the "Kitchen" spot constructs a triple portrait: that of the smoker, his victims, and possible rescuers in terms of their ethical, psychological, and social qualities. What is striking is that the conventional public health narrative that has traditionally targeted working-class men as risks to the well-being of their family households (through their purported drinking, venereal disease, and violent behavior) is now being extended to the middle-class household, the conventional setting of healthy bodily practices. This is now threatened by a new enemy within: the smoking father and husband. Indeed, in the DHS media campaign against secondhand smoke, fathers and husbands are the primary vectors of tobacco smoke and women, children, and fetuses their victims. This designation of secondhand smoke as predominantly "male" in origin plays off the logic of conventional domesticity, according to which men are frequently represented as "intruders" to the very extent they bring into the household "outside" influences from the world of work and the public sphere.

What is even more striking in this particular ad is that the immediate threat to the fetus is not the pregnant woman, whose lifestyle, habits, and diet had been targeted in the U.S. during the 1980s as inimical to the fetus's well-being (Terry 1989), but rather the husband. In spite of herself, the mother-to-be is "doing drugs in the womb," thanks to her spouse.[15] The supersession of the cultural drama that opposes the woman and her fetus by one of male aggression in the contemporary U.S. context could not but resonate not only with debates over domestic violence against women but also with the obsession with child abuse in the U.S. in which child abuse had come to mean physical or psychological sexual abuse perpetrated in the home mostly by men (Hacking 1991). In child day care centers, women were more often the accused (see Nathan and Snedeker 1995). The coding of secondhand smoke in the home as child abuse became common currency in public discourse and had been repeated by health officials, including then U.S. Surgeon General Joycelyn Elders (quoted in American Academy of Otolaryngology, n.d.).

In 1980s and 1990s U.S. discourse, the smoking fetus signified the greatest possible danger posed by secondhand smoke to the social body and invited prompt intervention. But by whom? At first glance, most likely not by Asian American husbands or fathers-to-be, if the spot is any indication. For, although the spot's dramatics clearly play off of male smokers' pangs of conscience, guilt is not evidenced within the frame of the ad. As smokers are wont to do in anti-smoking narratives, the man is utterly absorbed in his activity and in his smoking pleasure. Not once does he look up from his paper; nor is there the slightest glimmer of awareness of his wife's plight. He seems devoid of any sentiment or thought that might announce the possibility of being converted to the new health consciousness. This is not going to be a narrative of the redemption of the unwitting male smoker/aggressor but rather one of familial rescue by other parties.

But, again, by whom? Other viewers? But which ones? Now, the gender coding of secondhand smoke as male threat seems to correspond roughly to smoking practices among Asian and Pacific Islander immigrants recently arrived in the U.S. Secondhand smoke as male threat also suggests that the implied viewership is structured as predominantly female (especially Asian American and Pacific Islander women) and that it is the pregnant spouses and female relatives and friends who are being goaded to act in defense of the fetus and, by implication, of the family line. Moreover, Pacific Islander or Asian American viewers could quite possibly read the drama as one of the flagrant violation of Confucian, Christian, or Islamic ethics, which by century's end became part and parcel of new nationalist discourses, particularly among Asian diasporic communities in the Pacific Rim (Ong 1999, 214–239).

Yet, as I have said, the field of vision in these ads is not only a familial and community space but also the space of a much larger imagined, internally differentiated state (California) or national (U.S.) community. This is made all but inevitable by the fact that the spot was broadcast over mainstream TV and to large audiences in English as well as in Chinese, Vietnamese, and Korean. Thus, one has the inescapable impression that there is a second audience: the "general population," those (nonsmoking but perhaps also guilty smoking) "mainstream" women and men whose duly delegated health officials are authorized to act in their name and save them and other citizens at risk. What the ad produces, then, is a second, parallel, racialized drama of health and family, in which the gendered drama of male egoism and men's insensitivity to family members and their needs inadvertently slides into a Western narrative that depicts Asian men as cold and indifferent to life and denounces their patriarchal domination of wives and children. Using particular members of a particular ethnic group to carry a universal message health warning to all citizens or targeting a specific community or group via a mass audience medium like television carries considerable risks.

Structured in such a manner, this and other dramas would be a replay of the old colonial narrative, which, according to cultural critic Gayatri Spivak, can be summed up as the story of "white men [and women] saving brown women from brown men" (Spivak 1985, 121; see also Spivak 1998, 332, 341–342). However, in this second narrative the oppressed women are no longer irreproachable "victims," for, in terms of current stereotypes of Asians circulating in U.S. popular and learned discourses (Hamamoto 1994), the wife in this ad is *too* submissive, *too* dutiful toward the husband, for the well-being of her fetus. Predictably, she puts her husband's pleasure and comfort before her health and that of her fetus. Here, the gendered discourse that constructs women as potential threats to their fetuses returns in the form of a story of race and ethnicity, on the one hand, and non-Western, nonmodern womanhood, on the other.

Thus, by virtue of its structure and its broadcast to multiple audiences, the spot seems to cut two ways at once: it operates as a goad to Pacific Islanders and Asian Americans, especially women, to protect the family *and* as a drama for consumption by the general population that replicates racializing narratives about a nonmajority group. While these narratives seem to function for distinct audiences, it could be argued that they come together to form yet a *third* narrative, one of assimilation, which explicitly addresses political and health issues in terms of acculturation.

Asian Exceptionalism: Immigration, Citizenship, and Imperfect Modernity

While the household's nuclear structure marks the couple as well on the road of assimilation toward U.S. middle-class private life (no extended, multigenerational family household here), the husband's callous smoking and the wife's passivity signal to viewers that the couple has a ways to go before enjoying full membership in the community of knowledge and health of the general population of U.S. citizens. The community that awaits them will be one of a revised modernity. For the cigarette, an icon of modern urban living throughout the industrial and industrializing world for much of the twentieth century, has been demoted to being a symbol of imperfect citizenship and failed modernity (Brandt 1990; R. Klein 1993). Or, perhaps better still, the couple embodies an older, discredited modernity that tolerated smoking because it was unaware of the dangers smoking posed if not to smokers at least to their friends, family, and community. Thus, in the manner of public health campaigns described by public health researcher Victoria Grace, the spot assigns to the otherwise "model minority" (enterprising, hard-working, socially conservative) that Asian Americans were deemed to be in dominant U.S. discourses of the late twentieth century a "lack" that needs correcting (Takaki 1993, 416–418; Grace 1991). Paradoxically, it is a lack that stems from a crisis of consumption from within modernity itself (the pleasures and risks of smoking mass-produced cigarettes) and is tied to a still older acceptance of unremitting male dominance that is almost "premodern."

The ad initiates a complicated negotiation that at once dissociates smoking from an alluring U.S. modernity (to which smoking has been historically tied for more than a century) and links it to an older, non-U.S. community identity in order to lay out a proper path to future assimilation and good health. In this sense does the ad, like the DHS campaign overall, propel viewers toward a reworked, alternative U.S. modernity, which is, again, that of democratic but fragile family, community, and social relations conducted in a smoke-free California. Here, as with the fiercely anti-industry ad ("Industry Spokesman"), we meet up with a narrative that tells of a future that eschews an older modernity's lingering undemocratic, callous, and premodern practices.

In this regard, it should be noted that while much of the rhetorical power of the spot lies in the woman's status as victim, her complete passivity not only may play into stereotypes of Asians in mainstream U.S. and Californian culture but also may have compromised the ad's effectiveness among Pacific Islander and Asian American audiences. The CDC *Media Resource Book for Tobacco Control* suggests as much when it reports that this ad and the one featuring the Latino household focus groups were criticized by members of the targeted communities: "Because most people [i.e., TV viewers] did not see all four versions of these commercials [the other two featured an African American man in a car with his wife and a white smoker in his bedroom with his wife] there was some concern that they were portrayed as being uncaring" (CDC 1995–98, vol. II, 102).[16] Similarly, Asian American colleagues expressed incredulity that a housewife would be so passive with respect to her husband's behavior. However, when queried about the spot, members of the Asian Pacific Islander Tobacco Education Network (APITEN) expressed no reservations about its structure or tone.

Here, it would appear that even as the DHS attempted to address smoking practices in both mainstream and nonmajority communities, the spot's aggressiveness toward unacceptable smoking behavior ends up "othering" entire groups and inadvertently links the hazards of smoking among recently arrived immigrants to volatile debates over immigration and thus ties the threat smoking poses to family to a threat immigration poses to the entire social body.

This slippage from a health issue to a social one is almost built in structurally into the media campaign by virtue of its statewide character, the division of California into eight mainstream media markets by the advertising industry, marketers' segmentation practices that neglect underrepresented communities, and the use of the English-language spots in a wide broadcast/cable medium such as television. As we have seen, this kind of spot and its placement in mainstream media inevitably address several audiences simultaneously—what U.K. health promotion researcher Keith Tones terms the inevitable "blunderbuss" effect of mass media (Tones 1993, 129)—and are thus quite vulnerable to mainstream interpretations by the "general population" that isolate, label, and pathologize the specific communities addressed by the ads.

The "Kitchen" spot demonstrates two points of public health concern: the wife and the fetus as victims and the husband as vector of smoking and secondhand smoke. As such, in the context of debates over citizenship, indifference to women, children, and fetuses associates outdated modern practices (smoking) not only with the persistence of presumably nonmodern practices (non-Western patriarchy) in the present but also with the heated political question of social "risks" created by globalization's international flow of workers and more specifically Pacific Islander and Asian (and Latino) immigration to California and the evolving composition of the state's social body. At that time, debates over immigration in California gave birth to a great backlash against new immigrants, fueled by Democratic and Republican politicians alike, which climaxed in 1994 with the passage of Proposition 187 that denied public assistance, medical care, and schooling to illegal immigrants (Stall and Decker 1994; Calavita 1994; Chavez 1998). Secondhand smoke in this ad and others like it may be "male" and non-middle-class in origin, but it is also framed as "alien" and un-Californian and un-American. Like cocaine and other "foreign" bodies, it crosses national borders into the California and U.S. social body just as easily as it violates the sanctity of the home and, in public space, drifts into the nonsmoking sections.

Black Lungs: Framing African Americans

In U.S. social discourses, African Americans have been long classed among the risk groups of choice as objects of solicitude, discipline, and blame by various government agencies and private foundations. The national hysteria in the 1980s and 1990s over the War on Drugs, "crack babies," and intravenous drug users and over teenage pregnancy and welfare mothers (both are always constructed as black) only heightened African Americans' visibility as a "problem" and "problematic" community ("World without Fathers" 1993; "Scapegoating" 1989; Reeves and Campbell 1994; D. Gordon 1994). And in the late-twentieth-century U.S., in the context of globalization and newly displaced populations, African Americans, who for so long the stood at the center of U.S. urban modernity along with Latinos and descendants of immigrants from Europe, are now, along with other long-standing "residential minorities," portrayed as "stuck" on the sidelines of the new prosperity, relegated geographically to their depressed neighborhoods and culturally to their less-than-modern middle-class household arrangements (the so-called fatherless households) dating back to the Reconstruction era (1865–77) in the U.S. (Ong 1999, 9–10). Here, African American households do not embody either an unacceptable nonmodernity or an incomplete modernity in terms of some sort of progress toward assimilation to the U.S. mainstream but rather are framed as failed citizens of a failed modernity. As such, in U.S. parlance, African Americans stand as the very embodiment of "hard to reach" citizens, for whom a more traditional social pedagogy is warranted.

Of necessity, any health promotion campaign that addresses African Americans will have to negotiate these discursive and bureaucratic framings of black bodies and families that are part and parcel of contemporary racializing projects (Omi and Winant 1994, 55–56). The California statewide anti–secondhand smoke campaign was no exception. We've seen that during the period under study (1990–96) the DHS made relatively few ads directed specifically at the African American community, especially in community newspapers, which elicited protests from publishers and advocates alike. In the electronic and outdoor media, those ads that were produced and dealt with the hazards of secondhand smoke were uncommonly aggressive and broke the "surface tension" around the smoking issue dramatically. I have already examined one such billboard: it featured a young man with a lit cigarette between his lips and the caption, "Eric Jones put a contract on his family for $2.65."

In 1996 in California, there appeared large outdoor advertisements that once again played out questions of health-related knowledge, viewers' identity, community affiliation, ethics, and social landscape. The ads displayed a young African American girl in braids who gazes out of the billboard's frame at passersby. Her face is superimposed over the faded picture of her father, whose face, situated in the background above her right shoulder, is convulsed with laughter as he holds a barely visible lit cigarette. The title reads: "She has her momma's eyes and her daddy's lungs. Secondhand smoke kills." It is a reworking of an older HIV/AIDS education campaign poster sponsored by the Urban League, a leading African American community organization. This poster featured a photograph of smiling black baby in a pinstripe dress and white shoes sitting and facing the camera in a three- quarter pose. To her left is the text: "She has her father's eyes and her mother's AIDS. Before you get pregnant, find out if you need to be tested" (Gilman 1995, 151).

At first glance, this outdoor advertisement, produced by Carol H. Williams Advertising, enacts a health and ethical drama with familial, gender, and racial overtones similar to the "smoking fetus" TV spot: the physical well-being of a helpless family member has been sacrificed to the selfish pleasures of the smoking father. The daughter is one of the millions of U.S. children exposed to ETS in the home of which the EPA warned in 1992. As in the original Urban League poster, the ad's punchy verbal hook is a play on those everyday conversations between family and friends on family resemblances they observe in the physical features of infants. However, the clever play on words proposes a different level of knowledge from what is available through the reading of surface resemblances by those associated with the little girl's family. As we will see, for the ad to work, it must furnish passersby with scientific knowledge about ETS to complete their viewing experience.

The Ethics of Knowledge, Speech, and Public Health Paternalism

Here, unlike in the "Smokescreen" ad, to see is *not* to know, or rather, is *not to know enough*. The visual scientific demonstration does not go without saying, for without the words supplied by the title, the viewer cannot become fully the knower. The phrase, "She has her momma's eyes and her daddy's lungs," takes us viewers below the surface of the body to the "semi-pathological pre-illness state at-risk" (Armstrong 1995, 401) of the little girl's lungs that recent studies have discerned.

Thus, in terms of its visual structure, the billboard is much less "populist" in terms of the distribution of knowledge than either the "Smokescreen" ad or the "Kitchen" ad. Passersby's everyday sight is not sufficient; it requires help from the

"Momma's Eyes, Daddy's Lungs."
Photograph by Michael Allen Jones.

DHS to grasp the visuals, and thus viewers are placed more in the more traditional position of "beneficiaries" of state and voluntary organizations' expertise and assistance than of knowledgeable citizens or consumers favored by marketers and proponents of neoliberal social policies. Yet the ad does convert the deeper scientific knowledge it extends to the public into a "commonsense" perception that passersby can take away with them. The billboard does this through another sense of knowingness that at once warrants and speaks the scientific one: the dominant discourse on African American family households. This knowingness would be perhaps most operative in the ad's reception by nonblack passersby.

I argue that the "success" of the clever verbal hook that teaches new knowledge to passersby on the street depends on a second, visual hook—the father's image—which reproduces what the general public already "knows" about dysfunctional black families and absent fathers. This is the pathological household that would stand as the measure of African Americans' failure as a community of modern, self-regulating subjects. In contrast to the Asian American husband, who, while occupying the same visual plane as his coughing wife, was psychologically aloof and scandalously unaware of her suffering, here the father, who is relegated to the background plane in the visual field, is very much "present" to his daughter through his laughter but almost *too present* because he persists in his smoking pleasure to her detriment.

Now, the second hook is actually a two-step operation. On the one hand, it reverses what viewers already "know" and come to expect of African American family households—the "fact" of absent fathers and abandoned mothers—by having the father present; *and*, on the other, especially in the case of passersby from outside the community, it then reverses that reversal by nonetheless confirming the dominant narrative about black men as shiftless, irresponsible, given over to their pleasures (sex, drugs, and music), and prone to violence. Not to put too fine a point on it, in the smoking fetus ad we saw the reversal of a positive stereotype of the Asian American model minority in the image of incomplete modern citizenship; here, we see the reversal of a negative stereotype, but it is a false one: the reversal is first proffered and then withdrawn.

Let me explain. In the midst of the crushing learned and mainstream consensus concerning black family households, the ad's refreshing promise of a different story, that of a black father laughing and playing with his daughter in a middle-class home, is withdrawn in the very same gesture by which it is extended. His laughter, presumably that of a joyful father, is reduced to the expression of self-absorbed pleasure. He may be back home, but his pleasure is not her joy (she is not laughing). In the context of late-twentieth-century public obsession with child abuse in the U.S., his abuse of tobacco and of her body is, willy-nilly, sexualized. Hovering over her shoulder in the billboard, he stands as a direct menace to her well-being, all the more so in that the mother is nowhere to be seen, and if she's there (outside the frame), like the Asian American wife she is found wanting, for she, too, apparently puts up with her man's vices and offers her child no protection.

The double bind into which dominant narratives place black men is played out once again. African American men, if absent, constitute threats to themselves and the community, yet if present in the home, they are no better. This is what gives the verbal and visual hooks their nasty edge and betrays the legacy of a deeper paternalism in this ad stemming from traditional public health and advertising practices that assumed expertise and prerogatives in a top-down manner. It's as if, with respect to African American men, the ad simply abandoned the pedagogical mission of translating expertise about ETS to the public. For ultimately, what the ad seems to focus on most intently is the father's ethical and psychological features; what the spot makes visible is his pathological character—his laughter, like his smoking, creates her toxic environment and threatens to destroy her. What we "see," finally, is his convulsed black face, which speaks what we now "know"—her blackening lungs. His pathological interiority (his character, his misplaced pleasures) translates into her pathological body. Here, a practice—smoking as a health hazard—comes close to being converted into an identity—black maleness or simply blackness. The aggressive narrative threatens to defeat the ad's ability to reach African Americans, especially fathers, to whose sense of guilt for betraying the trust of their children the billboard makes a blunt appeal. However, unlike in the billboard that compares a young smoker to a gang member, when brought up in an interview with a member of the African American Tobacco Education Network (AATEN), surprisingly the ad elicited no apparent reservations and no explanation was given (Ethnic network director C 2000).

As in the "Kitchen" ad, the highly gendered story of male aggression (which ostensibly reflects different smoking rates between black men and women in California, although the gap is not large—25 percent versus 21.4 percent—compared with that of Pacific Islanders and Asian Americans) would seem to imply above all a female viewership (especially African American), namely, mothers, wives, partners, sisters, in-laws, and their friends. At the same time, the deployment of long-standing U.S. discourse on African American families and men would also seem to designate men and women of the "general population" as among the ad's implied audience. And in either case, once again, those excluded from the community of shared knowledge about secondhand smoke tend to be male smokers of color. For, if the laughing face ascribes to the father some sort of inner life of feeling and thought (and thus perhaps some possible awareness and sense of caring), it is wholly consumed by the selfish pleasures of the present moment. The daughter has no future, for the father cannot think one, and, for whatever reason, the mother cannot guarantee one. Once again, as in the smoking fetus ad, deficient family life is made to signify a health hazard.

The father, unlike the mother, other women, and other knowledgeable viewers, does not qualify as the "thinking, acting subject" presupposed by the emerging prudentialism of advanced liberal societies and to whom these health ads would appear to make their most direct appeal (Armstrong 1993, 407; O'Malley 1992, 257, 261; N. Rose 1993, 296). Presumably, the father is incapable of knowing and thus incapable of self-governance and discipline. He is one of liberal government's failed pedagogical subjects. As such, it would appear that he is not entitled to speak, let alone to name and identify himself, his thoughts, his desires, his subjectivity. The California DHS, the ad agencies, female passersby, and the general population have the privilege of doing it for him.[17]

Converting to Health: Giving Mainstream America a Second Chance

There is only one English-language TV ad, a thirty-second spot, in which the smoking husband gets to join to the community of knowledgeable viewers and self-regulating citizens. He gains admittance by undergoing an internal conversion to the new health sensibility and acquires thereby the privilege of speech denied the other men, but unfortunately it is too late. The ad, titled "Victim/Wife," drops the aggressive cleverness of the all-knowing third-person narration in favor of the intimacy of first-person melodrama while upping the anti–secondhand smoke ante yet again: it is the story of a white retiree of the all-American middle class (to which everyone presumably belongs or wants to belong), whose secondhand smoke killed his wife. First broadcast in 1994 and periodically aired up through 2001, it names smokers as so many suicidal murderers but suggests that if they are of European descent, they may be cast as *unwitting* killers belonging to mainstream America and thus capable of regret.

Our retiree gets to voice the discourse of guilt. Shot with full close-ups of his face and hands in the handheld camera style, the interview unfolds in a comfortable living room against a background track of soulful piano music:

> My wife was always getting on my case about smoking. She said, "It's bad for you, it makes the curtains smell." She even threatened to stop kissing me if I didn't quit. I said, "It's my lungs, it's my life." But I was wrong. I didn't quit. I had no idea the life I'd lose wasn't mine; it was hers. But she was my life. My wife was my life.

The final title flashes: "Secondhand smoke kills 53,000 people every year."

In the anti–secondhand smoke TV and billboard ads (including those not reviewed here), no other male smoker, and certainly no man of color who smokes, gets the benefit of the doubt and is extended the opportunity of speaking scientific fact through giving voice to "family." Here, the ad confronts TV viewers directly with a full-blown "thinking, acting" psychosocial subject, that of a smoker, who, by virtue of his age and habit, literally embodies the deadly smoking policy of the previous generation that was associated with the older, flawed modernity marked by the legacy of brutal industrialization, European immigration, the Great Depression, the Second World War, and the Cold War. As the short narration unfolds, the knowing subject recounts his path toward his new consciousness and the production of a new self. He has come to recognize himself in terms of "family" and the new health consciousness by naming his failure as a loving husband and knowing subject. Upon doing so, he has acquired a new body—if not one of health (we learn nothing about the state of his heart and lungs), then certainly one that reveals a depth, psychological in character, in the form of inner feeling and reflective thought. This is what individuates him, sets him apart from the other smokers, and leads him to quit his habit (Crawford 1980; Greco 1993).

As virtual witnesses of sorts, we are awash in the affect of his tears and trembling voice. An intimacy between viewers and his white face convulsed with regret is briefly established but it is not a dangerous one, for we recognize *him*: he has changed, and he is becoming one of "us." We extend him a therapeutic

"Victim/Wife."
Courtesy of California Department of Health Services.

second chance, for what we learn is what we already knew about people like himself—that he didn't mean to kill his wife—and we *want* to believe him. His vulnerability opens a space of sympathy and identification largely absent from the other anti–secondhand smoke ads: here, vulnerability cuts both ways and is both physical and affective. As nonsmokers, or even smokers, our bodies were always vulnerable to beings like his; now his inner life resembles our own. Moreover, nonsmoking viewers may identify not only with the victim (here, the dead wife) as in the other ads, but also, to a certain extent, with the aggressor; and, most important, *current* (male) smokers may see themselves in the crying widower and be induced to drop the habit.

The extent of the grieving man's conversion can be gauged by the fact that his wife's knowing reproaches, couched in the mundane, domestic conversations of an aging couple ("It's bad for you, it makes the curtains smell"), literally haunt him and fill his words as a speaking subject. Her knowledge has become his now; in death, her voice shapes and polices his thoughts, feelings, and pleasures. The fact that halfway through his tale he becomes his own melancholic voice-over while the camera focuses on him with his eyes closed permits viewers to fathom the depth of his new knowledge. The trade-off between her life and his pleasures was a poor one, indeed; in the neoliberal world of secondhand risks and relations, he miscalculated in the prudential exercise of his "preferences." A better guide to his choices than personal pleasure would have been family sentiment and scientific fact.[18]

Health, Marketing, and Modern Social Subjects

My analysis of these preceding ads marks an attempt to explore the rich, powerful rhetoric that the California anti–secondhand smoke media campaign has deployed in its appeal to multiple audiences. The play of knowledge, speech, and voice in melodramatic tales of family households under threat produces an apparatus of perception and a commonsense understanding of new scientific fact. What emerges is a fragile space of democratic community made up of singular and aggregate populations organized in terms of a social geography of health, disease, and risk. Within that epistemological and affective frame, the ads inscribe a new object of knowledge (secondhand smoke), invalidate existing anti-smoking policies, and reconfigure built space and social relations. The already-known stories and facts of common sense both validate and recruit Californians as caring citizens, virtual witnesses of emerging scientific fact, worthy learners of the new knowledge, and new agents of public health policy, and they powerfully position viewers as potential members of public health's ideal community of nonsmokers or those responsible smokers who wish to quit.

The ads present the privilege of knowledge and speech as open to many but not to all. In all the anti–secondhand smoke ads the only smoker who truly enjoys this privilege is the white middle-class retiree. This suggests, once again, that while the media campaign targeted for mobilization citizens of all communities, it nonetheless constructs as its ideal receptive audience a class of citizens that by virtue of its much-vaunted health and capacity for self-governance has normatively defined the general population over the years—that is to say, the professional/managerial class or, more inclusively, the class of knowledge producers and governors of conduct, which are predominantly of European descent and tend to view themselves as "healthy."

I have shown that as they address multiple audiences of California's diverse inhabitants and the peculiar health problems relating to recently arrived populations in an era of globalization, the more aggressive ads inadvertently tend to designate men in general and men of certain communities in particular (Asian Americans, African Americans, and Chicanos/Latinos) as vectors of secondhand smoke. The "hook" of the new common sense concerning secondhand smoke lies in the language of "family," what citizens are presumed to value—the normative middle-class family household—and to know and reject—the purportedly dysfunctional nature of communities of color and their private life, which are so many markers of an incomplete or imperfect modernity. Through family melodrama, the ads inadvertently reproduce the aggressive social discourses of the 1990s that tended to construct men of color as oppressive, uncaring, and out of control, as opposed to a more thoughtful and healthy (white) middle class. Thus, of the ads that feature white husbands or fathers as vectors of secondhand smoke, none portrays them as indifferent, selfish, or patriarchally oppressive to the same degree.

Take one last example: in 1994 the DHS widely aired an English-language TV spot called "Numbers." It portrays a middle-aged, white, upper-middle-class male professional (a lawyer or senior executive) with sympathetic understanding denied men of color who create secondhand smoke in other ads. The thirty-second spot follows him as he moves through different spaces in the course of his workday, distributing secondhand smoke as he goes to individuals identified by numbers. He is hardly an immigrant on his way to assimilation; nor is he a member of an economically and culturally "stuck" community relegated to the sidelines of a globalizing world. On the contrary, he glides effortlessly and gracefully through the recognizable spaces of the new service-based economy from car or van to the office, then to a restaurant, and, finally, to home (where he is greeted by his wife and daughter). The male voice-over, in a tone of concern and reproach, describes the smoking subject as "not a bad guy; on the contrary, he's a decent human being. His associates like and respect him, he's a caring father, and a good husband, but today the thirty-seven people he comes into contact with will die a little, because they were forced to breathe secondhand smoke from his cigarette." As the camera follows his arm as he reaches out to caress his daughter, the male voice-over concludes, "Secondhand smoke takes more than your breath away," followed by the title, "5,100 Californians die every year from breathing secondhand smoke."[19]

These cultural narratives of family and race, I believe, are essential to the rhetorical functioning of the aggressive anti-ETS ads, especially insofar as they also address not only specific communities but also a broader audience. My analyses suggest that ads that are perhaps appropriate or effective in particular community settings using community-based media (but that is far from clear for many of the ads analyzed here) may function quite differently when received by a general audience. From privileged objects of solicitude, communities of color come close to being named once again—as they were throughout the twentieth century since the Progressive Era—as the privileged *sources* of illness and social disorder. Thus, "whiteness"; the possibility of good family life and citizenship; middle-classness; a revamped modernity; and rational, healthy behavior tend to merge in these ads and stand for one another. In this discourse it would appear that the more individuals or groups embody some of these qualities, the more they come close to embodying the other ones as well.[20] Or, to use another

"Numbers."
Courtesy of California Department of Health Services.

language, aggregated together, these qualities stand for a particular norm but operate flexibly and inclusively insofar as the acquisition of on feature or quality implicitly promises the future acquisition of the others. By the same token, the apparatus of perception of health may also function as an apparatus of social intelligibility that tends to render nonmajority bodies and entire classes of citizens legible through the markers of risk factors such as smoking and imperfect family life. Here, smoking, dysfunctional families, flawed modernity, and social marginality come to mutually signify each other and, when aggregated, to designate particular populations.

Such may indeed be the liabilities of a statewide campaign that works to make populations available for public health intervention by using much more mass than community-based media and operates within certain social marketing assumptions in relative isolation from the communities it is trying to reach. From the preceding analyses, it would appear that the severity of the problems follows roughly the amount of influence the respective tobacco education ethnic networks enjoyed in the Tobacco Control Program: the weaker a community's influence, the more the play of questions of health, knowledge, ethics, and social identity is problematic in ads ostensibly addressing the community's needs. Still, it should be said that even community input provides no absolute guarantee and that mainstream discourses and dominant assumptions in tobacco control, public health, and advertising can sometimes override community issues and ethical reservations (Robinson 2000; Guttman 1997). Compounding matters, as always, is the role played by aggregate data—"economies of scale"—in policy circles, which trump particularizing objections (Boltanski and Thévenot 1991, 18–23). In this sense, one could ask whether, in the end, the statewide DHS ads—like earlier community intervention trials in Europe and North America, social marketing campaigns in the industrializing world, and marketing in general—don't speak more to citizens affiliated with marketers' and tobacco control advocates' own professional and social background than to others.

Transferring California: The Cases of Minnesota and Massachusetts

The interlocked issues of tobacco control education, liberal governance, knowledge, common sense, healthy behavior, and social identity specific to the California anti–secondhand smoke media campaign return us to one of the questions that opened this book: the transferability of media components to other states and even other national contexts. They remind us that this question entails not only the movement of health promotion strategies to other sites but also the circulation outside California of normative discourses of citizenship, family, and community.

To be sure, the media campaign, like the Tobacco Control Program overall, has roots that extend back in time and beyond the borders of the state. It stands as a particular moment in health promotion locally, nationally, and internationally. I have analyzed in particular how the media campaign marks the moment of the ascendancy of social marketing methods and approaches in U.S. tobacco control and demonstrates the successful negotiation of the changing U.S. and global public health context in which neoliberal thinking dominates policy circles and elite opinion in industrialized nations. Some of what became known as "Californian" in the media component of the campaign has indeed been widely shared within the U.S. tobacco control community itself: the use of paid advertising, its signature anti-industry approach, and the implementation of local smoking ordinances. However, as standard globalization theory reminds us, the flow of goods, workers, media, information, and vectors of disease not only suggests a homogenizing tendency in the present moment but also highlights differences and particularities. In California the local struggle against the transnational tobacco industry and the laws belonging to an older smoking culture has thrown into relief local and regional identities, which in turn entailed a discourse of healthy practices in terms of family, good citizenship, and community accountability that wasn't easily translated to other local contexts.

Within the U.S., state and regional conditions are indeed a factor in tobacco control. This situation, widely acknowledged by many researchers, officials, and advocates I have interviewed, appears to have been especially true in the case of the media campaign against secondhand smoke, which until the tobacco MSA of November 1998 had *not* been picked up by other state campaigns (California health official A 1999).[21] One reason was technical: the original contracts with members of the Screen Actors Guild made transfers prohibitively expensive, a situation that wasn't remedied until the CDC stepped in and created a bank of TV and radio tobacco control public service announcements in 1995 for U.S. distribution. Moreover, in interviews it was underscored that the limited circulation of the better-known anti-ETS ads also had something to do with secondhand smoke's intensely social character, which, although it has great potential to mobilize the multitudes as a health issue, also involves the complex, delicate nature of social and family relations that are specific to regions and communities.

A case in point was the Minnesota tobacco control program. At first glance, Minnesota, one of the least racially diverse states in the U.S. (96 percent white), would be an unlikely candidate for ads developed for a vastly different population. Indeed, although public health personnel from Minnesota directly participated in the conceptualization and launching of the California media campaign, during the 1990s the Minnesota program consciously avoided doing anti-ETS advertising and thus borrowed no ads of this kind from California. However, the reasons evoked did not seem to stem directly from the relative homogeneity of Minnesota residents. It was something at once broadly "cultural" and specifically related to the presumption of state government to intervene in family household relations. As one Minnesota public health official put it, "Minnesota is a different kind of culture. . . . I think when you get into secondhand smoke issues, you have to be real careful with them, otherwise you can end up blaming the victim, and I think [Minnesota] is much less confrontational [than California]." In particular, the use of narratives of the violation of family life by adult smokers as a means to promote awareness and discussion among smokers and nonsmokers would not transfer well: "I think we could have gotten into a lot of trouble. I mean, there were a lot of issues with pitting parents against children, and it would have been a rat's nest in Minnesota" (Minnesota health official 1997). In the view of this particular official, the acceptability and even success of anti-ETS ads in California also had to do with the fact that it's simply a "different milieu" and had benefited from the effects of local activism and municipal ordinances, which had prepared public opinion.

One state program that did borrow heavily from California was the Massachusetts Tobacco Control Program. Initiated by the passage of Question 1 in November 1992, which increased cigarette taxes by 25¢ per pack, the program generated the highest per capita tobacco health education funding of any state or country—about twice that of California. Modeled after the National Cancer Institute's ongoing ASSIST intervention study and California's Tobacco Control Program, it was credited with continuing the decline of adult prevalence rates in Massachusetts through the 1990s even as declines leveled off elsewhere in the U.S. except in California (Biener, Harris, and Hamilton 2000; Koh 1996; Hamilton and Harrold 1996). Although Massachusetts is a far less diverse state than California (roughly 92 percent white in the 1990s) with smaller recent immigration (a larger presence would lower overall prevalence rates), and although there are huge differences in population and geographical size, economies of scale, political structures, social climate, and attitudes, the program followed California's fairly closely in terms of media components (paid advertising), local ordinance implementation, and funding of local health department initiatives.[22] The media campaign replicated California's anti-industry and anti–secondhand smoke themes and had three broad targets: preventing youth initiation; shifting public opinion and social norms; and, one less stressed by the California media campaign, adult cessation (Connolly 1997; Advertising executive B 2000).

The kickoff of the media campaign put out paid advertising, some of which was expected to include borrowings from California. One idea that Massachusetts borrowed from the effort to pass Proposition 99 in 1988 but that the California campaign never really deployed until the late 1990s was the use of ethical testimony of former smokers and actual victims of smoking-related illnesses, including Janet Sachman, a former cigarette ad model (who first appeared in a French anti-smoking spot); they also employed public denunciations of tobacco industry lies by former industry researchers (Victor DeNoble) and lobbyists (Victor Crawford) that were produced specifically for the Massachusetts campaign. With respect to the anti-ETS spots, what is striking is that on the one hand, Massachusetts did not air any California anti–secondhand smoke ads, and on the other, while Massachusetts' own spots wove in anti-industry attacks and highlighted the hazards posed by ETS to young children, babies, and fetuses, no one spot features identifiable smokers, let alone smokers of specific non-English-speaking or nonwhite communities. Filling out the ethical and social features of smokers seemed to interest the campaign even less. Rather, ads most often dramatized potential victims who were speaking out against secondhand smoke.

Take, for example, a spot released in 1996 titled, "Unborn," which was a 1996 Clio Award winner and which was subsequently adopted by the tobacco control programs of other nations, including France and Japan. Shot in black and white, it features a woman in her thirties of European descent in a medium close-up who faces the camera in a three-quarter stance. She details the dangers of ETS as the camera crosscuts between stylized footage made to look like the negative print of shots of people inhaling smoke (they resemble animated x-rays): "Secondhand smoke contains nicotine. Nicotine is a dangerous drug. In a room where others are smoking it spreads, everywhere. And ends up in the body of every single person there [cut to an extreme close-up of a thumb-sucking fetus]. There are no exceptions." At which point we get a medium shot of the speaker as she glances down at her visibly swollen belly. In small print a tag then appears: "A Message from the Massachusetts Department of Public Health." The spot's effectiveness depends on the contrast between the woman's flat delivery of a typical public health lecture and the fast, cross-cutting, surreal visuals and final, dramatic shot.

In contrast to the Californian smoking fetus spot analyzed earlier ("Kitchen"), there is no male voice-over. Nor is there any mention of thousands of deaths due to passive smoke. It is the woman who speaks, and she does so in a calm voice, whose authority, while grounded in her pregnant body, moderates and contains the melodramatics. As viewers we are pulled in (the fetus appears helpless) but only so far (the mother/speaker already knows the danger); there is no narrative of endangerment followed by dramatic rescue implied here. Nowhere does the ad identify who the smokers are or suggest their social or psychological characteristics, nor does it indicate where they are located within local or national communities as in the California ads.[23] Thanks to the stylized footage that references an older scientific vision (the x-ray photograph),[24] viewers get visual confirmation of the speaker's discourse, and, as in the California "Smokescreen" ad, everyday speech and experience are wedded to recent scientific findings.[25] As such, the ad would seem to pitch itself as very much a public health message to smokers and nonsmokers alike; to pregnant women, their loved ones, friends, and coworkers; and to the general public. Still, when the Tobacco Control Program released the anti–secondhand smoke ads, health department officials and ad executives in charge of the account were caught off guard by a strong, hostile reaction from smokers, who called them "zealots."[26] They came to the realization that while they judged such spots necessary to stimulate change in public opinion, they courted the risk of appearing to ostracize smokers. Consequently, they decided to accompany the anti-ETS spots with other ones supportive of smokers' struggle to quit, including the animation, "Quitting Takes Practice," from the first wave of ads released in 1990 in California (Advertising executive B 2000; Connolly 1997). Gregory Connolly of the Massachusetts Department of Public Health summed up the strategy this way: "What we tried to do is, we do ETS but we juxtapose ETS with a love message to the adult smoker, so we bang them but at the same time [we] hold out an image of a smoker who's not scum, who's not killing someone, who wants to think about quitting. . . . I think just calling smokers killers over ETS is probably counter-productive. Government should be very cautious about doing that" (Connolly 1997).

That the early Massachusetts anti–secondhand smoke ads did not assign identities to smokers and refrained from accusing them of killing fellow citizens is more than merely suggestive. At the very least it would appear to avoid some of the potential divisiveness of exclusive focus on health issues, which worries some health advocates (Klaus 1993; Nathanson 1996; Robinson 2000). Yet, by the same token, a less richly detailed picture of smokers as ethical and psychological subjects attenuates the melodrama and engrosses audiences less in terms of their sense of self and of community and, when combined with the dry, didactic mode of address, establishes the traditional distance with the audience characteristic of older, top-down public health paternalistic stance that presents itself as the sole possessor of expertise and claims to speak for one and all. In light of U.S. public health's roots in the Progressive Era's obsession with identifying collective health problems with unhealthy "nonwhite" others (blacks, Native Americans, and Latinos but also, up to the Second World War, Irish and Southern and Eastern European immigrants), I do wonder, if either Massachusetts or Minnesota had had considerably larger immigrant and nonwhite communities resulting from recent, dramatic population shifts brought on by globalization and neoliberal policies, whether the local culture's reluctance to aggressively stigmatize smokers for polluting others around them wouldn't vanish. In a word, a relatively homogeneous (i.e., white), stable population with a strong sense of entitlement as citizens is at bottom willing to apply methods of governance to others that it—as descendants of nineteenth-century European immigrants—would no longer allow to be applied to itself.

Health Expertise and Hybrid Forms of Power

Of late, much ink has been spilt by writers attempting to analyze technologies of power and forms of governance in advanced liberal societies as the latter have undergone changes wrought by the campaign of neoliberals to weaken and dismantle the welfare state. Nikolas Rose writes that contemporary liberalism (in the classical sense of free markets and limited government, as embodied by the neoliberal and neoconservative programs) is

> dependent upon the proliferation of little regulatory instances across a territory and their multiplication, at a "molecular" level, through the interstices of our present experience. It is dependent, too, upon a particular relation between political subjects and expertise, in which the injunctions of the experts merge with our own projects for self-mastery and the enhancement of our lives. (N. Rose 1993, 298)

Such a project would seem to drive the California anti–secondhand smoke campaign. Through narratives of family under threat, it invites viewers to adopt the position of public health's knowing subject and that of the self-care movement (Crawford 1980, 379). Yet in these ads, self-governance (healthy behavior) in the name of family also entails governance of others (interventions in private and public life). Indeed, I have argued that the discursive power of "secondhand smoke" is that it authorizes strategies of normalization that target citizens not only as singular "thinking, acting" individuals by other individuals but also implicitly as aggregate members of a group by local, state, and federal government.

The logic of the secondhand smoke threat is thus at once far afield from and in secret relation with "the megalomaniacal and obsessive fantasy of a totally administered society" that, according to Rose, liberalism presumably had abandoned long ago in the nineteenth century (N. Rose 1993, 289). In this sense, the California anti–secondhand smoke media campaign is better viewed as an example of *hybrid* social discourse, which, according to U.K. sociologist Pat O'Malley, is a more helpful way to view contemporary concrete technologies of power, which are always an mix of punitive sovereignty, therapeutic discipline, and neoliberal self-government. This what U.S.-based anthropologist Aihwa Ong has termed, in the context of the regulation of labor relations, "graduated sovereignties" (O'Malley 1992, 257–258; Ong 1999, 215). Thus, public health's knowing subject has affinities with neoliberalism's (and neoconservatism's) knowledgeable, rational-choice actors (projected as mostly middle-class and white) capable of self-government and thus government of others, while other citizens (especially men of color), deemed ignorant and incapable of self-regulated conduct, are often pathologized as irrational, beyond therapeutic intervention, and thus, at least by implication, deserving only of punitive measures. However, unlike in the airless community of hard-core neoconservatism that favors punitive retribution, here the therapeutic benefit of the doubt persists for offenders, but it is reserved for strayed members of the "general population" alone. Not all bodies that circulate through the space of risky, secondhand relations constitutive of democratic community are constructed as vulnerable and educable to the same degree.

four
Revising Late Modernity

Smoking as Icon of Industrialism and the Cold War in Public Health and Media Culture

In one year, a typical one-pack-per-day smoker takes in 50,000 to 70,000 puffs through the burning column of a unique chemical factory which contains over 2,000 compounds. Many of those compounds are established carcinogens and appear in the particulate phase or "tar" of the smoke.

—*Smoking and Health: A Report of the Surgeon General* (1979)

The cigarette—the icon of our consumer culture, the symbol of pleasure and power, sexuality and individuality—had become suspect.

—Allan M. Brandt, "The Cigarette, Risk, and American Culture" (1990)

Flawed Modernity

In the course of examining California spots and billboards targeting secondhand smoke, I have traced how many of them aggressively "broke the surface tension" of public discourse in order to provoke debate and discussion of the hazards of tobacco smoke. They interwove questions of knowledge, family, and good citizenship and produced inadvertently, as it were, the outlines of a social map consisting of aggregated and particularized communities. Within this projected cartography, the ads tended to locate different groups and communities differently according to what would appear to be their amenability to amendment and reform. Through family dramas, the ads often framed men as sources of secondhand smoke and men of color as less able to grasp the tragic consequences of their smoking than their white counterparts, to whom, by contrast, are extended at times sympathy and the benefit of the doubt. The latter emerge as the only smokers who qualify as "acting, thinking subjects" favored by dominant U.S. health prevention and neoliberal discourses of the late twentieth century. I have argued that the sectarian contours of this social map are the product of tensions within U.S. public health between the democratic goals of the community empowerment movement, epidemiological studies, and the segmentation practices of social marketing; of the isolation of tobacco control officials, voluntary health organizations, marketers, and researchers from the communities they are committed to serving; and, ultimately, of racializing public health discourses on non-middle-class, immigrant, and former slave populations dating from the Progressive Era (1890–1920) that classified them as second-class citizens.

At the same time, these ads also tended to correlate this social map to a timeline, that of the one-way arrow of economic and cultural modernization. This was clearest in ads attempting to reach communities whose members include large numbers of recently arrived immigrants: in this case, incomplete citizenship meant partial assimilation to the cultural norms of American modernity[1] as, for example, the Asian American wife's passivity before her husband's cruelly indifferent smoking in the "Kitchen" spot and the Latino older brother's obliviousness to his younger sister's discomfort in the "Livingroom" spot.

However, what was striking is that if these groups along with African Americans are made to stand in, in so many words, for belated or incomplete modernities in the anti–secondhand smoke ads, the spots featuring the smoking and nonsmoking sections ("Smokescreen") and very middle-class white men ("Victim/ Wife," "Numbers") actually turn the tables on the very notion of American modernity itself: they discern a flaw in the original project of that modernity that tolerated or even identified with the wide consumption of cigarettes. That is surely one of the significations of the spot with the aging widower who tearfully recounts how his smoking killed his wife. He is of that great World War II generation that made the twentieth century the "American Century." His and his parents' generations were the first generations that were able by the 1940s to indulge in massive numbers in heavy, regular smoking of cigarettes in the U.S., Canada, and the U.K., thanks to the availability of abundant cigarette rations in their countries, which had been spared the worst devastation and hardships of the war. Their per capita consumption was not matched in other countries until much later in the 1960s, if at all.

Historians and observers of the tobacco industry have long noted that tobacco has been part and parcel of Euro-American modernity, from its very beginnings with the rise of mercantilist nation-states and the colonization of North and

South America and later Africa and Asia to the era of industrialization (J. Goodman 1993). The late nineteenth century witnessed the invention of machine-made cigarettes, one of the first mass consumer products (adopted in France, Japan, the U.S., China, and Europe in the 1920s and 1930s), and along with it the beginning of modern marketing practices and the rise of the first national markets for consumer goods (Brandt 1990; Schudson 1993; Kluger 1996, 3–79). Consequently, the beginnings of everyday consumption of singular cigarette brands in the U.S. across all states in the 1880s, for example, can be understood as having contributed to the construction of a national identity, of an imagined but very material community of "Americanness" (R. Foster 2002, 2, 7, 111; Fox and Lears 1983).

Thus, it should come as no surprise that in these anti-smoking ads, public health stories of health, citizenship, and community also turn out to be revisionist narratives of U.S. history that reassess the recent past in terms of contemporary social changes and project a transformed future. Moreover, during the early and mid-1990s these narratives emerged not only in the California tobacco control media campaign but also nationwide in mass media—primarily film and broadcast and cable television. My focus on electronic media follows that of U.S.-based anthropologist Arjun Appadurai, who claims a privileged role for electronic media in an era of globalization marked by renewed mass migrations, for "electronic mediation transforms preexisting worlds of communication and conduct" and offers "resources for the experiments of with self-making in all sorts of societies" (Appadurai 1996, 3). With respect to the culture of cigarettes, the role of the film industry in the promotion of the values, meanings, and styles attached to smoking (individualism, personal freedom, Americanness, pleasure, power, and so forth) is something of a commonplace in public discussions, and recent studies have documented the increase in smoking scenes in U.S. films during the 1990s (Hazan, Lipton, and Glantz 1994; Stockwell and Glantz 1997). However, less studied are the finer aspects of the interrelationship between popular culture and public health discourses, especially concerning tobacco control, which has undergone tremendous expansion in the U.S. since the late 1980s.

In this chapter I expand the focus of the preceding ones to link certain ads of the California campaign meant to "deglamorize" and "denormalize" smoking and to delegitimize the tobacco industry to the wider U.S. media culture. The California campaign arguably marks the ascendancy in health promotion of social marketing's conscious ambition to intervene through commercial advertising in the wider culture that had long been dominated by images and narratives circulated by Hollywood and independent cinema and by television serials. By looking at the form, content, and structure of counteradvertising, television serials, and films (Williamson 1978; Gilman 1995), I explore how together they reframed smoking as the tragic persistence in the postindustrial present of practices belonging to the recent but discredited past, namely that of industrial-based economies and the Cold War.[2] Drawing on several well-known California anti-smoking spots; new television series popular at that time (*The X-Files* and *Homicide: Life on the Street*); big-budget Hollywood films (including *Waterworld* and *Demolition Man*); independent or art-house cinema (including *Short Cuts, Kids, Clerks,* and *Smoke*); and new black, Latino/Chicano, and Asian American cinema (such as *Boyz n the Hood, Juice, To Sleep with Anger, American Me, My Family, The Wedding Banquet,* and *The Joy Luck Club*), I suggest that there was a shift in U.S. media culture in the early and mid-1990s in which smoking began to operate as an instrument of producing historical difference, reflection, and self-understanding. Wracked not only by oppressive and dirty factory work, urban decay, industrial pollution, and violent crime, but also by social inequities, unaccountable political authority and governance, secret power brokers, and the unspeakable terror of possible nuclear annihilation, the older era in many of these media narratives continued to haunt the present and compromised the future. Cigarette smoking was made to stand as the individual and social practice that jeopardizes citizens' chances for a stable and healthy future and that symbolizes all the risks, irrationality, and fear associated with the nightmare of the recent past, which refuses to go quietly in the night.

Sites of Media Production

Yet this shift in U.S. media culture did not take place in the same manner across all sites of production. For example, cable and broadcast television provided the most consistently anti-smoking narratives, while mainstream Hollywood productions, which had glamorized the culture of smoking for so long, did less so. Meanwhile, U.S. independent cinema, traditionally resistant to positive portrayals of model behavior and middle-class propriety, proved to be deeply ambivalent on the matter of smoking and seemed no longer able to embrace defiantly the pleasures of lighting up. Finally, it is striking that emerging black, Asian American, and Latino/Chicano cinema, whose community audiences were considered to be the privileged targets of tobacco industry marketing by anti-smoking advocates, offered an inconsistent engagement (positively or negatively) with the older culture of smoking as a social practice and a cultural icon, and in their films cigarettes played a comparatively limited role as indexes of modernity, national identity, or the recent historical past. I speculate that this may have something to do with their communities' position within U.S. society and their strife-torn histories of oppression, exclusion, resistance, and partial assimilation—that is to say, as one health researcher put it to me, the experience of "being in the culture but not of it" and thus not enjoying the privileged sense of entitlement and ownership that comes with first-class citizenship.

The "Lonely People" Spot: Modernist Pasts and Smoke-Free Futures

One of the most powerful California ads deglamorizing smoking aired in 1993. Titled "Lonely People," it takes direct aim at the older smoking culture dating from the 1930s and 1940s. Against the sound of a match striking and an eerie, melancholic jazz trumpet on the sound track, it features a series of short clips shot in slow motion of isolated, mostly middle-aged smokers indulging in their solitary habit in abstract, alienating urban spaces, from men's and women's restrooms to back alleys, rainy sidewalks, restaurant entrances, and hotel lobbies, to which smokers have been relegated by the new health ordinances. The final clip is a long shot of a single sports fan alone in a vast stadium hallway, who turns toward the off-screen bleachers as the crowd roars. The spot concludes with a white title against a black background, which reads, "There has always been a name for this: *addiction*," or alternatively, "If your closest friend is a cigarette, you need new friends. Call 1–800–NO-BUTTS," and with the tag: "Paid for by the California Department of Health Services."

The spot transforms the older culture of smoking described by Richard Klein (1993) and Lesley Stern (1999). That culture interwove cigarettes, on the one hand, with the fascinations and fears of urban, democratic social mixing in which citizens from different walks of life rubbed shoulders in crowded streets, restaurants, bars, clubs, and the workplace, and, on the other, with the thrill of the illicit pleasures of social and sexual independence less available in small-town life. In this ad, all that remains of the seduction of modern urban life are the cold ashes of alienation, the obverse side of modernist individualism, now stripped of its charm and, it should be added, of its youth. This seems to be the fate of smokers in the encroaching future of a smoke-free society. The ad brilliantly locates viewers in an imminent future by evoking settings of an earlier modernity—those public spaces of an old, mythical black-and-white urban America glamorized in film and advertising (in fact, all that is missing is a shot of a train station from the 1930s). This imminent future is thus a transformed, revamped past that clarifies and makes visible what was already there but unacknowledged years before: the truth of smoking and smokers. Here, smoking no longer involves opportunities for social pleasure, consolation, self-possession, or private self-reflection (Brandt 1990; R. Klein 1993, 17–21, 38–40; Gusfield 1993) but rather the tired reiteration of gestures of addiction. The fascination and fear of losing oneself to the enticements of a chance encounter or conver-

"Lonely People."
Courtesy of California Department of Health Services.

sation in a crowd are collapsed into the pitiful sight of mostly misshapen, middle-aged bodies "lost" to the meaningless repetition of smoking.[3]

Waterworld: Nicotine and Petrochemical Dependency

During the 1990s in Hollywood the most explicit link between the culture of smoking and a problematic industrial, urban modernity was to be found in Kevin Costner's big-budget, environmentalist film *Waterworld* (1995). In this science-fiction extravaganza—the most expensive film made to date at that time ($175 million)—the smoking subject is literally the carbon embodiment of a polluting, petrochemically based society that survives in a postapocalyptic world. There, inhabitants eke out an existence in an environment submerged by the melting of the polar ice caps. In this landless world, human society has reverted to anarchic tribalism, and the most precious commodity is soil; the greatest obsession is finding land. Some of them, like the hero Mariner (played by Costner), have mutated and sport gills and webbed feet and are much feared as less than human. However, in this transposition of the Australian postapocalyptic film *Mad Max 2: The Road Warrior*, set in the outback, to a landless world, still greater fear is aroused by the Smokers, a band of antisocial misfits who, cigarettes stuck between their lips, terrorize their neighbors with their smoke-spewing firearms, gas-driven Jet Skis, and flying machines—technologies they alone still possess. Dennis Hopper, playing up his image as Hollywood's aging bad boy and perennial counterculturalist, is Deacon, their shaven-headed, one-eyed leader.

What is their base of operations? What else but the rusting hulk of the former supertanker, the *Exxon Valdez*, which created the most destructive oil spill up to that time off the Alaskan coastline in 1989. Interestingly, the Smokers' Hell's Angels–like antics and Hopper's performance as their leader threaten to overwhelm Costner's flat style and the ponderous environmental drama (which ultimately was not a critical success and lost money at the box office). Perched atop his powered barge as he leads his forces into battle, Hopper, camping up his role with considerable humor and irony, strikes a pose midway between that of a chain-smoking Napoleon and a Roman Marc Antony. He is one of those Hollywood villains who smoke and whom audiences love to hate (R. Klein 1997). And as the *Exxon Valdez* sets out in search of land, viewers get to descend into the ship's steel bowels, which resemble nothing so much as a vast industrial forge. There, viewers are treated to the spectacle of hundreds of gang members pulling on steel oars as the smoking slave driver scans "Smoke!" (instead of "Stroke!"). The supertanker as slave-powered Roman galley articulates smoking as an addictive habit in terms of a violent, petrochemically dependent, industrial civilization and undemocratic political culture that did itself and nearly everyone else in before the beginning of the film's drama and that by the film's end is finally dispatched once and for all by Mariner, the loner-hero, who sends the Smokers and the *Exxon Valdez* to a watery grave.

The Smokers are so many lumpen proletarians and should look familiar to readers. In a sense they are the poor cousins—and lower-class customers—of the tobacco industry executives plotting the recruitment and death of future consumers in the California media campaign's "Industry Spokesman" spot, discussed previously. The ad's industry executives constitute a mix of autocratic aristocrats and working-class white ethnics of the industrial era. Together, both the executives and the Smokers are leftovers of an older society and embody all that was flawed in American modernity. As such, they relocate cigarette manufacturers, their representatives, and their deadly products outside of contemporary, globalized California temporally, economically, and even ethnically.

The "Factory" Spot: The Cold War and Devices of Mass Destruction

Still other anti-smoking TV ads and other media productions condemn cigarette smoking not only as the unsettling persistence in service-based economies of polluting practices of presumed bygone industrial societies but also as part

"Factory."
Courtesy of California Department of Health Services.

and parcel of corrupt power practices of the Cold War and illiberal modes of governance. Take, for example, another California TV spot that accompanies one of the boardroom executives on a plant inspection. It features a sequence of shots that follows a black stretch limousine with tinted windows as it clears security at an old plant gate.

The camera, alternating between exaggerated low and high angle shots, follows a silver-haired executive as he strides across the dirty shop floor to where he joins a group of traditionally clad workers manually sorting cigarettes and gingerly inspects a single cigarette as his face breaks into a self-satisfied smile. An accompanying male voice-over comments above the din of factory whistles and sirens: "The Cold War is over, yet one of the most threatening devices known to man is still being manufactured right here in America." Against a shot of an unmarked tractor-trailer exiting a town identified by a sign, "Greenfield, Pop. 7,411, Elev. 346," the voice-over concludes, "And the really frightening part is that it is aimed at men, women, and innocent children." The usual white-on-black California Department of Health Services (DHS) tag then closes the spot.

Here, to penetrate the secrets of the cigarette manufacturing process is to take a step back in time to another age: gone are the gleaming automatic machines spewing thousands of cigarettes per minute in an immaculate, laboratory-like setting, which was a common image in media reportage on the U.S. tobacco wars in the 1990s. Instead, the ad deliberately treats us with an older industrial process invoking anonymous workers toiling in a dirty, dark décor of clanging gates, whining elevators, and scurrying forklifts with no automatic machines in sight. There, away from the eyes of public scrutiny, a weapons factory continues its deadly operations with impunity even today. The ad transfers the universal horror at an earlier generation's development and deployment of nuclear weapons that primarily targeted civilian population centers to an entire culture of smoking and the industry that underwrote it. In so doing it gives the anti-tobacco industry narrative an unexpected Cold War inflection, for in a sense it now counterposes the industry to mythical preindustrial small-town America not simply as a polluting industry but as a political threat approaching that of Communists and New Deal liberals during the height of McCarthyism (L. A. Rose 1999). Interestingly, the ad mobilizes one U.S. Cold War narrative—the (Communist/liberal) Enemy Within—against another—the nuclear arms race—in order to discredit the tobacco industry.

Television's *The X-Files*, the National Security State, and Illiberal Governance

In the wider U.S. media culture, the most sustained anti-smoking productions of the 1990s were by television, led by Fox Television's immensely popular series *The X-Files* (1993–2002), which won numerous Emmy and Golden Globe awards over nine seasons. A provocative mix of spy, horror, science fiction, and melodramatic genres, it powerfully articulated questions of governance, health, and historical difference to an immense national and international audience and deserves extended analysis here. Virtually every week *The X-Files* intertwined narratives of betrayal of citizens' trust by high government and industry officials with the culture of smoking and gave them an explicit generational twist. There, unaccountable, absolute power, the balance of terror, and the possibility of mutually assured destruction live on in *The X-Files*' Cigarette Smoking Man, whose nickname, the Cancer Man, substitutes the disease for the pathogen. Member of a shadowy government within government, he is one of a group of gray-haired men in dark suits (called the "Syndicate") drawn from the old white Anglo-Saxon protestant (WASP)–dominated business and government elite who still holds the secret reins of the national security state as if the Cold War were still on (Knight 2002). Played by Canadian actor William Davis (who, it turns out, is a major anti-smoking activist in his country),[4] he is omnipresent but would be virtually invisible were it not for the telltale palls of smoke, ashes, and crushed cigarette butts and packs he leaves behind as he glides through dark rooms and the halls of power. Nameless, he is apparently without wife, children, family, or friends. A failed artist to boot, he's the abject modernist individual personified, whose lack of human ties frees him to decide effortlessly the fates of ordinary citizens, FBI agents, presidents, and aliens from other galaxies alike. At his command stand a contingent of expressionless agents and baby-faced soldiers who execute his orders wordlessly.

Each episode follows young FBI agents Fox Mulder (David Duchovny) and Dana Scully (Gillian Anderson) as they investigate cases that exceed the explanatory categories of forensics, criminology, and known science. Their work takes them into the realm of paranormal phenomena, which, if they remain unresolved by standard methods by local and national law enforcement, are reclassified under the FBI's "X-Files" and brought to the agents' attention. More often than not the trail of evidence (everything from unexplained deaths, mysterious disappearances, UFO sightings [dating from the inception of the Cold War in Roswell, New Mexico], and alien abductions to occurrences of witchcraft and vampirism, serial killings, and viral outbreaks) leads back to abuses of secret government-sponsored research, weapons development, and, eventually, a possible conspiracy by the Syndicate's aging Cold Warriors to collaborate in the ultimate betrayal—the alien colonization of Earth through the hybridization of human and alien genomes with the help of various governments. Here, Cold War *realpolitik* leads straight to the destruction of the very populations with whose protection the Cigarette Smoking Man and his colleagues have been presumably entrusted by the old state paternalism.

Liberal and well-intentioned to a fault, often the agents, especially Mulder, will intervene to defend indigenous peoples (Native Americans), former test subjects, the mad, and refugee aliens—those who stand on the margins of the national security state and who have suffered at the hands of U.S. government researchers and agents. These are the victims of liberal government's most illiberal practices. In this fashion, a double-thematics of "otherness" underwrites the series in the form of the paranormal and the nonhuman, on the one hand, and of unspeakable actions and policies by the Cold War generation, on the other. In the truest manner of spy, thriller, and horror genres, much of the series' suspense consists in not knowing what type of otherness is at issue in a particular case, and the suspicion that what is most other or alien (either good or bad) may be also be present in what is most familiar and intimate.

During the 1990s, *The X-Files* was the most consistently anti-smoking television drama in the U.S. The series both ridiculed cigarette smoking and exploited the larger culture's fascination with it. In an almost arbitrary fashion, smokers became the embodiment of bodies out of control. For example, in one 1994 episode a smoker, who is the survivor of secret research gone awry, involuntarily vaporizes hapless citizens who happen into his shadow. The show is not without humor: one of his victims is a Morley tobacco executive, who, oddly enough, turns out to be a nonsmoker ("Softlight," 13 May 1994). Another smoker, an undisciplined female employee of the Transcontinental Express delivery company, while on one of her many cigarette breaks in the restroom, is stung to death by killer bees ("Zero Sum," 27 April 1997). These are mutant bees that were developed to spread the smallpox virus and to wipe out the human race in order to ready Earth for the colonization by the hybrid species. Or as yet another example, after a particularly harsh divorce settlement, a husband, in a rage, carefully burns holes through the faces of his former wife and children in family photos with his cigarette ("Never Again," 2 February 1997). Shortly after going to a tattoo parlor in a seedy neighborhood, he begins to hear the taunting voice of his spouse and is driven to kill other women. Even the coldly calculating Cigarette Smoking Man, whose Morley cigarettes indicate a terrifying self-possession and detachment from the world around him, signifies a control over self and others gone berserk. And several years into the series, his body in turn becomes literally out of control: he learns that his body cells are multiplying at a geometric rate, and he cuts a deal with a refugee alien to cure him of his cancer in exchange for sparing the alien's life.

The most spectacular episode aired 17 November 1996. In "The Musings of the Cigarette Smoking Man," the villain is invested with every paranoid fantasy about the most traumatic domestic events in recent U.S. history during the Cold War. The episode begins with the creaking of a door opening and the close-up of two shoes in a darkened doorway. A rat scurries by as a smoking cigarette butt drops to the floor. The camera tilts up to reveal the looming figure of the Cancer Man as he scans the darkened room and lights another fag with a lighter bearing the inscription, "Trust no one." The Cancer Man then proceeds to set up an eavesdropping device that picks up a conversation between Mulder, Scully, and Frohike, their informant. The device effortlessly defeats Frohike's state-of-the-art countersurveillance measures. Meanwhile, the camera follows the shadowy movements of the Cancer Man as he sets up a tripod on which he mounts a rifle with a telescopic sight. As he trains the rifle on the door leading to Frohike's office, Frohike begins to recount the tale of the villain's life, a narrative that then fills the rest of the show's hour. What sort of story does Frohike deliver? The beginnings of the Cancer Man's career as a government agent and of his cigarette habit. The two coincided with the most spectacular U.S. assassination of the century: the murder of John F. Kennedy. If viewers thought Lee Harvey Oswald was the assassin, they were mistaken: it was the Cancer Man, who then set up Oswald to take the fall. His work done, he lit up his first cigarette that very afternoon. Five years later he will murder Martin Luther King Jr.

Indicting the Cold War Generation

From its inception, *The X-Files* mixed the detective and spy genres not only with horror and science fiction but also with family melodrama that sets a younger generation of Americans in opposition to its elders and their deeply flawed modernity. The scandals of Cold War government cover-ups of what it knows about UFOs and its botched human experiments affect Mulder and Scully directly: early on, viewers learn that Mulder witnessed the alien abduction of his younger sister and that Scully herself was abducted ("Ascension," 21 October 1994) and received an implant of possibly alien technology in her skull ("Nisei," 24 December 1995). At another moment, U.S. national security agents out to assassinate Scully mistakenly murder her sister ("Blessing Way," 22 September 1995), and later still, she will find herself inexplicably pregnant ("Requiem," 21 May 2000), possibly with a hybrid human-alien life form ("Per Manum," 18 February 2001). Moreover, their family ties to the national security state are close: Scully's father was a senior naval officer, and Mulder's was an employee of a national security agency and, as it turns out, a close collaborator with the Cigarette Smoking Man on various projects, including the development of a virulent strain of the smallpox virus that involved former Nazi scientists ("Paper Clip," 29 September 1995). As if that weren't enough, *The X-Files* throws the generational narrative into higher gear when Mulder discovers that the Cancer Man not only had his father murdered to silence him ("Anasazi," 13 April 1995) but most likely had an affair with Mulder's mother ("Talitha Cumi," 17 May 1996). Evidence accumulates that the master of deceit himself may indeed be Mulder's father ("Demons," 11 May 1997; "One Son," 14 February 1999).

The ambivalent fascination that the Cigarette Smoking Man exerts over most viewers—according to Richard Klein, the evil mastermind attracts viewers as much as he repels them (R. Klein 1997)—reaches a peak when we learn that he indeed has a family—a wife and a son—which he has ruthlessly exploited to further his own designs: he forced his wife to be a test subject in the hybridization program, something similar to which, when Mulder confronts him, he claims that Mulder's own father did in allowing aliens to abduct his sister Samantha ("Two Fathers" and "One Son," 7 and 14 February 1999). Meanwhile, the Cancer Man's own son turns out to be also an FBI agent (Agent Spender), whose task has been to befriend Mulder and keep tabs on his activities. A potentially deadly sibling rivalry ensues, but in yet another twist of the Oedipal logic, Agent Spender sabotages his own assignment and allows rebel aliens bent on stopping the hybridization and colonization programs to incinerate the Syndicate members and their families. At this point the Cancer Man, who has escaped, commits the ultimate paternal violence: he confronts his son and shoots him dead ("One Son").

In *The X-Files* the bankruptcy of the national security establishment, the abuses of public trust, the crisis of Cold War scientific certainties faced with paranormal phenomena, and the wrenching twists and turns of family melodrama converge to deliver an indictment of a entire generation and the industrial, Cold War society it managed and ran. The sheer indifference of those gray-haired men to the lives of fellow citizens and foes alike is relentlessly signified by cigarettes, which destroy both their users and those around them. If, according to the traditional ideal of government, the government of others always involves as its precondition the government of oneself (Foucault, 1985, 3–13), then smoking signifies the greatest violation of that principle. The Cold War logic of mutually assured destruction is identical to the logic of smoking cigarettes: in the words of an anti-smoking billboard in California featuring a lit cigarette, it is "a weapon that kills at both ends." And like cigarettes, there is nothing more collectively significant and entailing of others, yet nothing more intimate and personally involving, than the Cold War and its legacy of a deeply flawed industrial modernity.

Homicide: Life on the Street: Smoke-Free Detectives

Meanwhile, in other televised serials those remaining tokens of hard-boiled alienation whose cynicism and minimalist urban styles of not living, but simply surviving, had nonetheless had their own world-weary heroism—the gumshoes and dicks—began to go smoke-free. This occurred at a time when anti-smoking discourse spread across the ideological broadcast spectrum, as television series, liberal and conservative, like CBS's series *Murphy Brown* (1988–98) starring Candice Bergen and *Touched by an Angel* (1994–2003), regularly began to focus attacks on smoking and the tobacco industry. Witness NBC's *Homicide: Life on the Street* (1993–99), set in the Homicide Division of the Baltimore Police Department. Critically acclaimed for its gritty realism, innovative scripts, camera work, and superb acting, it features a group of intense homicide detectives of all ages and backgrounds, most of whom smoke at the start of the series. But beginning with the eighth episode of its opening 1993 season, macho cigarettes begin to disappear from the lips of detectives like Howard, Kellerman, Lewis, Felton, Bayliss, and Pembleton as they attempt to quit ("Smoke Gets into Your Eyes," 24 March 1993). Several seasons later Kellerman and Bayliss continue to smoke, but their habit is no longer woven into their tense negotiations with their surroundings or an intimate part of their personal style. The fags just hang there awkwardly from their lips.

U.S. Independent Cinema: New Ambivalence

Independent cinema in the U.S. was not immune to the new discourse on smoking emerging in the 1990s in big-budget movies and highly rated TV shows. Generally speaking, non-Hollywood films aren't known for promoting favorable representations of conventional middle-class behavior, but one can detect subtle shifts in terms of visual electronic media's culture of smoking.

Take, for example, Robert Altman's celebrated film *Short Cuts* (1993), situated in 1990s Los Angeles and featuring some of the most popular Hollywood actors of that time (Lily Tomlin, Matthew Modine, Tim Robbins, Andie MacDowell, Robert Downey Jr., and Lili Taylor). Based on a collection of Raymond Carver short stories of the same title set in contemporary Washington State, the film follows the interlocking lives of various white, largely middle-class characters in the course of a twenty-four-hour day that stretches from the opening eerie scene of helicopters spraying vast areas of Los Angeles at night against the medfly infestation to the daytime earthquake that closes the film. Throughout, we get to observe the characters and their family lives: every single household is dysfunctional, and, almost without exception, every adult smokes. In the older film and television culture, smoking operated as an aesthetic index of character and social background and visual anticipation or confirmation of turns in the plot and dialogue (R. Klein 1997). In *Short Cuts* smoking is so universal that it slips from being a particularizing dramatic and psychological device to an activity that blankets the film in its entirety. Its very banality matches the flatness of the film's overall affect—one of detached irony applied indifferently to all characters and situations. Thus, at first glance, smoking in *Short Cuts* seems to start out as a sly protest against the prevailing anti-smoking climate in California, yet by the film's end it approaches being a moral index for the listless anomie of daily life in late-twentieth-century postmodern Los Angeles.

Smoke and *Blue in the Face*: Smoking as Nostalgia for Urban Modernity

Or look at film director Wayne Wang and writer Paul Auster's two films released in 1995, *Smoke* and *Blue in the Face*. In these films the action centers on the comings and goings of patrons in a Brooklyn neighborhood cigar store. The center of neighborhood activity, the store is presided over by Auggie Wren (played by Harvey Keitel), the quintessential native New Yorker who belongs to the generation born around World War II and still embodies the older U.S. urban modernity of white ethnicity. Here, the pleasures and cares of smoking receive extended embodiment and articulation, and smokers are allotted space to voice their experience of the habit. These films are among the very few in the U.S. that explicitly do so in the mid-1990s.

What is so striking is that *Smoke* and *Blue in the Face* are able to evoke cigarette smoking positively but only in a nostalgic mode—in this case, nostalgia for old Brooklyn smoke shops and their declining working-class neighborhood culture of straight males who engage in cross-class and cross-race (black, white ethnic, Latino) sociality of shooting the breeze, sports talk, betting, business deals, local gossip, and disquisitions on women. The nostalgia shows through in the films' charm but also in the characters' shopworn tales common to modernist novels and film noir of masculine friendship and betrayal, lost fathers, middle-age crises, parasitic former wives, murdered loved ones, lost savings, and writer's block. It is an urban culture that is on its way out but with no hint as to what will replace it and the culture of smoking that it supported.

It is worth noting in passing that in the realm of print media, similar nostalgia suffuses Richard Klein's book *Cigarettes Are Sublime,* which appeared in 1993. An explicit paean to Euro–North American modernism (as embodied in poetry, novels, film, and photography) and the rise of the cigarette as a tool and symbol of individualism, risk taking, and sexual and aesthetic pleasure, the book had the goal of proposing, in a time when anti-smoking sentiment was sweeping America, that the best way to have smokers quit was for the culture at large to acknowledge the benefits and pleasures of smoking cigarettes and that their very dangerous nature constituted one of their fundamental attractions. This is why, Klein would later argue, anti-smoking campaigns will never work with youth in the U.S., for forbidden fruit always remains attractive by nature (R. Klein 1997).[5] As is the case for Wang and Auster, for Klein the richest acknowledgement of those pleasures is now located in the past, in the bygone Euro–North American cosmopolitan culture of modernist aesthetics and subjecthood, from Baudelaire's poetry, Svevo's novel *The Conscience of Zeno,* and Bizet's opera *Carmen* to Bogart, Bacall, and other icons of American film noir to, finally, the photos of Cartier-Bresson, Davidson, Benvenisti, Erwitt, Barbey, Hoepher, Capa, and Kalwar in Magnum Photos' portfolio (R. Klein 1993).

No Future: *Kids* and *Clerks*

Unexpectedly, in U.S. independent films focusing on white slacker, grunge, and skateboard youth cultures, cigarettes did not fare much better: teenagers and young adults, if they practiced the habit, were considered likely candidates for a sorry fate or simply fools. A case in point is Larry Clark's AIDS drama (cowritten by Harmony Korine) *Kids* (1995), a film that caused a sensation by virtue of its raw depiction of the speech and sex lives of young teenagers (twelve-to-fourteen-year-olds) hanging out in the summer in Manhattan's Washington Square Park with other skateboarders. The film follows the sex-obsessed protagonist Telly (Leo Fitzpatrick), who is ignorant of the fact that he is HIV positive, and his group of friends over the course of several days, as they smoke, do drugs, get drunk, beat people up, make love, and rape. Across the East River from *Smoke*'s beloved cigar store, a new generation of working- and middle-class kids pursue a dissipated and violent existence in a decaying postmodern landscape that is devoid of any promise.

Telly is something of an updated straight, younger version of "Patient Zero," the Canadian flight attendant Gaëtan Dugas, blamed for spreading HIV among U.S. gays by Centers for Disease Control and Prevention (CDC) officials and by reporter Randy Shilts, author of *And the Band Played On* (1987), one of the most influential accounts of the early years of the HIV/AIDS pandemic (Treichler 1989). In the HBO film version of Shilts' book of the same title broadcast in 1993, Dugas' "irresponsible" gay sexual behavior is marked by the camera's focus on his stylized smoking. Interestingly, in *Kids* it is not Telly who smokes but his mother. In a sequence that encapsulates well the film's gritty but moralizing narrative about out-of-control teenagers and lack of adult supervision, Telly returns home with his friend Casper in search of spending money. After quietly ransacking his parents' bedroom for cash, the two boys emerge to find Telly's mother smoking while nursing Telly's baby brother. In a sequence of shots the camera transfixes the breast, the cigarette, and Casper's lecherous gaze. Here, the adolescents' unconscious behavior is more than matched by the mother's; the cigarette and breast are framed as drug delivery devices that the sequence seems to suggest are the special attributes of an older, discredited generation that has let the world of unregulated private and public space of *Kids* come to pass.

Another example is director Kevin Smith's first full-length feature, *Clerks* (1995), which won prizes at the Sundance and Cannes film festivals. It, too, marked a shift in the culture of smoking in U.S. independent film. With considerable humor, the film affectionately recounts the ups and downs in the lives of two nonsmoking white college-age clerks, hard-working Dante (Brian O'Halloran) and his cynical, lazy friend Randal (Jeff Anderson). They work in adjoining convenience and video stores in depressed, small-town New Jersey. Their colorless existence in dead-end jobs lends itself to the deadpan irony of twenty-somethings with no prospects. Many, if not all, characters smoke, and, as one might expect in an independent film, *Clerks* indulges in funny comments on contemporary anti-smoking sentiment and the new restrictions on the sale of cigarettes to minors that convenience store clerks must enforce, an almost impossible task given the low pay and crushing boredom of their jobs. Thus, in one episode Randal, while minding the store for Dante, distractedly sells by mistake a pack of cigarettes to a five-year-old girl, and Dante winds up slapped with a $100 fine by a state health official. In another scene, one vehemently anti-smoking man berates fellow customers for smoking and starts a protest against the sale of tobacco, during which customers pelt Dante with cigarettes. Things come to a humorous halt only when the convenience store clerk's girlfriend douses the crowd with a fire extinguisher, and the irate customer turns out to be a chewing gum sales representative trying to undermine tobacco industry competition. However, here, independent cinema's mockery of conventional attitudes takes a turn and ends up satirizing young smokers as mindless creatures of their habit and vulnerable to the manipulations of others, in this case salesmen masquerading as anti-smokers. In contrast to them stand the film's two heroes, who, the sole repositories of smarts, must witness and suffer the foolish behavior of their less enlightened peers throughout the film.

Filmmakers of Color and the Culture of Smoking

The 1990s also marked the emergence of new black, Latino/Chicano, and Asian American film in the U.S., much of it independently produced.[6] Here, in terms of the culture of smoking, it is less a question of an unexpected ambivalence, as in the case of other independent films, than of a lack of sustained engagement with smoking as an issue in and of itself, let alone as a fundamental metaphor for contemporary social life or the passing world of generations belonging to another era. A good number films focused on the desperate situations communities faced in the 1990s as they struggled with not only dismal economic prospects for many of their non-middle-class youth but also the pressures of criminal drug trade, the rise of gang violence, and oppressive police tactics dating from the 1980s. In Los Angeles these conditions led to the riots in May 1992 following the acquittal of the police officers responsible for beating black motorist Rodney King (Massood 2003). As one might expect, in these films smoking is an index of wayward youth, such as the protagonist Strike's homeboys in Spike Lee's *Clockers* (1995) or the murderous Bishop (played by Tupac Shakur) and his friend Raheem (Khalil Kain) in Ernest R. Dickerson's *Juice* (1992). Even so, the smoking is rarely stylized or developed in African American films as in other contemporaneous independent films such as Quentin Tarantino's *Pulp Fiction* (1994).[7] And in other black films touching on similar material, smoking is either barely present as in Albert and Allen Hughes' *Menace II Society* (1993) or absent all together as in John Singleton's *Boyz n the Hood* (1991) and *Poetic Justice* (1993). Perhaps the purest reproduction of the older smoking culture in African American film is in a middle-class female setting: Forest Whitaker's *Waiting to Exhale* (1995), based on Terry McMillan's novel of the same name, about the twists and turns in the lives of four African American women who console one another and share friendship and, occasionally, cigarettes. But it stands as very much the exception rather than the rule. Smoking has no cultural role in other films featuring middle-class dramas such as Lee's *Jungle Fever* (1991), Singleton's *Higher Learning* (1995), and Charles Burnett's *To Sleep with Anger* (1990).[8]

Many Latino and Chicano films of this time recounted histories of immigration and economic struggle that involve generational narratives that move back in time to the 1930s, 1940s, and 1950s, and forward to the 1990s. Smoking works as a sign of gang membership or a brush with the law but inconsistently, as, for example, in Gregory Nava's *My Family* (1995): Chucho Sanchez (Esai Morales), the older brother involved in drug trafficking and later murdered by the police, never smokes, whereas his younger brother Jimmy (Jimmy Smits), jailed for armed robbery, does. And when Nava does *Selena* (1997) on the Tejano music star Selena (Jennifer Lopez) whose life was cut short by the head of her fan club who shot her, cigarettes do not operate as privileged signifiers of the irregular lifestyle of musicians. The most systematic association of smoking with the hard-bitten criminality of prisoners and former prisoners is found in Edward James Olmos' *American Me* (1992), which follows the story of a Chicano crime gang boss Santana (played by Olmos) from the 1940s through the 1970s. Here, moments of decisive action, self-reflection, and macho posturing among men are underscored by lighting and stubbing out cigarettes and blowing smoke.

Nowhere is the older culture of smoking consistently reproduced in African American and Latino films. By the same token, nowhere is the culture of smoking designated as such and made to stand in for an older generation or historical period or oppressive government practices as in anti-smoking spots, *Waterworld*, *Smoke*, or *The X-Files*. Revealing in this respect is the case of Burnett's *To Sleep with Anger* (1990), which recounts the disruption of a middle-class family's life by the sudden appearance of an old friend, Harry (played by Danny Glover), on their doorstep. He's a mesmerizing, seductive ghost out of their distant Southern past and brings with him the old folkways, superstitions, and vices of the black rural poor, who later migrated to West Coast urban centers during the Second World War in search of industrial work. Harry's presence causes family tensions to come to a boil. He may wield a switchblade, gamble, drink grain alcohol, and tell stories of murder and mayhem, but Harry doesn't smoke or chew tobacco; neither do his old cronies when they come by to see him and later when they get together on a stormy night to drink and play cards. In Burnett's film, smoking does not function as a tool of establishing historical difference, reflection, or self-understanding. Smoking doesn't tell a story of any kind.

Even in a properly historical drama like Julie Dash's lush *Daughters of the Dust* (1991), set in the Sea Islands off the Georgia and South Carolina coast in 1902, no one in the four-generation extended family visibly consumes tobacco except the matriarch, Nana, who quietly indulges in chew. In the drama of the family's migration to the mainland in search of a better future, the beckoning modernity is not signified by cigarettes.

Similarly—and in marked contrast to Hong Kong, Taiwanese, and Chinese films—in the early and mid-1990s films directed by Asian Americans on Asian American topics and Asian films staged in the U.S. do not replicate the allure and fascination of cigarettes stemming from the older, dominant culture of smoking dating back to the 1930s and 1940s. Nor is there much in the way of traces of cinematic depiction of smoking in ordinary people's lives as in Wayne Wang's breakthrough film *Chan Is Missing* dating from the early 1980s (1983). For example, Wang's film, *The Joy Luck Club* (1993), based on Amy Tan's novel of the same name, which delivers the involved story of four women and their scattered families that survived the upheavals of the Asia Pacific War and Chinese Revolution and immigrated to the U.S., contains no smoking whatsoever by older, younger, men, women, soldiers, courtesans, police, or refugees. None. And in Taiwanese director Ang Lee's English-language film *The Wedding Banquet* (1993), a comic story set in a gentrifying and globalizing New York about a gay Taiwanese-American couple and their attempt to negotiate the pressures of the Taiwanese character's parents for him to marry, the only major character who smokes is the gay groom's father, a retired senior officer of the Taiwanese army. And during the lavish wedding reception, smoking is seen but remains undramatized. At best, smoking would appear to index a generational difference marked by the experience of immigrating to the U.S. and assimilating to an alternate modernity presumably more tolerant of gay sexuality, but this is suggested at best, and the film doesn't develop these connotations of smoking in a systematic way.

Hollywood Backlash: *Demolition Man*

Even while cable and broadcast television seemed to be on the verge of going smoke-free and while independent and non-mainstream filmmaking underwent shifts or revealed in some cases an indifference to smoking as a social practice and cultural icon, in commercial Hollywood productions the first major signs of a negative reaction against anti-smoking sentiment and the promotion of healthy lifestyles appeared in the satirical film *Demolition Man* (1993), directed by Marco Brambilla. Like the anti-smoking spots, *Waterworld*, and *The X-Files*, it stages a drama of past industrial urban civilization and postindustrial present or future, but here a smoke-free twenty-first-century sci-fi future becomes the target of much humor. Starring Wesley Snipes, Sandra Bullock, and Sylvester Stallone (who once accepted $500,000 to smoke Brown and Williamson tobacco products in five films [Glantz et al. 1996, 365–367]), the film is set in postapocalyptic LA in the year 2032, twenty-two years after the "Big One" (an 8.9 earthquake) leveled southern California in 2010. From the ashes of greater Los Angeles a utopian society has arisen named Sanangeles Metroplex, a sparkling new city combining the areas of San Diego, Santa Barbara, and Los Angeles. Run by the authoritarian Dr. Raymond Cocteau (Nigel Hawthorne), Sanangeles is a peaceful, nonviolent, communitarian-based society devoid of crime where healthy, clean living is mandatory: the consumption of alcohol, cigarettes, caffeine, chocolate, and meat is banned; so are the use of foul language, protected or unprotected genital sex and kissing ("fluid transfers"), pregnancy, violent toys, graffiti, and pornography. Citizens walk around clad in flowing silk garments and wide hats; the civilian men have an effeminate air, and those at the top speak in affected (English) accents. It is as if rough-and-ready American society of yore had been "feminized" through the import of Asian-inspired New Age philosophy and sartorial fashions and English patrician manners. Opposed to the new social order is an ineffective movement of homeless and politically incorrect renegades who literally lead an autonomous existence underground but at the price of near-starvation.

All is well until two hypermasculine men from the twentieth century burst in upon the city. Simon Phoenix, a psychopathic killer played by Snipes, breaks out of cryo-prison, where he had been in suspended animation since 1997, and begins a murderous rampage. Helpless before the onslaught, the nonviolent police department appeals to have John Spartan (Stallone), the violent cop who had Phoenix put away, defrosted to stop him. The only people in Sanangeles Metroplex who understand and value Spartan are an aging black street cop, who remembers the old days, and Lenina Huxley (Bullock), a restless young recruit who is committed to the new values but who collects twentieth-century memorabilia and yearns for action.

Besides a violent, masculine nature, another twentieth-century characteristic shared by Phoenix and Spartan is smoking. In the film's opening sequence, a confrontation unfolds before viewers' eyes as the LAPD SWAT teams stage a military-style assault on a building where Phoenix and his fellow gang members have sequestered kidnapped civilians. Camping up his earlier role in Rambo films of the 1980s, Stallone as Spartan rushes to the rescue aboard a helicopter as explosions shake the area and tracer bullets streak through the sky. A title flashes: "Los Angeles, 1997." The second-largest U.S. city has been transformed into an urban Vietnam. In the face-off, Phoenix (after snorting cocaine) conspicuously lights a Marlboro (underscored by extreme close-ups) and taunts Spartan, standing unawares in a pool of gasoline, which Phoenix proceeds to ignite with a stylized toss of the cigarette. The ensuing conflagration and explosions cost the hostages their lives and Spartan his career as a police officer, but he gets his man.

As for Spartan's cigarette habit, we encounter it when he reenters the land of the living thirty-five years later after a stint in cryo-detention. The issue that initiates the humorous clashes between him and his twenty-first-century surroundings is smoking: immediately after he awakens from suspended animation and is informed that his wife and daughter died in the Big One and that a new world has arisen in place of the old Los Angeles, Spartan's first request to the perplexed guards is, "Give me a Marlboro!" "What is a Marlboro?" "A cigarette, just give me a cigarette, any cigarette!"

The remainder of the film follows through on the drama's initial premises: Spartan ends up, once again, breaking the rules in order to bring in Phoenix; he even allies himself with the underground to overthrow Dr. Cocteau when he discovers that Phoenix was deliberately set free by the powers that be in order to protect their secret racket; and Huxley and Spartan, in true Hollywood fashion, pursue a love interest in which the urges of the body win out over the rules of order. In a certain fashion, the film can be read as the revenge of the twentieth century on proponents of the twenty-first and its revamped modernity, and the return of masculine, non-middle-class behavior (black and white ethnic) to an all-too-middle-class (coded as Asian/European and effeminate) California. But what is striking in this satire of new middle-class health norms is that the cigarettes that made a dramatic entry early on in the film are never seen again. Even as the film makes fun of current narratives that wish to relegate smoking to the dirty and violent past, never after the opening sequences that establish character and plot does it picture on screen people casually smoking, even in the underground or after the overthrow of Dr. Cocteau. In the end it would appear that the drama of historical difference sets limits on the film's satirical high spirits: cigarettes do not cross over into the twenty-first century, even as the new political order stands poised to integrate the bodily pleasures of the preceding century. Cigarettes do not operate as ciphers of reborn pleasure and renewed American heterosexual masculinity. That role would have to await tobacco industry efforts several years later to capitalize on a shift of public sentiment away from the anti-tobacco trend.

By 1996, according to anti-smoking advocates, researchers, and public health officials, a full-blown backlash against anti-smoking campaigns was under way in the U.S. (Carol 2000; Glantz 2000); and in Hollywood there was a sharp rise in films prominently featuring smoking far exceeding current prevalence rates

(Hazan, Lipton, and Glantz 1994), to the point that First Lady Hillary Clinton openly admonished Hollywood for allowing it, and the *Washington Post* in 1997 began to list at the end of its movie reviews not only content ratings regarding the amount of sex, adult language, and violence but also the amount of smoking. Richard Klein argued in a lengthy *New York Times* article that the reversal of public opinion was an utterly predictable reaction to demonizing cigarettes to the point of making them once again attractive as forbidden fruit, especially to youth (R. Klein 1997); even public health researchers shared a similar position with respect to youth campaigns (Glantz 1996). Within California, blame was laid in part on the implementation in 1994, without adequate education of citizens, of Assembly Bill 13, the statewide ban (as opposed to locally developed ordinances) on smoking in workplaces, restaurants, and bar areas. In this view, the extension of the ban in 1998 to all non-owner-operated bars without preparing the public through a media campaign only made matters worse. I also add, following somewhat Klein's arguments, that the foundations of a backlash were perhaps laid in California and elsewhere by campaigns' strenuous deglamorization of smoking that emphasized the foul odor of stale smoke, filled ashtrays, and smokers' breath and mouths and by their apparent reluctance to deal with the issue of pleasure—by no means a simple task.

Whatever the case may be, R. J. Reynolds seized upon the persistent association between danger, illegitimate pleasure, smelly bodies, and smoking in tobacco control education to launch a counter "No Bull" campaign in 1996. One prominent billboard for Winston cigarettes without additives featured a large title in red against a white background, "Finally a butt you can kiss" with the tag, "Winston. No Bull." The ad turns the accusation of hypocrisy and misrepresentation of the tobacco industry concerning its products (and the questionable additives they contain) against the public health community by adopting a rough, no-nonsense language to suggest mockingly that there are many pleasures in life besides smoking that involve smelly body parts such as sex. Moreover, the ad cleverly turns the popular portrait of smoker-cum-stooge of the tobacco industry into that of the fiercely independent individual who may have to put up with the more powerful (the boss, to be sure, but *also* the tobacco industry) and take risks (dangerous habits) but does so knowingly in order to get what he or she wants. In case the ad's rough language left any doubts as to the social class (and perhaps gender) with whom the ad associates itself, on other Winston billboards and even matchbooks appears the photo of an imposing and barrel-chested all-American (presumably heterosexual) white male blue-collar worker, standing in overalls, arms defiantly folded, next to the tag, "Winston. No Bull." The ad elicits the perception that the older culture of smoking and the twentieth-century industrial world to which it belongs may be on the defensive, but it refuses to quietly fade away.

Alternate Modernities versus the Mainstream Culture of Smoking

This has been an all-too-brief foray into U.S. media culture of smoking in the early and mid-1990s and allows little room to appreciate the playfulness and irony driving many of the examples from popular culture. The goal was not a complete taxonomy and statistical sampling of television, film, and anti-smoking advertising. Rather, its point and purpose was to tease out new shifts in discourses that made smoking and its health hazards operate as at once symbols of the recent but discredited past, markers of historical difference, and a tool for reflection on U.S. modernity. As local, state, and national campaigns deglamorizing and denormalizing smoking gained in intensity, there was a moment in which the older media culture of smoking faltered: in major Hollywood productions it briefly became the object of scorn, while in television series it began to disappear all together. To be sure, by the mid-1980s smoking had largely exited from the small screen (Hazan, Lipton, and Glantz 1994) with the exception of broadcast films and TV series like *Cheers* (1982–93), set in a Boston bar. In the 1990s it made a comeback but in many cases only to be attacked (*Murphy Brown*, *Touched by an Angel*), marginalized (*The X-Files*), retired and consigned to irrelevancy (*Homicide*), or otherwise assigned to an outmoded past, either dangerous or regretted.

A movement between rejection of and nostalgia for the older smoking culture emerged also in U.S. independent cinema, traditionally a cultural haven for resistance to middle-class norms. Smoking remained an index of violent character (*Juice*, *American Me*) and social anomie or marginality (*Short Cuts*, *Clerks*, *Smoke*, *Waiting to Exhale*) but now often unattractively (*Menace II Society*, *Clockers*) and without purposeful style (*Kids*) when it wasn't absent altogether from films exploring similar or related themes (*Boyz n the Hood*, *Poetic Justice*, *Selena*).

Overall, it would appear that the deepest engagement with smoking as a social practice, cultural icon, and index of character and identity lies in films and television serials written and produced by and for members of mainstream U.S. culture. It is as if, in the end, many attempts to rethink dominant U.S. culture and problematize the present in terms of the past had as their obligatory passage point the culture of smoking. Following the work of other scholars, I have suggested several reasons for this, not least of which is the fact that manufactured cigarettes helped lay the foundation for national markets in the U.S., which in turn underwrote new patterns of consumption that, in the words of U.S. anthropologist Robert Foster, "materialized" a national American identity (R. Foster 2002, 113). The connection between everyday consumption and national culture was perhaps deepest for those who could see themselves as not simply in but also of the culture—groups of European descent assimilated into the norms of middle-class life. All of this leads me to speculate that cigarettes in the U.S. in the late twentieth century may have played a more powerful role defining and shaping mainstream culture than those cultures standing on the margins, if recent African American, Latino/Chicano, and Asian American film production is any indication. After all, these other communities' relation to twentieth-century U.S. modernity and American identity has arguably been complexly different and marked by particular histories of racism, exclusion, and limited assimilation. These histories may indicate the limits of theories of assimilation that seek to include marginal communities through the introduction of a shared timeline of modernization. In any case, it seems that the older U.S. culture of smoking does not make claims on the attention of these filmmakers of the same order it does for directors and producers stemming from the majority population. Reassessing the present and the past and imagining possible futures for one's community from the position of little experience in owning the dominant culture does not lead to an extended reconsideration of smoking as an icon of what it means to live in the U.S. today.

Part II

France and Japan: Alternate Trajectories

I think Europe is like a laboratory. Europe is fascinating to me because it shouldn't be so much different than the U.S. . . . But when it comes to smoking and tobacco use, it is very different than here. And especially when I hear what you've said now about France doing these things real early on. Why is it that they continue to smoke everywhere all the time? We were there, my family and some friends and I were there on a hiking trip when we went to England and Ireland a couple of years ago. We just couldn't believe it, and we had teenagers with us. . . . In fact, my sister and cousin kept pulling me back and saying it's their country, they can smoke if they like to! We'd go listen to music in the pubs and people were smoking and we kept asking them to open the windows. And I know the public health people are somewhat frustrated.

—Minnesota health official (interview, 1997)

In adopting different norms of hygiene have Europeans strictly listened to sanitary rationality or have they been caught up in a great cultural movement that prizes more highly the body and life? It is an open question. The struggle against tobacco use has certainly a side which remains hidden to its most active advocates. It belongs to the history of customs whose effects escape our actions.

—Serge Karsenty and Albert Hirsch, "Une lutte contre le tabagisme" (1992)

In the Meiji Era, Japan enacted a law, ahead of other nations, to ban smoking by minors. Lawmakers in those times had the foresight to protect the health of youngsters. Their successors, however, showed little interest in the harmful effects of smoking on youth and the general public.

—Kiroku Hanai, "Common Sense Up in Flames" (2004)

In part 2, I turn to the tobacco control campaigns of France and Japan. They are two hypercapitalist societies that stand outside of the British-American sphere of nations despite being deeply engaged with those nations politically, economically, militarily, and culturally throughout the twentieth century. The chapters that follow throw a another light on the underlying tensions of anti-smoking policies and liberal government in the context of late-twentieth-century globalization. They seek to demonstrate how many of the contradictions manifest in the implementation of the California campaign play out in the history of tobacco control efforts in France and Japan but differently so: between civil-society initiatives and state bureaucratic prerogatives, older paternalisms based on scientific and policy expertise and citizens' entitlement as self-acting subjects, inclusive and stigmatizing tendencies within public health practices of segmenting populations, the public media sphere and the goals of health promotion, and claims of exceptional and universal culture. With respect to the California campaign itself, given the book's historical frame (roughly 1975–2000), part 2 returns to the question of the transposability of California's media spots to other regional and national contexts but does not offer a study of how the California campaign was "globalized" in the years following the adoption of its media spots and community mobilization methods by the World Health Organization's (WHO's) Tobacco Free Initiative in 1999.

As with the California campaign begun in 1990, here, too, local constraints and global imperatives clash and interpenetrate in powerful ways. Both France and Japan have traditions of liberal governance of populations, which emerged out of the fulcrum of colonial rivalry, industrialization, regional and world war, and histories of the nation-state before undergoing new articulation during the wave of neoliberal thinking in international policy circles in the 1970s and 1980s. In both countries the rise of anti-smoking movements and organizations was shaped by the circulation of scientific studies conducted abroad and by particular domestic histories of tobacco consumption marked by the presence of long-standing state tobacco monopolies, which made citizen-consumers' access to cheap cigarettes into not only a virtual right and an integral part of national identity but also a service provided by the state itself. At the same time, other outside forces have played a decisive role. For example, the devastation wrought by the Second World War interrupted the steady growth of cigarette consumption in Japan and France and thus provided a different historical experience of tobacco-related illnesses and death from that of the U.S., Canada, and the U.K.; and the invasion of domestic markets by U.S. and U.K. cigarette brands in the 1980s colored anti-smoking rhetoric, especially in Japan, with its long history of resisting Western pressure.

Reviewing the specific practices of central and local government, public health policy and research, citizen activism, and health promotion campaigns thus provides an opportunity to rethink questions of French and Japanese as well as American particularisms in the arena of international public health. I am interested in how each site constitutes what I've called a "global singularity" that exceeds the simple determination of either internal dynamics or outside influences (the common staple of many theories of national exceptionalism and globalization). With respect to France and Japan, it will become apparent that U.S.

liberal practices of decentralized governance find an unexpected echo in the history of the France and Japan, which are known for their highly centralized governments but which often assigned responsibility for matters of welfare and public health to local government and private organizations. Yet, in another twist, questions of knowledge, ethics, citizenship, and competing modernities turn out to be not as fundamental as in the U.S. and California, and issues concerning globalization and public health are articulated differently in the two nation-states. For example, in terms of public discourse, in France the question of U.S. and French exceptionalism structures much debate not only in the media but also in public health circles: the agency closely tied to the state had fewer international contacts in tobacco control and proved more reluctant to adopt Anglo-American approaches to health promotion than did a more autonomous anti-smoking group. And in the case of Japan, the legacy of Western colonialism in late-twentieth-century globalization dominated anti-tobacco discourse even as anti-smoking activists willingly drew on U.S. epidemiological studies and the example of the American anti-smoking movement for inspiration. By contrast, in U.S. public health circles, exceptionalism is a less salient topic, and when it arises, the exceptionalism at issue is that of other countries. Finally, the population segmentation and evaluation techniques of social marketing, whose ascendancy in U.S. public health promotion was heralded by the California media campaign and a subject of much debate, play a smaller role in the underfunded anti-smoking initiatives in France and Japan.

Thus, part 2 completes my argument that tobacco control in the U.S., France, and Japan is better understood outside of some grand narrative of American globalization (with its single timeline toward which all peoples should converge) and local cultural resilience that power most discussions of globalization. Moreover, from the perspective of treating each site as a global singularity, the old fieldwork narrative of studying either "down" (of the "nonmodern" from the assumed guilty position of the culturally advanced) or "up" (of powerful moderns from an a priori position of authentic powerlessness) gives way to an approach that I've termed studying "across," which favors articulating contradictory material histories over evaluative, cultural judgments often favored in public discussion and policy debates. Assumptions about the workings of "culture" take second place to a focus on actual practices, past and present.

France

Unexceptional Exceptionalism?

Tobacco destroys the body, damages the mind, and stupefies nations.

—Honoré de Balzac, epigraph to the *Bulletin de la société contre les abus du tabac* (1868)

There is one illness one is never cured of and that is the desire to smoke.

—Attributed to Charles de Gaulle by Gaullist politician

In France nothing is being done. Trembling Public Health will await permission from the Ministry of Finance. With our 1,400 cigarettes per capita we are at a remove from our English and American friends. When our consumption joins up with theirs, another 400 to 500 billion old francs will go to the Treasury. Apparently that is what people are expecting. Obviously, that will multiply a little our current 6,000 annual deaths from lung cancer, but, in general we can count on their silence.

—Dr. Pierre Zivy, *Le Tabac* (1965)

Advertising executives must not take the place of educators.

—*L'Action politique* (1989)

"*Tabagiquement correct*": The Clash of French and U.S. Exceptionalisms

In 1997, when I first mentioned to U.S. interviewees for this book plans to expand my study to include France, my project was met with a mix of interest and skepticism. Many offered their impression of France, primarily that of a country of plentiful smokers and ineffective regulations. One senior U.S. tobacco control scientist cautioned me with a chuckle: "My initial observation of your choice was that comparing France to anything else is perhaps not a productive exercise. . . . The French do things so differently from anybody else on the planet that it is sometimes difficult to draw lessons from what they do" (Burns 1997). The researcher's remarks intrigued me, for they seem to suggest two things: first, the commonplace observation that meaningful knowledge is one that travels well and lends itself to generalization; and, second, that the singular practices of one nation-state or people don't always generate knowledge of this kind and often defeat attempts to transfer aspects of their experience to other national, regional, or global contexts. Put another way, certain singularities are comparable (the U.S., Singapore, Australia, Thailand, Canada, and the People's Republic of China were often mentioned to me in interviews) and others, like France and Japan, are not. If we recall the World Health Organization's (WHO's) decision to "globalize" California's media campaign and its approach to community action, France (or Japan) would stand as the antithesis of, say, Californian or U.S. "exceptionalism" (which is at once particular and universal), as an intractable singularity of little value to anyone else.

Now, what is striking is that French *opponents* of government tobacco control have been more than happy to exempt the "French way of life" from the claims of globalizing currents of public health, which they deem as too peculiarly "American" even as these policies attempt to universalize themselves in the guise of science (Pracontal 1998, 314, 333–335). Such views dominated most French reporting on anti-smoking measures with the exception of *Le Monde* and the Catholic *La Croix*, two dailies sympathetic to anti-smoking policies. Take, for example, a report aired on the France 2 television channel's evening news on 20 October 1996, four years after the passage and implementation of the Evin Law (*loi Evin*) restricting smoking in enclosed public areas and banning alcohol and tobacco advertising. Titled "Malraux Stamp Censored," the story decried the decision by government authorities in a time of tobacco-related political correctness ("*à l'heure du tabagiquement correct*") to eliminate André Malraux's cigarette from Gisèle Freund's famous 1935 photograph of him that was to be featured on the commemorative stamp issued to mark the transfer of the writer's ashes to the Panthéon. As the allusion to "political correctness" makes clear, France 2's news desk regarded the action taken by the French post office as the unacceptable importation of American methods to combat smoking that in the U.S. led to the similar mutilation of historical photographs of bluesman Robert Johnson for U.S. postage stamps ("Timbre Malraux censuré" 1996).[1]

In the cited U.S. and French examples, what seems to be at issue is a clash of practices framed as one between distinct nation-states. At the same time, there is also the transnational, tautological construction by Americans and French alike

André Malraux commemorative stamp.
Courtesy of Poste Française.

of the Frenchness of things French, a local singularity called French exceptionalism upon which there would appear to be some agreement. The point of disagreement between U.S. tobacco control researchers and French opponents would be over the status of the U.S. or Californian examples. In one view, U.S. (or, for that matter, Thai, Singaporean, or Chinese) tobacco control has much to offer the world; in the other, it does not.[2] As for French anti-smoking advocates, health officials, and researchers, their discourse oscillates between affirming cultural differences and identities and attenuating them as ideology and circumstance warranted. Indeed, one agency closely linked to the state, the French Committee for Health Education (Comité français d'éducation pour la santé, or CFES), rejected Anglo-American health promotion strategies while one with weaker government ties, the National Anti-Tobacco Committee (Comité national contre le tabagisme, or CNCT), actively argued for their adoption.

Universal Particulars: Governance, Nation-States, and Empires

These brief examples return us to standard theories of globalization whereby the worldwide circulation of knowledge, goods, labor, and capital underscores local identities and conditions as much as it dissolves them (Giddens 1990; Featherstone 1990; Hall 1991; Buell 1994; Appadurai 1996; Sakai 1997).[3] They also highlight the paradox of the mutual convertibility of local particularisms and global universalisms: French political philosopher Etienne Balibar and Japanese studies scholar Naoki Sakai suggest that they have been not so much opposed as mutually constitutive and that one is amenable to being translated into the other. The most salient case is that of nineteenth- and twentieth-century nation-states and their rival imperialisms that each accorded its particular culture both exceptional and universal status and relegated that of its rivals to the lesser rank of hopeless parochialism. Balibar extends this argument about the play of the singular and the universal to the *internal* dynamics of the nation-state as well: within traditions of popular sovereignty (including French Republicanism), any particular identity within the national community has often had as its precondition the self-recognition of citizens as French, Japanese, American, German, what have you. Moreover, the advent of globalization and the end of the Cold War in many regions of the world has tended to make borders seem more porous and national identities less stable; and as internal frontiers have multiplied, whole population segments of nation-states have been excluded, especially recent immigrants, who are present in ever greater numbers. These actions are part and parcel of illiberal or repressive policies endemic to liberal governance (Balibar 2001, 113–117, 176–177; Dean 2002).

Smoking and the Discourse of Culturalism

The Malraux stamp news story was just one of many in the French media that addressed the effects of new French government actions regulating the culture, marketing, and consumption of cigarettes in France. Woven into accounts were almost obligatory references to policies underpinning the anti-smoking climate in the U.S. and Canada. Moreover, especially around the time of the introduction, passage, and implementation of the anti-smoking and anti-alcohol *loi Evin* in 1991–93 by the Socialist government,[4] news stories filled the airwaves and print media with sensational headlines decrying something of a revolution in daily life in the U.S. in terms of a wave of puritanism and religious intolerance. To take several examples: “United States: Smokers in Hell” (“États-Unis: les fumeurs en enfer”) announced the conservative daily *Le Figaro* (2 June 1991); “When America Demonizes Tobacco; Prohibition Is Back” (“Quand l’Amérique diabolise le tabac; la prohibition est de retour”) added *Le Quotidien de Paris* (29 August 1991); “Tobacco Hell” (“L’Enfer du tabac”) reconfirmed TV channel Antenne 2 several years later (27 October 1994); not to be left out, the left weekly *Le Nouvel observateur* brought in a reference to Iranian fundamentalism: "Tobacco: The War of Brimstone: The Ayotallahs of Oxygen" ("Tabac: la guerre du feu: les ayatollahs de l'oxygène") (5 June 1991). For its part, the center-left, anti-smoking *Le Monde* would run headlines in a flatter vein: "United States: The Death of the Cigarette Butt" ("États-Unis: la fin du mégot") (21 March 1987) and "New Crusade" ("Nouvelle croisade") (Folléa 1994).

In these and other reports, audiences and readers were treated to stories emanating principally from California and New York City of company policies that exiled workers to balconies and sidewalks, new "McCarthyite" surveillance of employees' private habits leading to their firing for unacceptable levels of cotinine and cholesterol in their bodies ("Nous licensions" 1988; "Si tu fumes" 1991); other accounts told tales of the deterioration of a litigious culture in which neighbor sued neighbor over smoking in the home, others still of mercenary lawyers on the prowl for potential plaintiffs for lawsuits against tobacco companies ("Exemple USA" 1989; "Interdiction tabac" 1991; "Faut-il interdire le tabac?" 1995; "Smoking No Smoking" 1996; "Tabac: Clinton" 1996). Even the CFES, the semipublic agency in charge of state-sponsored anti-tobacco campaigns since 1976, was careful to distinguish its methods from aggressive U.S. and Canadian approaches in an internal review of its campaigns over a twenty-year period (1976–96). It cited long-standing cultural differences: "We are a far cry from the witch hunt against smokers experienced in North America and which opponents of the decrees have referenced numerous times as a major risk of a campaign getting out of control. . . . Moreover, French society, which is rather Latin, has regularly rejected directive, melodramatic or excessively moralizing messages" (Baudier et al. n.d, 6; all translations from the French throughout this chapter are my own unless indicated otherwise).

In France, beneath the sensational headlines the discourse of U.S. and French exceptionalisms ran strong. The Tocquevillian nightmare of the "tyranny of the majority" and American conformism crushing individual freedom returned in the form of passive acceptance by U.S. smokers of new restrictions symptomatic of America poised to enter the third millennium: "These smokers, constrained or forced to habituate themselves to this new life in turn-of-the-century America, remain stunningly silent" ("L'Enfer du tabac" 1994). An ethical portrait emerged of U.S. citizens and consumers that folded them back into a puritan past that underwrote the not only intrusive but also impersonal and sadistic regulation of individual behavior. For a reporter at *Le Figaro* witnessing the anti-smoking revolution in the U.S., French identity constituted a unique line of defense against it: "Faced with this broad offensive, resistance has trouble getting organized. One must be doubtless French, reputedly an arrogant people of complainers, in order to decry certain legislative excesses" ("Les États-Unis: les fumeurs" 1991). In a similar vein, guests on a late-night talk show *Durant la nuit* in 1992, devoted to debating the *loi Evin*'s restrictions on public smoking that took effect the preceding day, complained that legislation wasn't necessary to regulate conflicts, which the tradition of *la courtoisie française* was more than capable of resolving smoothly in personal interactions (*Durant la nuit* 1992). Like the Malraux stamp news story with which I began, reports attacked new French anti-smoking measures as the unwarranted extension of very American practices outside U.S. borders: "What's good for America is good for the universe. Quietly we are falling in line behind the Yanks as so many deserving neo-colonized populations," complained the *Quotidien de Paris*'s reporter (Peletier 1990). One writer went so far as to claim that the *loi Evin* was an example of "the poisoning of the country by the American model" ("*l'intoxication du pays par le modèle américain*"; quoted by former journalist Philippe Boucher, *Le Monde*, 30 June 1990).

As the 1990s wore on, the U.S. press returned the compliment and ran occasional stories deploring the persistence of unrestricted smoking in European and French bars, cafés, hotels, and workplaces in terms of residents' oblivious, almost innocent ignorance of the hazards of smoking: "As the U.S. Clamps Down, Europe Is Smokers' Paradise" ran one story (1997); in a more wistful vein, another report titled, "Paris Would Still Be Lovely without the Curls of Smoke" (Overholser 1999). More aggressively, the French were sometimes assigned a place of honor in terms of their willful indifference to matters of personal risk and presumably Gallic defiance of health codes and selfish disregard for the well-being of others: "Oblivious to Laws, the French Puff in Public" ("Oblivious" 2000).[5] Meanwhile, throughout the decade, informal reports filtered back from Europe as returning U.S. travelers complained about smoke-filled cafés and restaurants and regaled their listeners with stories of having to take all their packed clothing—whether it had been worn or not—directly to the cleaners upon their arrival back in the U.S. to rid it of the unpleasant smell of stale tobacco smoke. In a sense, a new subgenre of travel literature emerged in the 1990s on both sides of the Atlantic, turning on questions of health and smoking, in which smoking became one of the privileged markers—along with multiculturalism, the death penalty, welfare state practices, queer culture and gay rights, identity politics, feminism, anti-Semitism, and declared respect for international protocols and law—of long-reiterated cultural differences between France and the U.S. that were framed in terms of French and U.S. exceptionalism more often than not. However, in the press on both sides of the Atlantic, even while the discourse of French-U.S. exceptionalisms colored reporting on tobacco control, other international references cropped up, primarily to other rival economic and political modernities—the People's Republic of China, for its new anti-smoking measures ("La Chine" 1996), and Japan, presented as the last smokers' paradise ("Le Japon" 1994; "Tobacco under Fire" 1997).

Beyond Culturalist Logic

The heightening of national and regional differences through currents of globalization produces in the domain of public health debates what it does in other arenas of collective life—a polarization of identities. Today this is one of the unavoidable tensions of the fields of French studies in the U.S. and American studies in France. The rivalry between two models of liberal democratic modernity in the West over the last two centuries has given rise to what intellectual historian Jean-Philippe Mathy has called the "paradigm of discontinuity" in which the Frenchness of things French and the Americanness of things American are reiterated and opposed (Mathy 1993, 9; see also Mathy 2000; Lipset 1996). This sense of discontinuity between the U.S. and France arguably deepened during the 1990s and early twenty-first century in which, according to sociologist Eric Fassin, the U.S. has served as the privileged foil for French writers and intellectuals confronting contemporary social and cultural questions such as multiculturalism, public queer culture, and feminism that seem to entail a crisis of French identity in their eyes. The polemics surrounding the introduction of antismoking measures replicate some of the features of the crisis acerbated by the pressures of globalization and have fallen prey to a sterile discourse of "national character" whose culturalist logic powerfully shapes thinking about these issues (Fassin 1997, 136; Fassin 1999b; see also Kidd 2000; Duncan 2000). In these instances, the discourses of exceptionalism and nationalism return with a vengeance, if not in popular opinion, at least in the pronouncements of politicians, journalists, intellectuals, and some researchers. Fassin argues persuasively that what are often political struggles that cause the two countries' respective cultures to shift and change are redefined as timelessly cultural ones. National identity kicks in as a way to stabilize troubling questions of social change in the routines of daily life. Transformed, these struggles emerge exoticized by this discourse and ultimately risk becoming *unthinkable* (Fassin 1998, 63). I introduce contingent and heterogeneous elements to the history and practice of public health and tobacco control with the hope of rendering the present moment more amenable not only to thought but also to different forms of thinking.

France 1976: The Veil Law

In October 1976 French viewers of state-run television were greeted by a twenty-second spot depicting patients in a smoke-filled physician's waiting room. As three men and one woman puffed away anxiously waiting their turn, the fifth patient, a calm, handsome, long-haired man in his twenties, coughs and ironically remarks to the others' surprise, "Do you mind if I don't smoke?" ("*Ça vous dérange que je ne fume pas?*"). A male voice-over follows enunciating the campaign's slogan, "Let's live life to the fullest without tobacco" (literally, "with full lungs": "*Sans tabac, prenons la vie à plein poumons*"), as the campaign logo flashes across the screen (a still cartoon figure of a half-pigeon, half-human puffing up its chest while striding from right to left) with the name of the sponsor, the CFES.[6]

Fourteen years before the first anti–secondhand smoke ads in California, the spot alerts us to the discomforts and dangers of tobacco smoke. The linguistic vehicle of the health message is the ethical discourse of common courtesy. In this early ad, in putting social relations to work to protect his own comfort and health, the youth of the ad appears to be the French version of neoliberalism's "acting, thinking subject" (signified by his calm demeanor) of California's anti-smoking spots, the example of whose personal intervention would incite fellow citizens to act upon themselves (to not smoke in the presence of others) and upon others (to ask them to extinguish their cigarettes or not light up). By such actions, citizens would contribute to the quality of collective life whose responsibility has presumably been that of the French state and the French health care system. (Here, it is almost as if the state gave license to nonsmokers to act in the place of the physician and nurse who are absent from the ad's frame.)

Moreover, the ad transforms one of the dominant contemporary images of students and young workers who threatened the French state in May 1968 with radical reform into that of the bearers of the promise of some better future (a healthier environment and, perhaps, newly won consumership), suggested by his confident manner and youthful good looks, to whom the French state lends its authority. The ad bypasses the left-libertarianism ethos of May 1968 (one of whose slogans was, "It is forbidden to forbid" ["*il est interdit d'interdire*"]) through an individually based articulation of a social bond based on health. Here, the spot gestures toward rewriting the recent past (May 1968) and scripting a generational narrative about an imminent future that would serve as the setting of an alternative modernity in which public smoking enjoys diminished tolerance or status.

The anti–secondhand smoke TV spot was one of eight released by the CFES in 1976 as part of an anti-smoking effort launched the previous fall by Minster of Health Simone Veil under the market-oriented conservative government of President Valéry Giscard d'Estaing and Prime Minister Jacques Chirac.[7] Remarkably for its time, the media campaign contained two other spots that also focused on secondhand smoke: one that featured a husband smoking at the wheel in stopped traffic who complains about automobile exhaust and rolls up the window, to his wife's consternation; another in the form of a cartoon in which viewers watch as smoke from parents' cigarettes in the living room infiltrates their children's bedroom upstairs. (The following year a more modest print campaign titled, "Don't smoke, don't smoke out others" [*"Ne fumez plus, n'enfumez plus"*] focused on the protection of nonsmokers.) Other ads deployed other strategies widely adopted in the 1990s: they deglamorized smoking (comparing a smoking boyfriend's mouth to a smelly ashtray), targeted pregnant women who smoke and fathers who pass the habit to their sons by virtue of their example, and staged the difficulties and rewards of quitting.

The media campaign, which ran from 1 October to 30 November 1976, addressed principally youth, women, and health care professionals and was part of a larger action undertaken by Veil under the 1976 law named after her (*loi Veil*) that included a ban on radio, TV, and cinema ads for tobacco products; new health warnings on cigarette packs and in print ads; new health courses in junior and senior high schools; and the protection of nonsmokers through the extension of smoking restrictions in public places (nonsmoking sections in public transportation, banning smoking in hospitals, etc.). These measures reflected recommendations made by the French Academy of Medicine in 1972 and by the WHO in 1975 and were inspired by British smoking studies ("La Nouvelle campagne anti-tabac" 1976; Lemaire 1999, 8–14, 91–92; Le Net 1997). They also merged at a time when the government was taking new public health–related measures on highway safety.

Those in charge of the campaign viewed their ambition of mobilizing citizens as a daunting task, for from their perspective of French liberalism, the historical experience of a highly centralized state had all but killed the spirit of initiative in French civil society. It had produced largely submissive citizens prone to occasional outbursts against state paternalism. As Michel Le Net, the head of the CFES under Veil and specialist in social marketing (or what he termed "*la communication sociale*"), was later to write of the French populace, "In fact, we are more apt to receive the gospel rather than to fight for it. Everything leads you to think that it is up to the state to serve as guide and tutor to the passive citizens on our soil" (Le Net 1981, 39).

In the end, the ad campaign with its tiny budget (2.9 million francs, roughly US$600,000) enjoyed a modest success reaching large numbers of adults and adolescents. Unlike the California campaign that in 1990 began at a time in which adult consumption had been declining for ten years, Veil's efforts intervened when French tobacco sales had been growing at 5 percent a year for the preceding ten years (as French cigarettes became ever more cheap), but her campaign nonetheless succeeded in precipitating a temporary drop in overall tobacco consumption of 1.9 percent, even as consumption of imported cigarettes soared—a sign of things to come (Le Net 1981, 88; "Les Premiers résultats" 1977). But otherwise, the *loi Veil* remained dead letter: the powerful Ministry of Finance refused to raise French cigarette taxes (among the lowest in Europe);

the state-owned monopoly SEITA (Société nationale de l'exploitation industrielle des tabacs et allumettes) put up fierce resistance; foreign cigarette makers regularly employed loopholes to circumvent the advertising ban; schools did not enforce smoking restrictions; and the hospital ban, passed in 1977, was not applied until 1991. Many of these reforms would have to await the implementation in 1992–93 of the stricter *loi Evin* passed by the Socialist government (Hirsch and Karsenty 1992, 78–83, 110–112; Padioleau 1982, 49–77; "La Progression du tabagisme" 1978).

The French State and Public Health

The 1976 anti-smoking campaign presented the classic features of a government initiative undertaken by a (still) highly centralized French state. The *loi Veil* and the media spots were not the national government's response to pressures stemming from voluntary health organizations and anti-smoking groups (there were no pressures). Nor were they the outcome of lengthy public discussion on the hazards of cigarette smoking after the release of a public health study in France. There was none. The *loi Veil* didn't emerge from parliamentary debate, either. Rather, it grew out of the individual intervention by a respected member of the French Academy of Medicine and member of Chirac's Gaullist political party (Rassemblement pour la République, or RPR), Maurice Tubiana.[8] Appointed by Veil as chair of a government commission on cancer, Tubiana privately admonished the minister for smoking during public televised appearances, and in a later interview he convinced her to take action against the alarming increase in cigarette consumption in France (Hirsch and Karsenty 1992, 180–181; Nathanson 1996, 632 n10).

Although French consumers had taken up cigarettes in large numbers during the 1920s and 1930s at the same time consumers did in the U.K., the U.S., Japan, and China (J. Goodman 1993, 94–95),[9] shortages of necessities during the Second World War had reduced French consumption (even as it may have spread the habit, especially among the armed forces), as it did in other war-torn areas (continental Europe, Africa, East Asia), while the U.S., the U.K., Canada, and Australia, spared the severest deprivations of war, saw domestic cigarette consumption jump as much as 75 percent during the 1940s. For example, per capita U.S. consumption reached 3,500 cigarettes (predominantly light tobacco) by 1945 and peaked at 4,000 cigarettes in the early 1970s, whereas in France numbers never came close to those figures, peaking at 2,300 cigarettes (35 percent of which were dark tobacco) much later in the mid-1980s as heavy smoking rose dramatically (Reid 2002; Corrao et al. 2000, 284). As of 1995, even though a significantly smaller portion of the U.S. population smoked (28 percent) than in France (36.5 percent), U.S. per capita consumption remained higher (1,788 versus 1,515; Pracontal 1998, 271). The difference in dates in the rise and drop of *heavy* cigarette smoking in various countries due to war also led to an analogous difference of some fifteen to twenty years in the onset and decline of epidemics of lung cancer and other smoking-related diseases (Pierce, Thurmond, and Rosbrook 1992).[10] Also, lower rates of tobacco-related heart disease were partially correlated to French diet (garlic, olive oil, and red wine), a phenomenon known as the "French paradox," but these were offset in the case of men by high rates of stroke due to alcohol consumption (Ford 1992; Got 1997). By 1997, 39 percent of adult men smoked, compared with 27 percent of women; among teenage smokers, girls had a slight lead, and in the 18–24 age group, the prevalence rate was upward of 50 percent, the highest in the European Union. Male manual and low-level service workers and female managers also had high rates (Grizeau 1993; Grizeau and Baudier 1995; Grizeau and Arwidson 1997).[11] As of 1999, 77 percent of cigarettes sold in France were foreign brands (45 percent U.S.), and half of young smokers favored Philip Morris' Marlboro cigarettes. In Europe, trained professionals, especially women, tend to smoke more than their U.S., U.K., or Canadian counterparts, and only in the 1990s did France began to see a class difference in smoking rates, as men of older middle-class generations started quitting and rates among the unemployed youth rose (but some anti-smoking advocates claimed that such trends for social class were overstated; Boucher 1996, 1997). By the late 1990s, cited French deaths from tobacco-related illnesses reached 60,000 annually.

In the eyes of many observers, individual initiatives originating within university and political elites, conjoined with the absence of public debate and prior mobilization of politicians, public servants, and civil society, were the hallmarks of a top-down public health initiative issuing forth from a centralized state and diminished the effort's chances of success (Nathanson 1996, 623–624; Padioleau 1982, 72–76). Once again, state policy was reduced to the status of symbolic politics and nothing more. It was a judgment reprised in many conversations I had with French tobacco control advocates concerning initiatives by the French state in general.[12] All too often, the government in power remains content with passing the core legislation but omitting the necessary articles of implementation that in its eyes would be politically too costly. As a result, the centralized state can fill headlines announcing new laws for the benefit of citizens (who haven't been consulted) but with little or no follow-through. It is a state that makes speeches but doesn't act, arrogating for itself the privilege of discourse but not the responsibility for taking action. According to one leading advocate of automobile safety and anti-smoking measures, writing during the heyday of neoliberal thinking in French policy circles in the 1990s, this is the perfect mechanism for state and social hypocrisy in matters of preventive medicine and public health in an advanced liberal society, which at once favors those most able to survive while abandoning everyone else to their fate, and it encourages dangerous practices of consumption (the highest rates in Europe of automobile fatalities, alcoholism, and teenage smoking) while leaving their victims in the lurch. He writes:

> In place of the simplistic vision of the carefree and happy bon vivant who dies a little before others after having "got the most out of life," is the image, difficult to accept, of a large fraction of the population ensnared by a society that promotes danger and leaves the least advantaged by the roadside of prevention. Our society applies to prevention the same methods it applies to the economy: save the most apt to survive and ignore the others. This return to Darwinian selection is experienced as good by those who feel strong, but it corresponds to the abandonment of the ideals of solidarity and justice that differentiates the state of society from the state of nature. (Got 1992, 12)

Part of the power of such criticisms lies in a French Republican discourse that holds that health and access to medical care are fundamental rights of citizens. However, revisionist historians of public health point out that this wasn't secured in France until the 1930s and that while Republican ideology stretches back to the eighteenth century and the French Revolution, in matters of actual public health practices, many of its tenets had little meaningful influence during most of the nineteenth century. Moreover, when it came to public health, as in many other matters, France's long-standing practice of centralized government was actually deeply liberal in character dating from the French Revolution itself (respect for property rights, limited government, free markets for goods and labor, etc.) and fairly decentralized, insofar as responsibility for medical assistance, care of the poor, and management of hospitals was delegated to semi-autonomous organizations and to local government and public and private agencies up through the early twentieth century. This was all the more true in the rural areas where most of the French resided and worked (Ramsay 1994, 51–52; Murard and Zylberman 1996, 65).

During the Third Republic (1875–1940) up to the Second World War, as the state stepped up its activities, it continued to delegate responsibilities to local authorities and private or semi-autonomous agencies and respected the prerogatives of physicians.[13] The Second World War, reconstruction, and the maturation of consumer society under the Fifth Republic (1958–) would occasion greater state intervention and the expansion of medical benefits and health services, but the hybrid mix of state and private actions remained dominant. The preponderant influence of the medical world since 1958 helped break with what was left of a public health tradition of preventive medicine dating from the time of Louis Pasteur in favor of care and cure (Ramsay 1994, 70–98).

For a late-twentieth-century example of liberal governing practices, one need look no further than the Veil anti-smoking campaign, less in terms of its lack of follow-through by the state than in the delegation of policy implementation and the media campaign's narrative of citizenship. Veil may have publicly launched the media and legislative initiative with great fanfare, but the ad campaign was not carried out by any ministry of government; rather, it was conducted by the CFES, one of the many semi-autonomous agencies based on the 1901 law governing nonprofit organizations deemed "publicly useful" (*"d'utilité publique"*) but most of whose funding in many cases does not come from the state. As for the CFES, it is funded by the Caisse nationale des assurances-maladie (CNAM, the National Medical Insurance Fund), an agency founded in 1967 that pools and manages the large number of premiums and payments of the array of private and public medical insurance plans that cover most French citizens.[14]

Thus, to return to the TV spot featuring the young nonsmoker, his authority may derive from the French state, but it is a mitigated one. It could be argued that he stands somewhere between civil society (and its population segments, such as youth, women, the elderly, the poor, children, and others) and the state (all citizens, or the "general interest"). And he literally embodies the two aspects of the nation-state's policies that both singularize and aggregate the social body: youth, an all-too-particular and narrow population segment deemed a threat in the aftermath of May 1968, emerges as a potential agent of change for the general welfare of French citizenry.

French Public Health and Smoking: *Le Rapport des cinq sages*

In the eyes of French public health advocates, the state of affairs in late-twentieth-century France left much to be desired. While the system of health care consisting of near-universal coverage and an extensive network of publicly run hospitals provided wide access to the latest in modern surgery and advanced therapies, propelling France ahead of most industrialized nations with respect to longevity (second-highest in the world after Japan) and infant mortality rates, it neglected the infrastructure of *preventive* medicine, a situation that was glaringly underscored by the HIV/AIDS pandemic (Setbon 1993, 41–47; Morelle 1996). In response to the scandal implicating high government officials in the contamination of the nation's blood supply by HIV during the 1980s, Minister of Health Claude Evin commissioned an investigation, and the final report on AIDS and public policy strongly recommended new state measures to secure the protection of the French population, many of which were adopted by the Socialist government (Got 1989, 101–104; Got 1992, 144–147).[15] The report revived the struggle between the centralizing and liberal tendencies of the state.

It was in the shadow of this debate that the tobacco control *loi Evin* was born.

The law benefited from state authorities' long-standing concern with alcoholism and recent public worry over another environmental hazard: the dangers posed by the presence of asbestos in construction materials. As with the *loi Veil* fifteen years earlier, the input of medical experts prompted the current health minister to propose a new set of measures to combat smoking that cost France 50,000 lives a year through tobacco -related diseases, primarily lung cancer. The experts' intervention took the form of a commissioned report to Evin in 1989 signed by five eminent specialists including Claude Got, author of the report on AIDS, and Maurice Tubiana. They represented a wide political spectrum, and most of them were heart, lung, or cancer specialists with extensive experience in public health matters from automobile safety and asbestos pollution to anti-smoking advocacy, including activism within the CNCT.[16]

Popularly known as the *Report of the Five Wise Men* (*Le Rapport des cinq sages*), it was titled, *Political Action in the Area of Public Health and Prevention* (*L'Action politique dans le domaine de la santé publique et de la prévention*; see Nau 1989). It was the last in a series of reports in the 1980s on France's health system (*Le Système de santé* 1983; *La Santé en France* 1985). Broadening the earlier reports' focus on budgetary constraints and the inadequate funding of physicians' training and of public health research, the report delivered a sweeping overview of French public health infrastructure. It situated the mission of public health squarely within the contemporary French context of a highly successful system of individually based primary care and the neoliberal consensus about containing exponential increases in health expenditures. In keeping with the emerging emphasis in international public health discussions on prevention and health promotion in addition to primary care at that time,[17] the report suggested that what public health offers are the advantages of preventive medicine's focus on collective practices that are not amenable to individually based solutions restricted to the sick population alone:

> By public health is meant reflection and action concerning the health problems of human groups, their health needs, and the means of implementation in order to promote, protect, and, if necessary, restore public health. . . .
>
> Public health action concerns groups or populations of persons—sick or not—or even all of society; it always has a collective or societal dimension.
>
> Its effective areas are prevention; organization of the health care system; evaluation of practices, techniques, and institutions; surveillance of major risks by tracking [*observation*]; and epidemiological intervention.
>
> Defined this way, public health exceeds the powers of intervention of public authorities in public health. But these powers are construed [*par construction*] as part of public health. (*L'Action politique* 1989, A1–1)

The document takes the position that aggregate problems require aggregate solutions. According to the authors, the struggle to bring health care expenditures under control requires intelligent guidance by an overall public health policy and by focusing on measures to reduce behavior that is dangerous to health, such as malnutrition, accidents, smoking, and excessive consumption of alcohol (*L'Action politique* 1989, 5). The report gestures toward certain actors and approaches and not others: the infrastructure of the health care system (the CNAM's members, perhaps industry and labor collective bargaining entities, public health agencies) but little in the way of nongovernmental organizations (NGOs) and citizen groups. There is no call to mobilize citizens in terms of their obligation to self-care, which was already a mainstay of neoliberal health policy reform in the U.S., the U.K., and elsewhere in the 1970s (Knowles 1977a, 1977b). As such, the report to the Socialist health and welfare minister stands as almost a classically French Republican, statist effort to redress the balance in public policy discussions weighted at that time toward neoliberal measures ostensibly favoring the initiative of members of civil society and the private sector.

In many ways, this document stands as an attempt to finesse strategically the pressures of limited budgets in a neoliberal climate and to turn them to the advantage of a state-sponsored public health agenda in the name of fiscal responsibility. Here, health care is both expanded to include "upstream" practices of prevention that target the healthy and sick alike and made subordinate to the broader vision of public health. In turn, "health" receives a more inclusive definition: it "conserves the state of health" of French citizens (*L'Action politique* 1989, 3). As such, health involves not a punctual activity (sickness and cure, trauma and recovery) but a continuous one (avoiding illness and trauma), yet it is a definition that eschews the very broad conception of health proposed by the WHO as early as 1946 based on the notion of "complete mental, physical, and social well being."[18] Finally, state intervention is invoked but within recognized limits. As spelled out in the document, the state's role is one of coordination and evaluation of national public health policies, collecting and pooling data dispersed throughout state agencies and research bodies, training public health physician-inspectors, and health education targeting essentially physicians and grade-school children (*L'Action politique* 1989, 4–12). However, exactly what "exceeds" the powers of state intervention is never made clear by the document, nor is what is entailed by information campaigns designed to educate the population other than that it is the sole responsibility of the state. The report does not engage directly with the noneconomic dimensions of neoliberal policy agendas that claimed to promote citizen or community empowerment and stressed citizens' ability to act upon themselves and each other through forms of community policing and self-help to remake themselves into responsible, self-governing citizen-subjects.

France's public health infrastructure is deemed quite poorly situated, in terms of preventive practices, to meet the report's goals. The authors don't hesitate to make one telling comparison with France's neighbors to the north and across the English Channel and the Atlantic: "To implement an ambitious policy in this area requires increasing the means of action in public health, which are underdeveloped in France when compared to Scandinavian or Anglo-Saxon countries" (*L'Action politique* 1989, 4). These international references return again and again in the French anti-tobacco literature and in interviews I conducted (Hirsch and Karsenty 1992, 108–110; Mélihan-Chenein 1997, 1999; Karsenty 1997; Hirsch 1997; Slama 1997; Dubois 1999; Got 1997). While in the area of health care and welfare policy France at that time seemed to continue its long-standing habit of quietly borrowing and appropriating concepts and practices from other nation-states, in public health matters such as preventive medicine and epidemiology, tobacco control advocates stress again and again France's relatively inwardly turned stance since the Second World War, taking little official notice of developments outside its borders, with few personnel linguistically competent enough to be able to participate in English language–dominated conferences (Ramsay 1994; Murard and Zylberman 1996; Hirsch 1997; Karsenty 1997; Slama 1997; Nau 1997, 1999; Folléa 1997; Le Net 1981, 23, 114, 148, 159). Published accounts of French public health confirm this. For example, well into the 1980s and early 1990s, few French government health policies explicitly refer to current reports and studies released by the WHO, which was felt to be an organization dominated by Scandinavian and Anglo-Saxon countries and their peculiar communitarian orientation (presumably rooted in Protestant-dominated cultures [Marrot 1995, 158–159, 198]).

It is important to underscore the fact that this isolation in matters of prevention-oriented public health and medicine was perceived by anti-smoking activists and researchers to be even greater in tobacco control, in which British and American dominance was long-standing (Hirsch 1997; Dubois 1999; Meurisse 1999), especially in the domain of epidemiological research and citizen activism. Students of public health generally credit smoking studies with the rise of the discipline of statistical epidemiology (Brandt 1990; Beaglehole and Bonita 1997). These began early before the Second World War in the U.K. and matured in the 1950s with the work by Richard Doll, a British physician, and in the U.S. with studies by German émigré Ernest Wydner. In France there was a single important early study, begun in 1954, which borrowed statistical and working methods from the U.K. and the U.S. and was published in 1958 in French (Denoix, Schwartz et al. 1958) and in 1961 in English by Pierre Denoix and Daniel Schwartz, who was the director of the French monopoly SEITA's Laboratory in Biological Research in Bergerac. Although the research was considered one of the best retrospective studies at that time, it had relatively mild impact on practicing physicians in France. This was so partly because French physicians themselves, compared with their U.S., U.K., or Canadian colleagues, had less experience during the 1950s with lung cancer among their patients due to the later onset of heavy smoking in France, and partly because medical research itself was isolated from practicing physicians and, in France as elsewhere, had yet to integrate statistical approaches into its toolbox. Moreover, the results were not picked up by the French press and received little notice outside of medical circles. SEITA did not bury the report but was not pleased by it, either. Shortly afterward, Schwartz left SEITA for the Gustave Roussy Cancer Institute (which Denoix directed), where he pursued his cancer studies but not on tobacco (Meurisse 1999; Zivy 1965, 210–215). No other French research team picked up where he left off. The gap between scientific and public debate in France and in the U.S. and the U.K. and the indifference of the French state deeply troubled some French physicians at that time. One acidly commented:

> In France nothing is being done. Trembling Public Health will await permission from the Ministry of Finance. With our 1,400 cigarettes per capita we are at a remove from our English and American friends. When our consumption joins up with theirs, another 400 to 500 billion old francs will go to the Treasury. Apparently that is what people are expecting. Obviously, that will multiply a little our current 6,000 annual deaths from lung cancer, but, in general we can count on their silence. (Zivy 1965, 246)

Concretely, the *Report of the Five Wise Men* proposed a National Council of Public Health with its own funding and staff to oversee public health practices, suggest policies, and remit an annual report; to endow the Ministry of Health with a health and social statistics agency, capable of intervening like the U.S. Centers for Disease Control and Prevention (CDC) in epidemiological emergencies; to provide funding explicitly earmarked for epidemiological studies to be conducted by INSERM (Institut national de la santé et de la recherche médicale), France's principal medical research institute; to create schools of public health oriented toward training personnel for work on the ground;[19] and to launch new health promotion initiatives against alcohol abuse and smoking (*L'Action politique* 1989, 32–35). The new initiatives built upon a report submitted to the health minister in 1987 on alcohol and smoking and reprised and toughened articles of the *loi Veil*: a 100 percent increase in cigarette taxes to discourage smoking by youth and the disadvantaged, the protection of nonsmokers in enclosed public spaces (transportation, hospitals, schools, workplace), and a ban on all advertising in all media of alcohol and tobacco products. Regarding the last measure, the authors stated laconically, "Advertising executives must not take the place of educators" (*L'Action politique* 1989, 13). These would become the key articles of the 1991 *loi Evin*. They positioned the state in a struggle with its own tobacco monopoly, the alcoholic beverage industry, and ad agencies over not only the bodies but also the minds of French youth.

The report had other effects: the High Committee on Public Health (Haut comité de santé publique) was assigned expanded responsibilities including the publication of an annual report, and an Institute of Sanitary Control (Institut de veille sanitaire) was established. However, many of the recommendations did not receive legislative support, and the road to implementation, especially of the *loi Evin*, was a bumpy one. The High Committee lacked substantial funding and authority, prompting two members, also coauthors of the report, to resign in public protest in 1992 (Got 1997; Dubois 1999; Nau and Nouchi 1992). The proposal to create new schools of public health was tabled, and tobacco control continued to occupy a minor place on medical research agendas. As late as 1997 only one out of eleven INSERM research commissions included tobacco control, and a conference sponsored by INSERM on risk and public health in 1995 gave little attention to tobacco (Slama 1997). Regarding the *loi Evin*, the ban on alcoholic beverage advertising was gutted through industry lobbying, cigarette taxes were increased by only 15 percent per year, no authority (other than a reluctant police force) was invested with the task of enforcing restrictions on smoking, and once again, as with the *loi Veil*, provisions concerning educational institutions were imperfectly implemented (Hirsch 1997; *La Loi relative* 1999, 45–83).

Smoking and the French Tobacco Monopoly

The case of Daniel Schwartz's early epidemiological work on lung cancer as a SEITA functionary brings into focus the intersection of cigarette consumption and the state-owned tobacco monopoly in French history. In the minds of the report's five authors, these overlapping histories are what authorize their own anti-smoking recommendations:

> The industrial use of the cigarette in France is a recent introduction (creation of the state monopoly in 1814 [*sic*], of SEITA in 1926). This behavior is not part of our cultural heritage [*patrimoine culturel*]. Taking into account the dramatic consequences tobacco use has on health, it is no longer possible to give free rein to cigarette makers in utter disregard for public health. Rather, the government must once again seize the initiative, in order to bring about in due time a cigarette-free France. (*L'Action politique* 1989, A3–1)

In other words, the widespread consumption of cigarettes is a matter not of culture but of government policy. What state officials helped bring into being, new legislation can remove.

The association between tobacco and the prestige and authority of the French state dates from the sixteenth century, when, according to traditional historical accounts, Jean Nicot, France's ambassador to Portugal, introduced tobacco to

France in 1560 by sending some plants to queen Marie de Medicis that brought her relief from migraines. The tobacco monopoly actually began in 1674 under Louis XIV and continued until 1995, interrupted only during the French Revolution, when, in keeping with its revolutionary commitment to free commercial and labor markets, the French Assembly dismantled it. In 1926 it was reorganized as SEITA, whose profits were to retire the huge public debt dating from the First World War. Only much later, in the 1980s and 1990s, did foreign cigarette makers gain free entry into the French domestic market.[20] The domestic monopoly came to an end with its complete privatization in 1995, and four years later it metamorphosed into a multinational concern when it merged with the Spanish tobacco company Tabacalera to form Altadis in 1999. In one stroke the new company became Western Europe's third-leading cigarette distributor and the world leader in the cigar market.

In the nineteenth and twentieth centuries, war and military service played a major role in propagating the consumption of cigarettes and together with the fact of the state-owned monopoly eventually lent smoking the aura of a patriotic act—something that the Gauloises Caporal brand, launched in 1910, capitalized on with its winged Celtic helmet, which lent it a martial, populist aura. Cigarettes entered Europe first in Spain, and some historians credit Napoleon's invasion of Spain with introducing cigarettes to France via veterans returning to their homeland (hence the other brand launched in 1910, Gitanes, or gypsies); others place the date around 1830.[21] Cigarettes' physical qualities presumably offered much to modernizing Europe and reflected the times in which they arose: "Light, fast, practical, cigarettes have the frivolity of modern life, and like the latter, they are burned up quickly; cigarettes are in its image, and thus modern life adopted them" (Zivy 1965, 89; *Encyclopédie du tabac* 1975, 222–228, 492–297; R. Klein 1993; *Graphismes et créations* 1996, 19–34). Until the 1970s, state control over the production and sale of tobacco products, together with the commonplace male experience of initiation to smoking in the army while performing obligatory military service, made smoking not only a patriotic act but also a rite of passage into modern adulthood (Hirsch and Karsenty 1992, 75).

French cinema, which dominated world production and distribution until the First World War, did much to propagate the culture of smoking at home and abroad. It exploited the associations of exotic passion, stylishness, war, and French populist sentiment present in cigarette advertising up to the Second World War in such figures as Arletty, Michèle Morgan, Louis Jouvet, Jean Gabin, and Fernandel. After the Liberation, a new generation began to embody the pleasures and glamour of smoking (Yves Montand and Simone Signoret, then Jeanne Moreau, Brigitte Bardot, Jean-Paul Belmondo, Alain Delon, Jean-Pierre Léaud, Michel Piccoli), but their example—especially those associated with New

Wave cinema—was inflected by the inimitable models of U.S. actors (Greta Garbo, Humphrey Bogart, Lauren Bacall, Bette Davis, Orson Welles, Rita Hayworth, Barbara Stanwyck, James Dean, and Marlon Brando), those icons of New World modernity who exhibited remarkable self-confidence and smoked interminably. Writer and feminist Annie Leclerc, who came of age in the 1950s, expresses eloquently the deep association between cigarettes and the cinema during her teenage years. Ruminating on her first cigarette and the aura of (American, masculine) invulnerability with which it endowed its users, she writes:

> I remember now that cigarette's first taste, its wicked, rough taste of a lie.
>
> It's not an innocuous cigarette. How many films would lose their savor, how many characters would lose their intensity, if you took away their cigarette?
>
> It's the "Humphrey Bogart" cigarette. The cop's, the reporter's, the street tough's cigarette—the "mature" cigarette. The always military, colonial, imperial cigarette. The politician's, scholar's, male or female activist's cigarette. A phantom of power, called up, inhaled, smoked for such a long time that it ended up taking shape, becoming flesh. (Leclerc 1979, 107–108)

Members of the intellectual bohemia and the world of fashion—famous smokers all (Sartre, Beauvoir, Camus, Boris Vian, Coco Chanel)—were equally fascinated with U.S. culture (literature, jazz, cinema) even as they worked to redefine postwar French life as it underwent decolonization and edged toward the consumer age. And once tariffs and government-fixed prices were lifted on foreign brands in the 1970s and 1980s—effectively breaking SEITA's monopoly on domestic cigarette sales—with U.S. brands leading the way the market share of foreign cigarettes rose from 3 percent in 1970 to 53 percent by the time the *loi Evin* went into effect in 1992–93. By the 1990s, with the advent of economic and cultural globalization, Marlboro had become the brand of choice of more than 50 percent of young French smokers, 18–24 years old (Hirsch 1995, 135). And, as is often the case with the decline of a state tobacco monopoly, tobacco advertising exploded.

Anti-smoking Sentiment

To date little historical research has been done on anti-smoking attitudes and government ordinances regulating the consumption of tobacco products in public in France prior to the late twentieth century. The few studies that do exist suggest that France has had its share of prominent doubters of the value of Jean Nicot's weed as well as its poets and promoters. Already in the seventeenth century Louis XIV's dislike of pipe smoke favored the spread of the more discreet snuff at court (which he also duly banned), and a century later Napoleon expressed strong distaste for tobacco (R. Klein 1993, 84; Zivy 1965, 153–154).

Nineteenth-century poet and playwright Victor Hugo claimed that smoking weakened his ability to work, critic Edmond de Goncourt declared that his writer's block lifted after he stopped smoking, and Honoré de Balzac, a fond smoker of water pipes (thanks to George Sand), wittily decried the deleterious effects of tobacco on masculine character and sexual performance in his *Traité des excitants modernes* (1838), and elsewhere declared that "tobacco destroys the body, damages the mind, and stupefies nations" (Zivy 1965, 163).

Meanwhile, in the nineteenth century, as nicotine became an object of scientific inquiry (first isolated by French scientists) and Claude Bernard began to conduct early experiments on its effects, the French Anti-Tobacco Abuse Association (Association française contre l'abus du tabac, or AFCAT), the ancestor of the current CNCT, was founded in 1868. Early adherents of the association included four members of the Imperial Academy of Medicine, popular novelist Alexandre Dumas (*fils*), as well as patrons drawn from high society. In 1878 the future pillar of French science, Louis Pasteur, became a member. Balzac's lapidary pronouncement served as the epigraph to the organization's mission statement, a document that claimed that the greater the consumption of tobacco, the greater were its nefarious effects on health, including mental illness; general paralysis; and cancerous infections of the lips, mouth, and stomach. It even named tobacco smoke as a cause of discomfort to nonsmokers. Early on, AFCAT would add the fight against alcohol abuse to its goals, and after the Franco-Prussian war it held both substances responsible for France's current weakness and political instability.

The French defeat at the hands of the Prussians at Sedan and the insurrectionary Commune (1870–71) did give a crucial boost to the idea of disastrous effects of alcohol and tobacco on the health of the French masses. One physician even declared that French troops were swept aside by the invaders because "many of those ravaged by the narcotic plant must have felt their minds too empty, their lungs too breathless, their legs too wasted, their arms too weak to pick up their rifle and march toward the enemy on the days of the invasion" (quoted by Nourrisson 1988, 541). The French anti-tobacco organization continued to flourish until the end of the century and held the first International Conference against Tobacco Abuse during the Universal Exposition in Paris in 1889 under its new name, the French Society against Tobacco Abuse (Société française contre l'abus du tabac). However, whereas the French anti-alcohol movement remained strong, during the ensuing years leading up to the First World War, the membership of the French Society against Tobacco Abuse fell off, and the movement became moribund as the old leadership died out (Zivy 1965, 20, 80–85; Nourrisson 1988, 540–542; Mélihan-Chenein 1997).[22]

In terms of daily practices, it would appear that pipes were not commonly smoked in the street, whereas cigars were tolerated. Until the widespread adoption of cigarettes in the 1920s, in France moderate smoking prevailed, and (middle-class) men smoked primarily after meals and on Sundays. One of the earliest reported bans on public smoking was passed in 1826 by the town of Saint-Claude; it forbade pipe smoking in its streets. A century later in 1924, strollers were forbidden to smoke in forests, and in 1942 smoking was first banned in theaters not because it represented a fire hazard but because cotinine showed up in the urine of nonsmoking patrons. By the 1960s nonsmoking was the rule in theaters, cinemas, concerts, music halls, and circuses, and nonsmoking sections were established in public transportation. Good manners still dictated that people not smoke in other people's homes without first being invited and that one never smoke in the guest bedroom (Zivy 1965, 20, 93–94, 243–245).

The Anti-smoking Campaigns in a Neoliberal Environment

It was in these overlapping public health, political, and cultural contexts that the anti-smoking laws intervened. Polls conducted before and after the implementation of the *loi Veil* in 1977 and the *loi Evin* in 1992–93 reported strong public support for regulating the marketing and consumption of tobacco and new restrictions on smoking in enclosed public areas. And when a new Conservative government came to power in 1993 and some of its members tried to undo the *loi Evin*, public opinion prevented it. Both laws gave a new lease on life to public health agencies and "publicly useful" voluntary health organizations, including two whose important activities I examine in some detail: the CFES, replaced in 2002 by the Institut national de prévention et d'éducation pour la santé, or INPES, which ran the Veil anti-smoking campaign; and the descendant of the French Society against Tobacco Abuse, the CNCT, which assumed its current name in 1968 and was assigned in 1977 the task of suing violators of the *loi Veil* in civil actions. As a semi-autonomous state agency and a state-supported private advocacy group, INPES and the CNCT stand as two different forms of organizations whose strategies, modes of collaboration, and complex relation to the French state bear the marks of all the tensions between the centralizing and liberal tendencies of French government. They are interesting cases of an ambiguous and contradictory mode of governance that operates through entities that stand somewhere between the state and civil society.

Other NGOs included the League against Tobacco Smoke in Public—Non-Smokers' Rights (Ligue contre la fumée du tabac en public—Droit des non-fumeurs) founded in 1973 and based in Colmar (Haut-Rhin [Alsace]); the Anti-Cancer League (Ligue contre le cancer), founded in 1918, which enjoyed ties to the International Anti-Cancer Union (based in Geneva since 1933 as the Union internationale contre le cancer, or UICC); the National Committee on

Respiratory Illnesses and Tuberculosis (Comité national contre les maladies respiratoires et la tuberculose, or CNMRT), founded in 1916, which cosponsored France's first anti-tuberculosis campaign in 1917 spearheaded by the Rockefeller Foundation; and, finally, the Cancer Research Association (Association pour la recherche sur le cancer, or ARC), a research organization founded in 1962 (Pinell 1992).

While public support remained strong, many other factors hampered the anti-tobacco campaigns, including the hostility of the Ministry of Finance (which controlled SEITA); the lack of interest of trade unions, which were powerfully represented on the CNAM board that financed health campaigns (they feared that new workplace regulations would pit worker against worker to the advantage of management); the CNAM's own focus on problems arising from spiraling unemployment rates (which rose from 4 percent in the 1970s to over 12 percent in the 1990s, reaching 25 percent for workers under age twenty-five); the limited nature of anti-tobacco NGOs' own activities; and, last but not least, the poor state of French statistics on tobacco use because of underfunding (more about which below).

As noted earlier, the Veil campaign originally targeted youth, women (especially pregnant women), and health care professionals and attempted to mobilize citizens as citizens and family members around the issues of secondhand smoke, smoking as an unclean habit, protection of the unborn and small children, the transmission of the smoking habit to the younger generation, and the difficulties of quitting. However, between the initial *loi Veil* campaign in 1976 and the mid-1990s, the CFES moved away from the ambitious scope of its efforts to focus mainly on adult cessation and prevention of youth uptake of smoking. Reflecting the priorities of its funding agencies (the welfare and health ministries and the CNAM), until 1997 the CFES had only one staff member assigned to tobacco control (who, moreover, brought no previous experience in tobacco control to the job). And when it came time to launch the *loi Evin*, the CFES, despite vigorous lobbying by activists, declined to start an information campaign in the public media sphere to publicize and explain to the larger public the reasons for the new regulations and what they entailed. In the arena of broadcast and cable media, it fell to news organizations and talk shows to promote public discussion. The CFES contented itself with distributing small brochures summarizing the law. Whatever beginnings of a mobilization in favor of a direct partnership in governance between health authorities and citizens that had been laid down in 1976 had all but vanished by 1991–92, and what remained were the modest activities of the publicly sanctioned organizations like the CFES, the CNCT, and the Anti-Cancer League that did little in the way of energetic recruitment of citizens against the hazards of smoking.

This lack of follow-through with the public is all the more striking in that neoliberal policy agendas were in ascendance during this entire period. In France, as elsewhere, they stressed, at first, budget cuts and reductions of enterprises' financial burdens pertaining to the welfare state and then later, under both Socialist (1981–86, 1988–93, 1993–2002) and Conservative (1986–88, 1993–97, 2002–) governments, privatization of state enterprises and services, tax cuts, and the deregulation of financial and labor markets. During the 1970s and 1980s, economic stagnation, the 1972–73 oil crisis, inflation, and layoffs of workers fueled the launch of these agendas. In the arena of political administration proper, the Socialists, responding to social movements, which proliferated in the aftermath of May 1968 and demanded more democratic governance, initiated the decentralization of the French state (including public health services) and the revitalization of regional and municipal government.

At the same time, as these agendas gained ground, the economic crisis caused the French welfare state—what sociologist Jacques Donzelot has termed "the social"—to falter. With the rise of long-term unemployment and dim job prospects for youth entering the labor force, a new class of "socially excluded" ("*les exclus*") located in urban areas emerged, and the dual commitment of the state and French society to social assistance (*assistance sociale*) of citizens in perpetual need and to social protection (*protection sociale*) of everyone else broke down.[23] What began to replace the language and promises of the *État-providence* was a discourse that took the radical democratic demand of contemporary social movements for "participation" and translated it into the language of involvement (*implication*) of citizens in "projects" involving "collaboration" guaranteed by contractual "obligations" of self-management and the realization of citizens' "potential" in the creation of "communities of responsibility." Thus, for example, inhabitants of subsidized or low-rent housing (*habitations à loyer modéré*, or HLMs) were asked to become managers of their own buildings and neighborhoods (Donzelot 1992, 19–36; see also Donzelot 1984). This bears some resemblance to "community policing" projects that sprouted up in the U.K., the U.S., Australia, and Canada at that time. In these instances, the state not only stands as a symbolic space of belonging for citizens, it also recruits and mobilizes them (Donzelot and Roman 1992, 11).

The State-Sponsored CFES Campaigns: Youth at Risk

However, the very modest French anti-smoking campaigns sponsored by the CFES after 1977 (1978, 1979, 1983, 1988, 1991, 1994–95) did not do much to encourage citizens' activity in general—as favored by neoliberal thinking, but also from a different perspective by radical democracy movements. Nor did they perform the Republican gesture of vigorously acting and speaking on behalf of all citizens in the name of the "general interest." Moreover, in their spots they made no mention of the links between smoking and tobacco-related illnesses and deaths and of the hazards of smoking in enclosed public spaces. Take, for example, the one group that the CFES did target regularly for recruitment: French youth. In the context of the management of any population, by virtue of their age, youth are the natural pedagogical subjects of the state and objects of its paternalism, and they stand as the symbol of the future of the nation. Yet in anti-smoking counteradvertising, French youth found their mission as a public health actor curtailed as compared with that of the young man in the 1976 ad. No longer does youth actively embody the dynamic possibility of a smoke-free future or the promise of an alternative modernity. Rather, in many of the spots the role of youth was limited to resisting the social culture of smoking among peers or to standing as passive models of healthy behavior for peers and adults. Thus, there was the 1978 media campaign, which employed the tag, "One crushed-out cigarette is that much more freedom" (it works better in the original French: "*Une cigarette écrasée, c'est un peu de liberté gagnée*"). It featured spots dramatizing "cool" teenagers in social situations (girl meets boy, boys and girls [or just boys] hanging out together) in which, against the background of a rock sound track, one teenager silently tenders a cigarette to another, who with a smile promptly crushes or breaks it in two and tosses it away without comment while a female voice-over repeats the tag and the CFES's name appears at the bottom of the screen. Interestingly, in these spots boys and girls alike both proffer and refuse cigarettes. Still, the attempt to address the problems of peer pressure and the desire to please in situations involving friendship or erotic interest remains awkward at best, for stiff and didactic acting betrays the paternalistic style of traditional public health messages and contradicts their appeal to teenagers as mature ethical subjects capable of acting upon themselves and each other in order to "free" themselves of the cigarette habit and social pressures. Similar ads were issued in the 1979 campaign (theme: "One less cigarette means a little more life"; "*Une cigarette en moins, un peu de vie en plus*"), and the 1988 campaign (titled, "Tobacco is no longer what it was"; "*Le tabac, c'est plus ça*"). Just what tobacco used to be is never specified in the latter campaign, nor is there any gesture toward a sustained historical or generational argument.

There was one spot that deployed a commonplace of twentieth-century health campaigns: the figure of the child as bearer and transmitter of public health consciousness and modern values (often learned at school). It targeted youth smoking by linking children's initiation to parents' poor example as unconscious hypocrites. The ad featured a father and son who watch circus clowns mocking parents' contradictory words and actions concerning tobacco. When they leave, the father pulls out a cigarette, and when the little boy frowns in disapproval, the father hesitates as the voice-over of a mature man comments, "One out of two children of parents who smoke smoke, too" ("*Des parents qui fument, un enfant sur deux fument aussi*"). Although both have watched the humorous skit, only the little boy has learned anything, and as a worthy pedagogical subject he becomes in turn the tutor to his own father in matters of health.

In 1994–95 a campaign did stage youth as symbols of health and a healthy future but only passively so. The CFES sponsored three humorous spots on the deleterious effects of smoking on physical performance by pitting teenagers against adults. One featured a man desperately trying to play goalie opposite teenagers (who are off-camera) who have given up the habit; another caricatures an average middle-aged French couple (*Français moyens*) driven to distraction by the incessant drumming of the kid next door (also off-camera) who has stopped smoking; and the third stages a dignified, retired bourgeois couple vainly trying to read in their living room while the sounds of a squeaking mattress of a young couple making love in the apartment above them grow louder. Resignedly, the wife comments, "I liked it better when the kids upstairs used to smoke. They didn't go on for so long" ("*Les jeunes du dessus, je préférais quand ils fumaient. Ça durait moins longtemps*"). The last shot is a medium close-up of the offending couple's noisy box-springs, with the campaign's title superposed, "Energy isn't meant to go up in smoke" ("*L'énergie, c'est pas fait pour partir en fumée*").[24]

A much milder and less ambitious reprise of the 1976 (anti–secondhand smoke) ad with the youth, these ads transform the image of teenagers as a problem (and more humorously as a source of irritation) for society into one of pedagogical examples of healthy habits presumably for their peers and perhaps even for adults but without dramatizing adolescents as active agents (in all three ads they remain off-screen) of the new health norms and without tackling the socially vexed issue of secondhand smoke. Moreover, just as the 1988 campaign never did make clear "what tobacco no longer was" and the general culture that underwrote it, so, too, these spots did not spell out what the smoke-free future would look like for which youth would serve as possible vectors. The CFES refrained from grappling with the deep cultural associations smoking enjoyed in the larger culture beyond some vague notion of glamour and cool. With few exceptions, this reticence was echoed by the cinematic and literary production in France at that time.

There was one spot last spot worth mentioning that created something of a stir when it came out in 1991, the year the French National Assembly passed the *loi Evin* and the WHO designated the CFES as a collaboration center of the WHO Regional Bureau for Europe in tobacco control. The televised ad was a direct attack on the global tobacco industry and its cultural images, which the CFES, in a temporary shift in policy, deemed to be ripe for criticism ("Fumer"

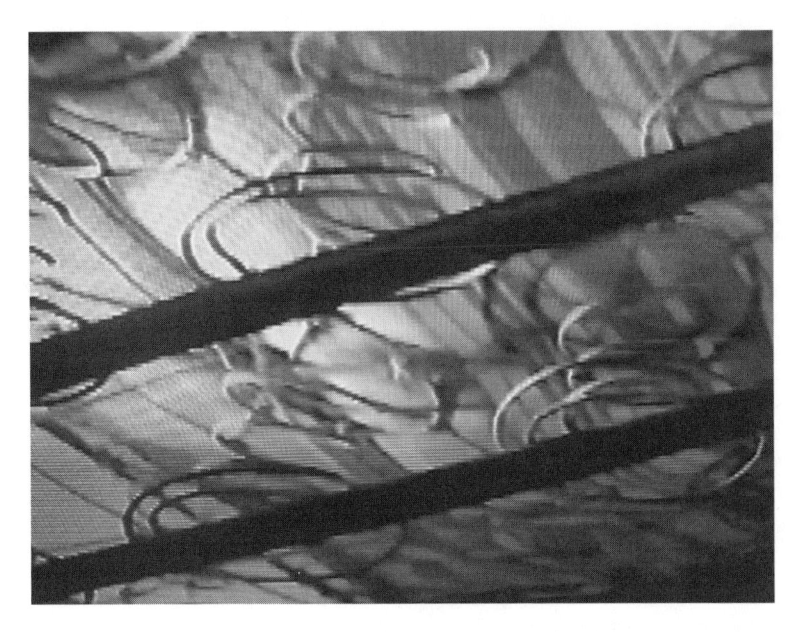

"Energy isn't meant to go up in smoke."
Courtesy of the Institut national de prévention et d'éducation pour la santé.

1991, 15). It was a collage of what looked like clips from Marlboro ads picturing sweeping vistas of cowboys riding herd in Monument Valley against the Marlboro theme on the sound track. A series of medium close-ups follows a digitally altered Marlboro Man (his face is a blur) as he squats by the fire warming his hands, his day done. At this point, his manly voice breaks in to declare, "It's not my nature to smoke" ("*Fumer, c'est pas ma nature*"). An example of what later came to be called "culture jamming" in the English-speaking world (N. Klein 2000), it was an eye-catching if somewhat clumsy reappropriation of Marlboro's cowboy logo (posters were more successful—the cowboys' visages were not erased). Sued by Philip Morris, the CFES prevailed in court and continued airing the ad ("Polémique" 1991). However, during the 1990s it was to be the other semi-public organization, the CNCT, which hewed to this approach most consistently and vigorously.

Aggregating and Segmenting Populations

Overall, the CFES media campaigns staged mildly dramatic situations in fairly private and ordinary settings (vaguely middle-class friends and family) structured by discourses of age and perhaps gender. They were not very marked by class, let alone by other social categories, as was often the case in media campaigns in California: all of the actors are of European descent and speak in standard French accents. This segmentation practice was also reflected in a range of French tobacco control data-gathering practices at that time.

What is striking to any reader of Californian and U.S. public health statistics is the near-complete absence of regional, linguistic, racial/ethnic, and immigrant communities among the "vulnerable populations" in France, a country with a long tradition of immigration and a definition of citizenship based, like the U.S., on place of birth (*droit du sol*) rather than on blood ties (*droit du sang*) as in Germany (until recently) or Japan.[25] However, the dominant narratives were far from identical; according to Eric Fassin and Martin Shain (Fassin 1999b; Shain 1999), French discourse stresses a relation between citizens and the state, while U.S. discourse foregrounds citizens grouped by national origin or ethnicity in their negotiations with government. This is the legacy of parallel but divergent histories. On the one hand, the French state deeply involved itself over the past century in promoting immigration policies targeting certain national groups (Poles, Italians, Yugoslavs, North Africans, Portuguese) to meet its labor needs in mines, industry, and agriculture and provided newly arrived immigrants with services (housing, employment, medical care, etc.) while addressing them *as individual subjects*. On the other, the U.S. government did not closely coordinate immigration from Europe with national labor needs (Asian immigration was another matter); nor did it actively recruit workers from Europe by nationality. And once they gained admittance into the U.S., the government provided them with little in the way of services. Faced with government indifference and private-sector discriminatory practices, immigrants in the U.S. obtained what few services they could procure for themselves through political and economic bargaining that was often based on national or ethnic origins. The legacy of these different histories structures the way in which respective statistic-gathering practices construe French and American populations.

When queried about the lack of race/ethnicity data in French smoking studies, anti-smoking advocates gave two responses. To begin with, the resources to segment the French population more finely are simply unavailable, and—more to the point—even if the necessary means existed, it would be illegal for the French state to do so. Since the 1870s, French practice has forbidden the collection of statistics by the state by race, ethnicity, and religion in metropolitan France. The ban was reinforced by the European experience of Nazi genocidal policies directed toward Jews, homosexuals, socialists, communists, gypsies, the handicapped, and the mentally ill, and it is supported by contemporary discourses of citizenship, social belonging, and solidarity that have deep roots in French Republican nation-state ideology and social democracy (Blum 2002). So to single out North African immigrant communities or the lesbian and gay community for a study of smoking behavior, let alone as targets of state-sponsored health campaigns specifically tailored to them, not only would contravene the law but also would prove to be very disconcerting to many and would court accusations of stigmatization from members of the community and human rights organizations in France. Explained one researcher-activist:

> The community-based approach stands in contradiction with the Republican tradition. But you have to see that even inadequate epidemiological data demonstrate that certain populations smoke more, drink more, die more in automobile accidents, etc. So the question is whether we are going to take on a particular problem concerning a particular fragment of the population, or are we interested in smoothing away inequalities such that people drink less and so forth. I'd say there's a tension there. The state, public authorities will do everything in order for the Republican way of doing things—which is the foundation of the nation and homogenization of the nation—to take its course but clearly professionals, associations, etc.—in short, civil society—struggle against this because they realize that it's too slow. (Hirsch 1997)

Exacerbating the issue were two things: first, high rates of French unemployment, in response to which arose a French social solidarity movement against "social exclusion"—the marginalization and disenfranchisement of the disadvantaged (*populations défavorisées*), including the chronically unemployed, working-class youth, the homeless (*les sans-abris*), undocumented workers (*les sans-papiers*), and second- and third-generation French of Arab descent. And second, there was the far-right National Front party's platform that blamed the AIDS pandemic in France on the presence of North African immigrants and gays. Thus, aggregating communities and subgroups of French residents, other than by sex, age, and economic status, and singularizing them by the state health bureaucracy as done in the U.S. in terms of "special needs" would be unacceptable.[26] As the chief medical reporter for *Le Monde* put it, "You're not going to tell me that it's more dangerous for blacks to smoke than for whites?" (Nau 1997). Still, according to the researcher-activist, community groups in France were starting to come together and call for community-based social programs and hoping that perhaps one day this would extend to public health (Hirsch 1997).

The absence of community targeting of health messages did meet with dissent, however, especially concerning the HIV/AIDS pandemic. In the eyes of some activists, the CFES's recent health promotion initiatives on AIDS, which did not tailor their health messages to the communication requirements of specific communities, amounted to criminal negligence on the part of public health authorities (French AIDS activist 1994). At that time state-sponsored prevention strategies promoted general condom use and free voluntary HIV screening in the mainstream public media sphere while refraining from moralizing discourse linking AIDS and homosexuality as was done in Italy, its Latin neighbor (Steffen 1996, 59–60). To address particular communities faced with the threat of HIV/AIDS in France, there arose many NGOs, including AIDES, founded in 1984, and ACT-UP, founded in 1989—inspired by U.S. NGOs (Edelmann 1993)—which stepped in with prevention campaigns designed by and for different groups, provided patient services, and engaged in political lobbying and public protests.

The only state-sponsored health promotion literature in France that did target a specific group besides women and adolescents at that time was nutritional brochures for the disadvantaged. This situation may well hold for much of Europe as well, with the exception of Germany, which has some public health initiatives targeting German residents of Turkish descent. Even publicly addressing specific adult audiences by socioeconomic status was explicitly rejected by one high French health official, who told the researcher-activist quoted above that if you took away from French workers their bottle of wine, their car, and their pack of Gauloises, nothing would remain. However, according to this anti-smoking advocate, such an attitude stemmed from what he termed an "elitist populism" that was rampant among leaders and intellectuals of the French Left and that was paternalistic at bottom. Here we witness something of a clash between the guardians of public health on the one hand (the anti-smoking activists) and those journalists and politicians acting as gatekeepers of French workers in policy circles and public debate on the other. In one case, the "general interest" required measures that would discourage smoking among workers (higher taxes, new workplace health codes, etc.); in the other, there was a resistance to the claims of the "general interest" over the lives of one segment of the population, stemming in part perhaps from the historical experience of intrusive public and private initiatives that sought to oversee and discipline working-class customs and daily life (Donzelot 1992; Joseph, Fritsch, and Battegay 1977; Murard and Zylberman 1976).

This general reticence to address subgroups may help explain why youth enjoy the privilege of being the target of choice in the public health policies of population management whose discourse refuses to singularize subgroups because of the risk of stigmatizing them: "youth" cuts across virtually all other population segments and in terms of individual biographies is a temporary category. In this sense, perhaps, could a singularity such as youth come to represent both symbolically and pragmatically the "general interest" of the population residing in France.

The meager data on tobacco-related diseases and behavior among the French made deriving meaningful information about various segments of the population for health promotion campaigns a difficult task. The title of an article published by the CFES allowed as much: "No survey, no campaign" (Rotily and Bregeault 1994).[27] Moreover, epidemiological studies of smoking garnered little interest and support because pulmonary specialists remained preoccupied through the 1970s with their old nemesis, tuberculosis, especially among immigrant workers; citizens' experience of lung cancer until the 1980s remained rare; and there was a taboo in France on public discussion of cause of death (Mélihan-Chenein 1997; Hirsch 1997; Karsenty 1997; Slama 1997; Meurisse 1999). For a long time, what French studies that did exist tended to focus on disease alone and not on behavior. One set of data came from the National Institute of Statistics and Economic Studies (Institut national de la statistique et des études économiques, or INSEE), the equivalent of the U.S. Census Bureau and Bureau of Labor Statistics. However, many tobacco control advocates deemed the INSEE studies of limited value, for they did not focus on health per se and were published only every ten years.

The other set were surveys conducted by the CFES.[28] These did take into account smoking-related behavior (initiation, number of cigarettes, cessation) by age, sex, and socioeconomic variables (education level, religion, occupation, marital status), but many of the authors of the *Report of the Five Wise Men* did not take them seriously, citing the studies' small sample sizes and lack of rigor.[29] As for secondhand smoke, studies conducted in Japan and the U.S. and cited by the WHO did signal the health hazards of secondhand smoke to the international community (Hirayama 1981; U.S. Dept. of Health and Human Services 1986), but up to the late 1990s French epidemiologists in tobacco control, who were trained in the British school of Richard Doll and Richard Peto, who did not consider it a serious health issue, refrained from investigating the topic.

Faced with the paucity of French-based data, French tobacco control advocates drew on international sources. This was true not only for the early Veil campaign, as noted previously, but also in the 1990s. Although seen by antismoking advocates as closely tied to the French state and thus less open to foreign research and policy trends, the CFES, at least in its brochures and campaign literature in the 1990s, did cite the WHO; the European Union's own campaign, "Europe against Cancer"; and studies conducted by U.S.- or U.K.-based researchers (see, for example, "Fumer" 1991; "Les Jeunes et le tabac" 1994). Still other researchers and activists (some associated with the CNCT) insisted that such foreign data on smoking behavior and health promotion campaigns were fully adaptable to the needs of French tobacco control and thus sufficed; to conduct local studies in France would be a waste of time and money (Sasco 1992). However, when it came to persuading public opinion of the dangers of smoking, especially the hazards posed by secondhand smoke, many felt that foreign data simply would not work.

Vulnerable Populations and Unruly Subjects in France

Insofar as statistical studies contribute to identifying population segments in terms of risk and disease factors and thus make them available to public health interventions, several groups did emerge from the tobacco control data in the 1980s and 1990s. We have met them before: adolescents (ages 12–18); young adults (ages 16–34); general practitioners, whose rates continue to match those of the general population; and women, especially the young, the pregnant, and those with university diplomas. All of them were deemed particularly at risk since their smoking rates were holding steady or increasing, while rates for older men as a group had begun to decline ("Fumer" 1991; Grizeau n.d.; Baudier et al. n.d.; *Lutte contre le tabagisme* 1996).[30]

What is fascinating is the picture that emerged from conversations with researchers and advocates, especially when asked to look at the future. Many privately admitted that smoking prevalence would never drop below 10–20 percent of the French population—something analogous to which I never heard expressed by U.S.-based advocates—and some thought that this group would include the deeply addicted and a good number of those living on the margins of French society—the indigent, the permanently unemployed or unemployable, and delinquent elements. The only campaigns that might prove effective with some of them would be aggressive "scare" campaigns. Several expressed theories of social cycles of the adoption and abandonment of popular behavior such as smoking in modern democratic societies: the elites (the wealthy, the famous) and margins (the poor, bohemians, artists) are often the first to adopt a practice and will remain with it after the mainstream has adopted and then later begun to drop it. Parallel to these cycles was one based on gender, according to which, in male-dominated societies, men would take the lead in both initiation and cessation (Boucher 1996; Got 1997; Hirsch 1997; Karsenty 1997; Slama 1997).

While the discourse of social differences in smoking generally remained private (with perhaps the exception of disadvantaged youth), it occasionally slipped out into the public media sphere. When it did so, it took the form of the figure of the older working-class woman as object of solicitude and blame in televised talk show discussions. A France 3 afternoon television show, *C'est pas juste*, in an early broadcast devoted a full hour in January 1989 to the problems of quitting. Titled "Halte au tabac," it staged the struggles with smoking of a mother from Malakoff, a Parisian working-class district, whose habit her daughter finds oppressive. Gathered around the table are the mother, her daughter, several medical experts, and a very young talk show host. Throughout the show, the mother is framed as a patient and weak ethical subject. She is the object of extensive moral and clinical commentary that refers to her frequently in the third person, even as she shares with those on the set the difficulties she has had in quitting ("Halte au tabac" 1989). The overall tone is one of condescension and is underscored by the camera work of shot/reverse shots of a smiling physician and a forlorn, uncomfortable middle-aged woman. Near the close of the show, turning to her, one medical expert remonstrates, "It's a question of inner maturation. It a little like that with every decision. Otherwise, it's what you'd call a drunkard's oath." The woman replies, as if on cue, "Me, I'm like a child. I have a weak character. I need to be helped and monitored for a long while."

Six years later, in November 1995, the mise-en-scène is reprised in the broadcast of TF1's late-night talk show, *J'y crois, j'y crois pas.* Titled "Faut-il interdire le tabac?" the show features a large group of activists, researchers, writers, artists, and politicians. At the center of these eminent guests sits a gravely ill woman in a hospital gown, who, even after several amputations, continues to smoke ("Faut-il interdire?" 1995). In this medical theater (reminiscent of nineteenth-century clinical demonstrations), she is not the constant center of discussion, but when attention does turn to her, the tone and manner are patronizing, as when one physician says to her and his colleagues, "She has become enslaved to tobacco." It is not hard to detect in these two examples the legacy of public health discourses whose roots lie in nineteenth-century social philanthropy directed at the poor and working classes and, more recently, in aspects of the welfare state and social medicine whose practicing experts targeted incorrigible populations for social assistance and discipline, if not social protection.

The Semi-Public National Anti-Tobacco Committee (CNCT)

When the state endowed the semi-private CNCT with the public mission of enforcing new bans on tobacco advertising and public smoking by authorizing it to initiate lawsuits against violators, the CNCT adopted an aggressive stance that eventually put it on a collision course with the more timid CFES and the Anti-Cancer League. It targeted SEITA and other tobacco companies and attempted to disrupt public discourse on smoking and tobacco-related illnesses. Still, in the early years before the *loi Evin* (1978–1990) the CNCT filed only forty suits. This corresponded to a period when much citizen activism from the anti-nuclear and ecology to the women's and gay movements fell off markedly after the success of the Left in the 1977 municipal elections and the advent of the first Socialist national government in twenty-five years in 1981, while the far-right-wing anti-immigration movement grew, led by Jean-Marie Le Pen and the National Front (Duyvendak 1995). The exception were the HIV/AIDS organizations that sprang up between 1984 and 1989. In 1991, just after the *loi Evin* was debated and passed, things changed radically when a group from the League against Tobacco Smoke in Public—Non-Smokers' Rights joined the CNCT. Accusing the old guard of a narrow medical focus on cessation while neglecting the protection of nonsmokers, they won a majority in the annual elections to the governing board (then comprising two physicians, a teacher, a teacher of physical education, a journalist, an activist, and a trade unionist). At that time the CNCT claimed to have approximately 1,600 members.

The new board promptly named Philippe Boucher as its director, a post he occupied until 1998. The French general director of health asked the CNCT to increase its legal activities, which it promptly did. These actions, together with the CNCT's own media promotion, would help radicalize public debate on smoking in France. Between 1991 and 1997 the CNCT undertook more than 200 legal actions, primarily civil suits against two groups: on the one hand, publicly and privately owned media, sponsors, and advertisers for circumventing or violating France's ban on tobacco advertising and promotion and for improperly displaying health warnings; on the other, public transportation authorities for not providing appropriate signs and nonsmoking sections or areas. With a staff of six (a majority of whom were graduates of France's Fondation Nationale des Sciences Politiques ["Sciences-Po"], a *grande école*) plus two lawyers on retainer, the CNCT was often successful in the lower courts, and violators were slapped with various injunctions and punitive damages, all of which garnered a great deal of media attention to the organization's activities and the cause of nonsmoking and smokers' rights (Nau 1991). An independent government audit of the CNCT would later credit the group with the disappearance of direct and indirect tobacco advertising in France (*Rapport de contrôle* 1998, 12). One other tactic the anti-smoking group employed to great effect was to settle out of court for substantial sums (to subsidize its activities) and for free space in television, radio, or print media, which it would then fill with its own spots or counteradvertising. The most spectacular case occurred in 1992—the so-called *Affaire* Williams Renault, which embroiled a nexus of national and global actors consisting of a U.S. tobacco company, a French television network, a French NGO, and the French state. The CNCT sued R.J. Reynolds, TF1 (France's principal private television network), and the state-owned Renault automobile company for promoting Camel cigarettes through prominently placed logos and signs during the prestigious Formula 1 Grand Prix de France broadcast. The suit created a national uproar. The controversy created divisions within the government, pitting the Ministry of Health against the Ministry of Finance and Ministry of Youth and Sports, and it weakened support for implementing the new *loi Evin*. It came close to forcing cancellation of the race the following year, since most Formula 1 stables were sponsored by the tobacco industry. Ultimately the CNCT relented, and in 1993, in exchange for dropping its suit, it accepted a budget allocation of 100 million francs (c. US$16.5 million) specifically earmarked for public health activities to the state Public Health Network and an additional 2 million francs (c. US$330,000) from TF1 in 1997. Later, taking advantage of France's stringent product safety legislation, on behalf of sick smokers the CNCT also filed suits against cigarette manufacturers for fraudulent misrepresentation of the health hazards of smoking (Boucher 1996, 1997; Hirsch 1997; Mélihan-Chenein 1997, 1999; Caballero 1999).

The Williams Renault case underscored the intrinsically ambiguous and contradictory mode of governance that characterized the CNCT's role in enforcing the law of the land. Emanating from civil society, the organization was endorsed by public authorities and made an instrument of state policy, which the government itself was reluctant to enforce directly. Dubbed the "secular arm of the state" (*"le bras séculier de l'État"*) by the audit's authors (*Rapport de contrôle* 1998, 7), the CNCT, as a semi-public agency, was understood to operate as a private "interest group" but one that transcended the narrow world of politics. As such, it had the daunting task of taking on state-owned industries (SEITA, Renault, SNCF [railways], and public transportation), state-supported events (Grand Prix), and powerful private-sector lobbies (the media and the tobacco industry). And it did so from a position of compromised autonomy stemming from its dependency on state subsidies for its legal actions and funds from the CNAM (the state-run agency that oversees the pool of health insurance programs) for preventive work. A case in point was the fate of the 100 million franc subsidy for public health actions: the CNCT saw very little of the funds, most of which were snapped up, with the tacit consent of state health officials, by researchers not working on public health issues at all. Another example were the actions of the CNAM, which in 1993 reduced its subsidy because in its view the CNCT had not done enough preventive work and withdrew its representative from the CNCT's board (*Rapport de contrôle* 1998; Slama 1997, 1999). The CNCT may well be the "secular arm of the state," but in the eyes of its Philippe Boucher, it has served rather as an alibi for the French state's own inactivity in tobacco control. Boucher's view seems to imply that the liberal project of "governing at a distance" through civil society's own organizations amounts to not governing at all (Boucher 1996).

Taking Liberties: The CNCT's Publicity Campaign

The CNCT's official adversarial mission and the activist background of its director and staff gave it a freer hand than the other hybrid, semi-public organization, the CFES, whose limited activities were closely monitored by the CNAM and state officials who provided almost all its funding. This was evident even in the CNCT's cessation and health promotion actions. The group set up an information source on France's Minitel computer service (Tabatel) and a telephone hotline with the CFES for smokers seeking to quit (Tabaphone); and it published *Tabac et santé*, a monthly newsletter relating tobacco control news in France and abroad. It distributed brochures and posters, disseminated anti-smoking ads in the print and electronic media, and started a European video lending library for schools and citizen groups. Yet, even with funds obtained through legal actions, the CNCT had little in the way of resources for media interventions and anti-smoking media campaigns, and its primary goal remained one of stirring up controversy and debate from time to time rather than setting agendas and shifting norms and behavior through full-blown exposure campaigns based on formative studies and follow-up evaluation. Still, the authors of the CNCT outside audit claimed that the lawsuits in and of themselves and the publicity they generated did constitute important acts of prevention and health education (*Rapport de contrôle* 1998).

The CNCT's media events resembled examples of discreet media advocacy more than anything else, tied as they were to international conventions, World No Tobacco Day, the implementation of the *loi Evin*, and other developments. Modest though they were in their scope, the CNCT media interventions took anti-smoking discourse in France in new directions: attacks on the tobacco industry's marketing practices targeting youth, the use of images of the sick and physically repulsive pictures of diseased organs, and the adoption of non-French materials. This approach challenged the parameters of then acceptable public discourse and put the CNCT at odds with the CFES. However, the CNCT did little in the way of publicizing the hazards of secondhand smoke other than through its legal actions.

For example, at the time of the Ninth World Conference on Tobacco and Health held in Paris in October 1994, the CNCT performed its own version of culture-jamming and brought out a set of posters caricaturing the famous gypsy and Gaulois logos of SEITA's leading brands, Gitanes and Gauloises. The posters reproduced faithfully the style, colors, and layout of the cigarette logos but with minor distortions, and they added a title. Thus, the dancing gypsy woman's seductive castanets become threatening shellfish claws, and the Gaulois helmet is transformed into a death's-head with the title, "Advertising kills" ("*La pub tue*"). The altered logos play off deeply ambivalent cultural registers concerning sexuality and war to appeal to smokers' own divided sentiments regarding the pleasures and dangers of their habit and, by extension, the manipulative practices of the tobacco industry. The posters cleanly invert values used to promote Gitanes and Gauloises cigarettes (French gallantry and battlefield heroism): exotic women as seductive possibility become a repulsive castration threat, and the martial glory of the common soldier is changed into the wanton slaughter of hapless recruits. SEITA immediately sued the CNCT but lost its case (Peyrot 1991). This gave the CNCT the opportunity to mock openly the state tobacco monopoly (but prudently not by name and not as a state-owned enterprise) in a tag placed beneath the first in a second printing of the posters: "Since the tobacco industry wanted to prevent you from seeing this poster, the National Anti-Tobacco Committee has undertaken to reprint it."

Other spots and posters took on the taboo on public representation of illness and physical handicaps in mainstream French media and culture. In one spot aired on TF1 in November–December 1992 (just as restrictions on public smoking took effect), the CNCT employed the approach, later adopted by the Massachusetts Department of Public Health, of recording tobacco-related cancer patients. They interviewed former U.S. model Janet Sachman, a lung cancer survivor, who regretted her appearances in cigarette ads for Lucky Strike and denounced the tobacco industry's marketing tactics targeting youth. The ad begins with a set of shots of glamorous stills from her youth alternating with barely perceptible extreme close-ups of her elder face and mouth as the sound track reproduces the troubling sounds of someone's labored breathing. The spot then switches to close-ups of Sachman speaking with difficulty to an interviewer:

> I want people to know what happened to me from smoking. The most important thing anyone can do is not to smoke. We must discourage advertising encouraging children to smoke. I owe it to the world. I might have encouraged a few people to smoke. Now I am trying to tell them what happened to me.

The French subtitles edit her remarks for clarity and succinctness, rendering her message even more forceful.[31] It was the first time in France that a link had been publicly drawn between advertising and tobacco-related diseases (Mélihan-Chenein 1997). The spot later won the top prize for a commercial in San Francisco.[32]

Another small poster took direct aim at the lack of public discussion of cause of death in the media and among friends and neighbors stemming from French laws protecting both physician/patient confidentiality and physicians' traditional right to withhold such information from not only family and loved ones but also patients at their discretion. Replicating the tactic of the famous 1985 U.S. spot of Yul Brynner, who issued a warning to young people from beyond the grave, the poster featured a letter coauthored by Jean-Paul Signoret, a man dying from lung cancer, and his female companion next to a still photo from a home video of him lighting birthday candles. Alongside a "postscriptum-postmortem" from his partner announcing his death, which appeared in red lettering, there was a "call for witnesses" ("*appel à témoignages*") addressed to those suffering from tobacco-related diseases to come forth and tell their stories.

The CNCT's borrowing of practices from anti-smoking campaigns outside

"Advertising kills."

Courtesy of the Comité national contre le tabagisme.

of France matched the liberties it took with SEITA's commercial logos and acceptable public discourse. It freely appropriated posters, spots, and film clips from the U.S., Catalonia, and Sweden and reflected a commonly shared opinion expressed by the CNCT members and tobacco control activists working in France that in an era of global products from Coca-Cola and Nike shoes to cigarettes, there is no reason why health communication materials borrowed from abroad shouldn't work in France (Boucher 1997; Mélihan-Chenein 1997; Slama 1997; Karsenty 1997; Hirsch 1997; Dubois 1999). As one younger activist put it, "It's true that the French will tell you all the time, that's America, that's not France. But the French are so Americanized! Why is it that they can smoke American cigarettes but should not adopt American anti-tobacco policies? Is it just cigarettes that are American? Cancer is, too" (Mélihan-Chenein 1997).

Referencing the threat posed by industrial pollution to nature and humans alike, in 1993 the CNCT reissued a medium-size American Cancer Society poster featuring a drawing of a suit-clad man walking toward the viewer but whose head has been transformed into one tall cigarette emitting heavy smoke like a smokestack and the title, "Smoking pollutes in nature and in the street. Each cigarette butt pollutes" ("*Fumer pollue dans la nature, dans la rue. Chaque mégot pollue*"). Tags identified the CNCT and the American Cancer Society as sponsors, and another title appealed directly to smokers: "Smokers, try being tidy: pick up your trash instead of leaving it in nature or the street" ("*Fumeurs, essayez la propreté: récuperez vos déchets au lieu de les abandonner dans la nature ou dans la rue*"). The poster moves to link environmental issues with tobacco control and gestures toward reforming society by inciting smokers to discipline themselves; there is no call for smokers to police each other and even less for nonsmokers to intervene with smokers.

The following year the CNCT opened the Ninth World Conference in Paris with a reworked version of a famous public service announcement (PSA) and poster originally released by the Minnesota Department of Health deglamorizing smoking as a cool activity. It featured close-ups of actual barnyard animals smoking against a blues sound track with the male voice-over (or wording in the poster) in colloquial French punning on the multiple meanings of the French word "*bête*" (foolish, stupid, animal): "With a cigarette you really look foolish" ("*Avec une cigarette . . . t'as vraiment l'air bête*"). It roughly translated the Minnesota wording, "It just looks ridiculous when you do it" and retained some of the original humor (which in the Minnesota case actually was said to be highly popular with all age groups, especially children; Minnesota health official 1997) while opting for a more aggressive and familiar mode of address. The contrast between sound and image tracks tried to subvert the success of R. J. Reynolds' Joe Camel cartoon figure (first launched in France in 1984, then in the U.S. in 1987) by playing off the silliness of actual animals smoking and the traditional association between cigarette smoking and the cool glamour of blues and jazz culture. The ad and poster were something of a hit in France as well (the image was even issued as a T-shirt).

However, there were limits to its practice of adopting non-French anti-smoking materials. For example, the CNCT did not rush to adopt the spots from the California campaigns in the 1990s whose hard-hitting style attracted international attention. Their contacts with Californian health officials were intermittent at best, and the question of royalties involved in the Screen Actors Guild contracts for the ads discouraged their appropriation abroad until the release of a small number of them to the CDC's Media Campaign Resource Center in 1995 for public distribution. Still, the CNCT did not adopt any California ads until 1997–98, and even then they chose to broadcast yet another dramatic testimonial by a suffering smoker. This was the famous "Debbie" spot featuring a middle-aged woman speaking to the camera of her addiction to nicotine as she smokes through a hole in her neck. Significantly, the anti–secondhand smoke ads were not picked up. This did avoid singling out and possibly stigmatizing any group (men, women, pregnant women, youth, etc.) and was in keeping with some researcher-activists' sense of the dominant risk perception in France: the idea that industries and even government could pose a threat to citizens' well-being but not friends and family through their actions and habits. The greatest example of this was the high public mobilization in France against asbestos in building materials as opposed to smoking (Karsenty 1997).

It was precisely to challenge this sense of safety among loved ones who smoked that the Anti-Cancer League issued several spots in 1994 at the time of the international conference in Paris. Of all the "publicly useful" anti-smoking groups, it was the only one to sponsor televised spots drawing attention to the dangers of secondhand smoke. This may have had to do with its greater independence from the state in terms of its funding; it also had been the recipient of no specific mandate from health authorities concerning its mission. The league aired three ads. They featured a little blond boy visibly distressed in three situations. In one he worries about his mother's smoking and runs up to her, exclaiming, "Mom, you don't smoke, do you?" ("*Maman, tu ne fumes pas*?"). In another he wakes up from a nightmare of a room full of adults smoking, provoked by the sounds of a party in his home, and cries out to his mother. When she tries to comfort him, he accuses her, "Mom, you burned me" ("*Maman, tu m'as brûlé*"). In the third, looking abandoned as he watches his mother busy herself while smoking, he finally pleads with her, "Mom, don't smoke anymore" ("*Maman, ne fume plus*"). All three conclude with the title, "Let's not smoke up our children's lives" ("*N'enfumons pas la vie de nos enfants*"). As in the CFES ad featuring the father and son at the circus, the child in these spots serves as the bearer of a new public health consciousness and norms, but here the ads directly target the mother, the traditional caregiver of children, to whom the boy speaks in a tone of betrayal as a family victim.

Collaborations and Exceptionalisms

Differences in articulating tobacco control issues to the public heightened tensions between the anti-smoking organizations, especially between the CFES and the CNCT. This was very clear concerning the question of the transposability of foreign-made spots and methods to the French context. Here, the discourse of French exceptionalism was deployed in opposite ways, for national differences mattered differently to each organization. Thus, as mentioned earlier, on the one hand the CFES, virtually a department within the welfare and health ministry, availed itself of foreign statistical studies but rejected the Anglo-American approaches as too melodramatic and moralizing for Latin France; they were also considered too divisive and risked, in a reprise of the rhetoric of the opponents of the *loi Evin*, setting off a war between smokers and nonsmokers. This view was held by both older and younger members of the CFES staff throughout the 1990s. Another example was how the CFES redid the poster proposed by the WHO for World No Tobacco Day (31 May) in 1999. The CFES readily altered the theme and logo to give it a national "stamp" likely to appeal to a French audience. The WHO logo for that year was an ashtray with a red rose placed in it. Working with the French ad agency Publicis, the CFES developed an alternative image, that of a picture of Earth from outer space, with the tag "CFES." No clearer articulation of the convertibility of the national and the universal could be found. The local French "stamp" on the World No Tobacco Day logo was to give it a universal theme symbolized by an image of Earth taken from outer space. By contrast, university researchers and activists in the orbit of the CNCT saw little problem with adopting foreign health promotion materials and aggressive, "non-French" ways of addressing the public. An interesting variation in this "internationalist" stance was that of a leading researcher in pharmacology. His was a modernizing view according to which France was in the throes of a long historical evolution, dating from the nineteenth century, of cultural discourses and practices, regarding an increasing vigilance against possible violations of the integrity of the body, which it shared with other leading industrial or industrializing nations. The rise of this "neutral" body was very far along in the U.S. However, since in his own view it was a matter of getting crucial information across to the public in the context of complying with a new law, it shouldn't involve so much "telling people what to do," as in so many health campaigns. Social norms were already shifting toward nonsmoking in public (Karsenty 1997; see also Hirsch and Karsenty 1992, 10).

Divergent approaches were apparent regarding international contacts and exchanges, which played different roles in the two organizations. Members of the CNCT actively attended international tobacco control conferences outside of France to discuss the latest concrete strategies being developed to combat the marketing and consumption of tobacco products, whereas until the late 1990s it would appear that this was rarely the case for CFES functionaries. If the CFES is closely identified with the state, it is fair to say that the CNCT is more affiliated with the international anti-smoking organizations, including the International Union against Respiratory Ailments and Tuberculosis and the International Union against Cancer, where it finds support. These differences surely stem in part from the backgrounds of their respective staffers and board members, for until recently the CFES could not match the international training and contacts, especially in the U.S. and the U.K., of researchers and staff associated with the CNCT.[33] The discourse of French exceptionalism did surface in conversations with anti-smoking advocates outside of the CFES, but, as one might expect, negatively so; in contrast to northern Europe and Anglo-American countries, they lamented the lack of a widespread anti-smoking movement in France, the absence of an aggressive French equivalent to the American Cancer Society (which actively funds research, health promotion, and professional lobbying), and the presence of a wide gulf, based on status and hierarchy, that separated activists from the medical elite, which contributed to the inability of citizen-driven organizations (like the League against Tobacco Smoke in Public—Non-Smokers' Rights) to present their positions in a convincing, professional manner to the wider public. In the best Tocquevillian manner, they attributed these shortcomings to differences between Protestant- and Catholic-dominated cultures originating in Northern and Southern Europe (Boucher 1996; Hirsch 1997; Mélihan-Chenein 1997).[34] Although their proposed measures received strong support in public opinion polls, the groups' base remained narrow and their numbers small, effectively limiting their ability to convert poll numbers into political clout. This is confirmed by simply consulting the membership of various commissions, oversight committees, and expert advisory boards that sprang up in the 1990s. The range of names of nongovernmental anti-smoking advocates is not wide.

Throughout the early and mid-1990s collaboration between the two organizations proved difficult at best. For a time, a representative of each organization sat on each other's board and shared membership on the tobacco control Oversight Committee (Comité de pilotage) coordinating various groups' efforts and on the Alliance for Health—Anti-Tobacco Coalition (Alliance pour la santé—Coalition contre le tabac, founded in 1991). While recognizing that the CNAM tightly controlled the CFES's policies, anti-smoking advocates inside and outside the CNCT characterized the CFES as timid bureaucrats who embodied the libertarian legacy of May 1968 that stressed a hands-off attitude when it came to regulation of private behavior (especially in public) and who underscored the mutual "solidarity" of citizens as expressed through the actions of the state (broad welfare protections and social medicine).[35] The advocates did not hesitate to go public with their criticisms of health and welfare ministry officials and the medical establishment in professional and media venues. One anti-smoking activist scathingly remarked, "In France we don't do public health, we do prudish health" ("*On ne fait pas de la santé publique, on fait de la santé pudique*"). In turn, functionaries at the CFES expressed exasperation with the activists' and researchers' pressure, calling them at times, in private, extremists and even "ayatollahs" that were eager to foment conflict between smokers and nonsmokers.

1996: A Turning Point

A shift in government attitudes began with the return of Simone Veil to her old post as health minister in 1993 with the Conservative victory in legislative elections. Internally the government was divided, for while some Conservative politicians attempted to abrogate the *loi Evin*, the Health Ministry increased CFES funding levels, which in turn led to improved collaboration between it and the CNCT, the professionalization of its anti-smoking campaigns through recourse to high-powered commercial advertising agencies, and fresh strategies that addressed the structure of the public media sphere dominated by marketing's traditional segmentation practices and altered how it addressed its audience in terms other than that of "vulnerable populations." Veil's ministry sponsored the Ninth World Conference on Tobacco and Health, which was already scheduled to be held in 1994 in Paris. Beginning in 1996, the CFES hired young staffers trained in health communication and statistics collection and established a working group to formulate the goals and methods of a three-year campaign (*Lutte contre le tabagisme* 1996). Meanwhile, in 1995 the CFES finally received an advisory board of experts called for long ago by the *Report of the Five Wise Men* in 1989. These consultative bodies included an array of public health actors including researchers, officials from the health ministry and the CNAM, and anti-smoking advocates. By the late 1990s relations between the CFES and the CNCT seemed to improve. It would appear that this was due to their active collaboration on well-subsidized, concrete projects and to new relations established between younger staff members of both organizations.

Meanwhile, the CNCT and the CFES pursued their respective tacks and commitments. The CNCT continued its mandated mission of filing suits against violators of the Evin statutes but focusing more on the tobacco industry's responsibility in encouraging smoking through inadequate warning labels and in the face of its own scientific findings. As for the CFES, it directed its energies not at the hazards of smoking (assumed to be well-known by everyone) but at overcoming the challenges of quitting (addressed to both smokers and their loved ones) and at the image of a new subject: *young ex-smokers* who symbolize quitting, not smoking, as a sign of maturity. Both spots devoted to these goals try to shift the narrative of quitting from a lonely face-off between the isolated smoker, his or her habit, other smokers and peers, and scientific fact to the reassuring interactions between a smoker and his or her friend or spouse. Here, in the intimacy of the young couple, one can detect echoes of contemporary discourses of "social solidarity" with people facing "exclusion" from French society. Regarding the questions of the dangers of secondhand smoke and smoking while pregnant, unlike in California, these issues were deemed inappropriate for the spots and billboards of the mainstream public media sphere (too reductive a venue) and better addressed though brochures available to women in gynecologists' offices and well-documented kits meant to educate the print media (which proved to be very popular with reporters). Later on, the CFES worked to distinguish the issue of female smokers from that of pregnant smokers (Grizeau 1997; Speisser 1997, 1999).

The CFES now carried out its task through better-funded campaigns working closely with major advertising agencies such as Publicis, France's largest and most prestigious firm. This involved both formative and cumulative forms of evaluation, which were subcontracted out to a third party such as Sofres, a leading polling organization. Here, it is worth noting several things. First, readers may recall how vexed the collaborative and evaluation processes were for the vastly larger California campaigns in the 1990s. Lines of authority, competence, and oversight were subjects of continual dispute, and the ad agencies enjoyed a determining role in that state–private sector joint effort. In the CFES and Publicis collaboration, in the minds of both the advertising people and CFES functionaries, it was very clear that it was the state agency that dominated the process. Not only did the CFES have the last word on promotion materials, but also it actively intervened in their development at every stage. The ad agency was clearly working not as an equal partner but rather at the behest of the state. This was the understanding and practice of both parties. At the same time, the advisory board of tobacco control advocates from outside the CFES worked in concert with the state agency to develop goals and campaign strategies, after which a public call for proposals was issued to the advertising industry (Speisser 1997, 1999; Maruani 1999). Even as allies of the tobacco industry worked to weaken the *loi Evin* under that same Conservative government that would later (in 1995) pass legislation to privatize SEITA, it would appear that third-party evaluation did not become a flash point of conflict between the CFES, the ad agency, the CNAM, the expert advisory board, and the evaluators.

Second, there seemed little manifest discomfort, at least within the CFES, with entrusting to a private-sector company an ethically sensitive mission related to guaranteeing public welfare. However, it would appear that this was not so much because the CFES did not share the *Report of the Five Wise Men*'s repugnance, noted earlier, for advertising's corrupting sway over public culture (tobacco industry marketing) as because the Republican state had long had dealings with the private sector—through state-owned enterprises like SEITA and French television and radio (until the early 1980s) and services (such as phone service and postal delivery)—in which it enjoyed the position of dominant partner.[36] Finally, the CFES attempted to come to grips with the public media sphere that was a mix of Paris-based public and private cable and broadcast companies and independent radio stations located throughout France by shifting principal media venues toward independent local radio, which arguably played a role in youth culture unparalleled anywhere else in Western Europe or North America. They provided local stations with CDs containing short clips of discussions with former smokers and presentations of topics as a prelude to live round-table debates open to radio audiences (Speisser 1999; Fertas 1999).

The CFES also set up partnerships with national radio and television networks, women's magazines (*Marie-Claire, Cosmopolitaine,* etc.) for special articles, and Family Planning for brochures to be distributed in the offices of gynecologists and general practitioners as well as in Family Planning's local offices. The Family Planning literature, designed by Publicis Étoile, was quite striking for readers used to standard anti-smoking literature. The brochures endowed female smokers with complex ethical features not generally found in anti-smoking representations of smoking. In particular, they focused on the nexus between dominant notions of feminine identity (self-affirmation, access to pleasure, sexual autonomy) and cigarette smoking, a connection underscored with the cover photo depicting an elegantly poised woman's hand in soft focus holding a lit cigarette. Inside, the text reviewed the opportunities and pressures women experience in their daily lives and the meaningfulness and utility of cigarettes before underscoring the dangers of smoking, especially while taking the pill, and ways to quit ("Femmes et tabac" 1998). The photo's sensuousness suggests something of a break with the spare if humorous style of most anti-smoking material put out by the CFES and, by means of its visuals, acknowledges the actual pleasures of smoking, which have been so effectively exploited by industry marketing but dismissed out of hand by deglamorizing anti-tobacco ads in France and elsewhere. In short, the brochures' cover appeals to women not as a "vulnerable population" but as embodied social actors and active consumers.

However, a subsequent attempt to extend this new approach to the public media sphere in the form of state-sponsored counteradvertising ran into major opposition. The CFES may have had the last word with the ad agencies, but

Brochure cover of "Femmes et tabac."
Courtesy of the Institut national de prévention et d'éducation pour la santé.

administratively all decisions by the CFES required final approval from the CNAM, its major funding source. In 1999 the Communication Department at the CFES collaborated with a leading women's magazine on a cessation campaign to combat the wave of pictures in magazines featuring models smoking. Together they actually reworked a fashion photograph of a model in panties holding a cigarette between her fingers with the title, "Dare wearing panties without a cigarette, unless you want to look vulgar" ("*Osez le slip, sans la cigarette, sinon ça fait vulgaire*"). In effect, they were using fashionable glamour to deglamorize smoking and to intervene in lifestyle advertising by detaching cigarettes as a consumer object from other bodily pleasures, but the playfully seductive photograph did not please their superiors, who rejected the concept out of hand, deeming it too vulgar. The CNAM did not make clear to the CFES whether it meant that the photo was too erotic or too fashion-oriented. Frustrated CFES employees later claimed that the CNAM's health promotion approach was based on a very dated image of French youth that pictured them as identical to the reading audience of newspapers and books put out by a progressive Left Bank Catholic publishing house. It was the clearest case of a clash between younger functionaries more comfortable with working within the style and values of advertising agencies and the consumer culture of pleasure they promote, and older public health officials identified with more austere values of progressive Catholic ideology and the French Republican state.

Unexceptional Exceptionalism

At one point during an interview in Paris on anti-smoking measures, *Le Monde*'s chief medical reporter exclaimed, "In general the French have two ways of viewing the United States: 'Americans are crazy' and 'It's inevitable. It will happen to us inevitably'" (Nau 1999). In this view, popular French discourse concerning tobacco control in the U.S. and France swings wildly between an affirmation of radical cultural difference and a resignation to the erasure of that difference by the relentless march of modernization led by the American juggernaut. Both attitudes would seem to stem from a sense of threatened sovereignty not only over economic, political, and cultural matters but also over the very stuff of daily life in an era of intensified global circulation of ideas, policies, goods, services, labor, and capital.

However, the preceding pages deliver a different picture that muddies the clarity offered by either position. For example, the 1976 *loi Veil*, considered quite advanced for its day, challenges any simple narrative of modernization under U.S. aegis. Yet at the same time the law's imperfect implementation places a cloud over any retrospective French claims of modernizing leadership in matters of tobacco control. Observers inside and outside of France have been tempted to impute this lack of follow-through to the nature of the centralized Republican state that is long on discourse but short on action and to the lack of robust civil-society organizations that can lobby politicians effectively and hold them accountable. However, even U.S. anti-smoking advocates will admit that over the years, despite high cancer rates, considerable citizen mobilization, and a wealth of scientific findings, the U.S. government did relatively little in effective tobacco control beyond sponsoring epidemiological studies, health warnings, and smoking bans on airplanes until the late 1990s, which is why it fell to activists and state and local governments to take decisive action in Arizona, Minnesota, California, Massachusetts, Florida, and New York City in the way of anti-smoking programs, sharp increases in cigarette taxes, bans on secondhand smoke, restricting youth access, and lawsuits against the tobacco industry to recoup the medical costs incurred by smoking.

Moreover, the apparent shared history of cigarette smoking across France, the U.S., the U.K., Japan, China, and Europe—adopted by consumers in the 1920s and 1930s—broke up with the advent of the Asia Pacific War and the Second World War (1931–45), which vastly reduced cigarette consumption among populations devastated by war even as it may have spread the habit among both civilians and those serving in the armed forces. The wide variance in current national experiences of tobacco-related diseases today arguably is the direct legacy of that fractured history. From this perspective, the recourse to the language of "lag" or, alternatively, "Americanization" to describe French tobacco control policies in the 1990s may have not so much to do with the unfathomable nature of French culture or the force of American hegemony as with the contingencies of a history that isn't shared with the U.S., Canada, or the U.K. but is more common to most other countries in continental Europe and East Asia.

Furthermore, close scrutiny of state governance in France attenuates any notion of a strict French identity. For example, we have seen a strong liberal strain commonly associated with the U.K. and the U.S. in actual practices of the French centralized state throughout much of the nineteenth and twentieth centuries that allotted responsibility for public health and citizen welfare to local and regional government and private organizations. In the realm of public health, French practices often had roots in practices of their European neighbors or even the U.S. Meanwhile, what little documentation that has come to light about French anti-smoking sentiment hints at a history stretching back to the early nineteenth century that includes French-based initiatives in regulating the consumption of tobacco.

So I argue that any understanding of the specificity of French actions in tobacco control is located outside of the narrative logic of intractable cultural uniqueness on the one hand and of a story of the one-way arrow of modernization that magically submits all cultures to its homogenizing (American) values on the other. Another narrative emerges, one that escapes the temptation of a false particularism (indelible Frenchness) and its foil, a false universalism (modernization, globalization). By the same token, the customary *frisson* of strangeness that radiates from the anthropological object often produced by the first two narratives is absent here as well. Regarding anti-smoking efforts, what is left is France's unexceptional exceptionalism, the messiness of current policies and initiatives stemming from contingent historical practices and events and from the given structure of the public health sector (Setbon 1993, 410).

If the French experience is any indication, the circulation of the experience of smoking and tobacco-related illness and of public health data, expertise, methods, and training from one sector to another and from one national site to another is uncertain even in an era of neoliberal ascendancy in international policy circles. Public health matters circulate smoothly here, in fits and starts there. Thus, the 1976 *loi Veil* and the accompanying anti-smoking campaign were the work of a Conservative government that was open to new developments in health policy and to new approaches to communication and management techniques found in the private sector and in other countries (mobilization of citizens against secondhand smoke through advertising; U.S.-based decision-making sciences inspired by behavioralism), whereas the successive Socialist governments retreated to a more discreet approach to the hazards of smoking until the adoption of the *loi Evin* in 1991 that in many ways simply revived earlier policies enunciated in 1976. And even the full implementation of the new law came up against, on the one hand, the Socialist and Republican tradition that favored social welfare measures and national medical coverage in the name of social solidarity over more aggressive preventive public health interventions and, on the other, the reluctance of trade unions (a group with little or no input into tobacco control in either the U.S. or Japan) to introduce yet more divisive workplace regulations in the way of anti–secondhand smoke rules. In the thinking of the CFES, the state agency in charge of health campaigns, the examples of more aggressive approaches in other nation-states were negative at best, and this was reflected in the timid nature of the agency's own information campaigns, much to the chagrin of the private, state-sponsored nonsmokers' rights group, the CNCT. Even so, the CNCT's own preventive activities concerning secondhand smoke relied on court cases and the publicity they generated, not advertising's direct intervention in the public media sphere.

Finally, in August 2000 it was quite striking to watch one of France's leading anti-smoking advocates, Gérard Dubois, holder of a doctorate in public health from the Johns Hopkins University and author of an article in France denouncing the government's backwardness in tobacco control relative to other countries, stand up in the audience at the Eleventh World Conference on Tobacco or Health in Chicago and forcefully remind a panel on the future of tobacco control that all of their speculations and projections were based on the assumptions and power of Anglo-American common law, which was not the foundation of the legal systems of many of the countries in attendance. His stance remained a modernizing one but one that was much more mindful of positive national and regional differences than is usually the case among anti-smoking advocates.

It is worth recalling that much of the culturalist discourse on French exceptionalism has been produced by intellectuals and journalists, groups famous for their many smokers, and that, as Jean-Philippe Mathy and Eric Fassin have pointed out, they are often more nationalistic than their compatriots and often do not reflect wider public opinion (Mathy 1993; Fassin 1999a). So during debate on the *loi Evin* in 1991–92, while polls showed a large majority of public opinion in favor of it, most journalists declaimed against the new legislation. Thus, it came to me as something of a shock to learn that in the news department of thirty journalists, where *Le Monde*'s medical reporters worked between 1994 and 1997, the number of smokers dropped from fifteen to just three. Anecdotal evidence seemed to indicate that France was undergoing a quiet revolution. Yet more surprises were in store. In September 1999 the leading left daily, *Libération*, dropped its abiding hostility to anti-smoking actions and headlined a weekend edition, "The Smoke Merchants' Plot" ("Le Complot des marchands de fumée"). The paper devoted its first five pages to decrying the tobacco industry's suppression of evidence concerning the hazards of tobacco revealed in documents unearthed by the lawsuit filed by the State of Minnesota and Blue Cross and Blue Shield against cigarette manufacturers ("Le Complot" 1999). And four years later, in January 2004, newspapers were convulsed by the news that between 1999 and 2003 the sale and consumption of cigarettes in France had dropped 12 percent, which translated into a decrease in French smokers age fifteen and older from 34.5 percent to 30.4 percent. The report by INPES (which replaced the CFES in 2002) also revealed that the greatest drop in smokers (18 percent) occurred among the two groups least inclined to quit smoking—women and youth: from 30.8 percent to 25.3 percent (women) and from 44.5 percent to 36.4 percent (youth). The rate among male smokers fell from 38.3 percent to 35.7 percent. More French were quitting, and fewer were taking up the habit; the main reasons cited were health concerns and the high price of cigarettes, which in 2003 averaged €4.50 or US$5.50 per pack (*Enquête* 2004). Something indeed is afoot in France.

Japan

In the Shadow of Colonialism and Japan Tobacco

In a world that is turning against smoking, Japan is a land that time forgot.
—Dan Rather, *CBS Evening News* ("Tobacco under Fire" 1997)

The enormous Marlboro cowboy that towers over the rooftop of a building on Aoyama Avenue in Tokyo, ads on television (after 11pm), the railway smoking cars that are smoke dens, or conversely, the small "revolution" provoked by the broadcast over private television networks of an ad warning against the dangers of tobacco on 1 May 1994 are revealing: Japan, like for that matter all of Asia, is the last paradise of smokers.
—"Le Japon: le dernier paradis des fumeurs?" *Le Monde* (1994)

I've seen a lot of progress in Thailand, Taiwan, Korea, Malaysia, and Singapore, and many of those countries have very long-standing traditions, but their progress in the fight against smoking is ahead of our culture, [they are] even more stringent than in the United States. It is not true that we just copy from the U.S.—we draw possible methods from other Asian countries.
—Bungaku Watanabe, TOPIC (interview, 1999)

Anti-smoking campaigns in California and France are the product of practices past and present: forms of liberal government of populations; epidemiologists' and marketers' methods of segmenting populations; the activities of local advocates, health organizations, and government officials; and the circulation of public health expertise, policies, and personnel within and between nation-states. The actual history of these recent anti-tobacco efforts in California and France breaks with the logic of both cultural uniqueness and a common project of modernization that underwrite many reports in the media and policy decisions in government circles. Thus, for example, while California's Tobacco Control Program drew on tactics developed in other states and countries, it created a set of highly publicized policies, segmentation practices, and media campaigns that in the end enjoyed limited exportability to other regions in the U.S. and to other nations. And although in France the deprivations of the Second World War postponed the onset of heavy, regular cigarette smoking and consequently the epidemic of tobacco-related diseases, the national government began to warn citizens early on about the hazards of secondhand smoke and effectively banned all tobacco advertising years ahead of many nations.

The case of Japan sharpens these questions, because Japan is a highly industrialized non-Western nation-state and, as such, has been widely perceived as the paradoxical embodiment of hypermodernity—a fascinating and disturbing electronic-based consumer society—and of backward, feudal values. In response to Japan's quick postwar recovery, proponents of Western-based modernization theory had to make an exception to their core belief that local culture constituted an obstacle to the forward march of modernization; in the late 1950s they began to argue that at least in the case of Japan, local cultural practices laid the foundation of rapid industrialization. By the 1980s, scholars claimed that the peculiar version of Japanese Confucianism dominant in the Tokugawa era (1600–1868), which stressed the worldly values of hard work, self-discipline, savings, and the fruits of one's labor, created the necessary conditions for the country's entry into the modern industrial era (Bellah 1985; Lipset 1996, 211–263).[1]

With respect to this paradoxical mix of modern and premodern, Japan shares much with the perception of California by European intellectuals—such as Jean Baudrillard—who view California as the first truly modern society that has decisively broken with the values of Old Europe and by the same token constitutes a truly primitive culture in the thrall of the magical power of mass-mediated simulacra (Baudrillard 1988; Mathy 1993). What allows the comparison of Japan and California is their extreme "modernity," which is in turn closely tied to their ex-centric status, at least when both are viewed from Europe. But a crucial difference remains: California is a former European settlement colony that has presumably escaped Europe's orbit, whereas Japan stands as one of the few nations not only to have resisted successfully Western conquest and colonization in the nineteenth century but also to have challenged Western nations as a rival military, colonial, and economic power. Japan's example decenters not only many traditional Western assumptions about "modernization" but also contemporary theories that view globalization as a simple process of Westernization or Americanization (Buell 1994, 40–71; Iwabuchi 2002). This may account for why Japan has been the object of deeply ambivalent scrutiny by Euro–North

Americans (and Asians) and the target of outside pressure concerning everything from its culture, economic policies, and consumption practices to its management techniques and modes of government. Japan persists as the modern anthropological object par excellence radiating strangeness and uniqueness and as the focus of international policy initiatives.

In Japan, local constraints and global imperatives arguably intersect in particularly powerful ways, and, as we will see, the obstacles and tensions shaping contemporary tobacco control, public health management of populations, and globalization emerge there forcefully. Thus, I look at the interplay of different actors and forces: the overwhelming presence of Japan Tobacco, the state monopoly privatized in 1985, and of the powerful Ministry of Finance (MOF), which retains a controlling interest in it; the galvanizing of anti-smoking groups by the actions of both the Japanese state, responsible for the welfare of its citizens but whose official policy promotes smoking domestically and regionally, and U.S. and U.K. tobacco companies, which forced open the Japanese cigarette market to their products in the 1980s; the culture of smoking characterized by a large gap in smoking rates between men and women (62.5 percent and 12.6 percent, respectively, in the late 1980s) that has led to the privileging of women and children as risk groups by tobacco control advocates whose fate often serves as the ethical frame of the battle in the public media sphere over the hazards of smoking; and the creation of a women's anti-smoking organization that is unique among industrialized nations.[2]

Asia, Women, and Global Tobacco Control

These intertwining issues were dramatically played out in public in November 1999 when Gro Harlem Brundtland, director-general of the World Health Organization (WHO), brought her organization's Tobacco Free Initiative (TFI) to Japan and Asia. The occasion was the WHO International Conference on Tobacco and Health, "Making a Difference to Tobacco and Health: Avoiding the Epidemic in Women and Youth," which gathered anti-smoking advocates and representatives of major international women's nongovernmental organizations (NGOs) in Kobe City (Hyogo Prefecture).[3] The main purpose of the colloquium was to recruit women's organizations into the TFI and to begin building tobacco control "capacity" among them. It also had the goal of pressuring the Japanese government to strengthen its anti-smoking policies through tax increases, improved health warnings, and other measures. In her keynote address that opened the conference, Brundtland outlined to the audience the TFI (the establishment of the Framework Convention on Tobacco Control regulating tobacco through taxes, advertising bans, etc.) and highlighted the special role women had to play in the worldwide struggle against the tobacco epidemic that left four million dead every year:

> It is time to strengthen such initiatives by expanding our outreach to NGOs and women's organizations. Women leaders can help set our priorities straight and are natural allies in health development. Prevention is a health practice that women know well. Women are the first line of family health workers; they tend to the health care of children and the elderly. Women are natural tobacco control advocates. (Brundtland 1999)

She urged women's organizations to join the fight all the more so in that women around the world are targets of tobacco industry marketing strategies that appeal to women's desire for autonomy and modern selfhood:

> Here in Japan we see Western cigarette brands marketed as a kind of "liberation" tool. We see cigarette companies calling on young Japanese women to assert themselves, shed their inhibitions and smoke. Last week a coalition of minority organizations in the United States demanded that a famous cigarette company withdraw cigarette ads that are seen to target black, Hispanic, and Asian American communities. The ads include glossy images of minority women including a geisha smoking a brand destined for women. (Brundtland 1999)

Here, she is referring to the protest staged in California by anti-smoking community groups against Philip Morris' Virginia Slims "Find Your Own Voice" print media campaign, which accused the company of deploying patronizing ethnic stereotypes in its series of ads that associated the affirmation of ethnic identity and female independence with smoking and acculturation (APITEN 1999b; see chapter 2 of the current volume). According to Brundtland, Asian women have an unusual opportunity to lead the way in tobacco control:

> In our efforts to limit the global tobacco epidemic, Asia and especially Asian women, will play a central role. Here, smoking rates are still low but they will not remain that way if the tobacco industry gets its way. In fact, there are worrying indications of increased smoking rates among the young. (Brundtland 1999)

Indeed, regarding Japan, in a later panel a health researcher from the Ministry of Health and Welfare (MHW) confirmed Brundtland's remarks in presenting the results of the second national survey on smoking: the smoking rate for men age 15 and older was 52.8 percent and for women, 13.4 percent. The rate for minors (15–19 years old) was 19 percent for males and 4.3 for females, but both had increased steadily through the 1990s, and that for young adults in their twenties and thirties was much higher—almost 60 percent for men and 20 percent for women. Still other studies had revealed that for women in low-level service occupations, smoking rates reached up to 40 percent. Smoking-related deaths are estimated to be between 95,000 and 114,000 annually.[4]

Brundtland's speech articulates the basis for a multiple global response and stands as a good example of how the struggle over the marketing and consumption of cigarettes constructs at-risk populations both transnationally and locally. On the one hand, her battle plan follows closely both the tobacco industry's own contemporary marketing strategies that segment populations into particular transnational aggregates based on age, gender, culture, and ethnicity—youth, women, women of Asian or African descent, and so forth—and community-based anti-smoking groups' very local struggles with those same marketing practices (as, for example, in California). Echoing TFI project manager Derek Yach's speech to community activists earlier that same month in Lake Tahoe, California, recounted previously, Brundtland attempted to bring together local and global efforts through the appeal to "communities of shared values"—here, among women (Yach and Bettcher 2000, 212). On the other hand, Brundtland's remarks point to the constitution of risk groups in tobacco control not only across political boundaries (women, women of Asian descent, etc.) but also in the more traditional terms of well-defined nations and regions (Japan, China, Korea, Malaysia, Thailand, Hong Kong, Singapore, Philippines, etc., and the overall Asia–Western Pacific region). Entire regions and nations as well as transboundary ethnic- and gender-based groups are understood to be at risk. It is within this complex public health landscape that the WHO director-general invited women's NGOs to take concerted action.

Recruiting Women's NGOs

The conference convened female health and government officials and representatives of major international NGOs in order to introduce many of them for the first time to current research on hazards confronting women who smoke or are subjected to the smoking of others. Panels followed by workshops were devoted to a wide variety of topics: the Japanese National Survey; the health impact of tobacco use; environmental tobacco smoke; media, fashion, and promotion; access and affordability; addiction; cessation methods; gender sensitivity of tobacco control policies; and regional and international strategies. Organizations included the Asia-Japan Women's Resource Center; Fundación para Estudio e Investigación de la Mujer (Argentina); Forum for Women, Law, and Development (Nepal); Women's Environmental and Development Organization (WEDO); and the African Women's Development and Communication Network (FEMNET).[5]

In its efforts the WHO assigned itself no simple task: up till then there had been little substantial contact between anti-smoking advocates and women's groups, either locally or internationally. It faced a classic problem of public health outreach. The gulf between the two groups was quite large, and building bridges across it was fraught with palpable tensions. As a result, the Kobe conference's organization and discussions replicated some of the difficulties we've met before, stemming from California tobacco control's isolation from the populations it was publicly committed to serving: distrust between interested parties, intermittent organizational paternalism and claims to exclusive ownership of issues by sponsoring agencies, presumptive definitions of community identity, and so forth, even as anti-smoking advocates tried to make contact with communities and organizations. For example, in its attempt to mobilize segments of international civil society, the conference appealed to representatives of women's NGOs in their personal and professional capacities as women and activists in women's causes (reproductive rights, human rights, labor, consumer rights, environmentalism, nutrition, violence, prostitution and the entertainment industry, family welfare, etc.), who see themselves as belonging already to multiple communities (based on gender but also on ethnicity, culture, language, national origin, and religion) that often transcend the boundaries of the nation-state. In Brundtland's speech and other presentations by anti-smoking advocates, the emphasis fell on region, culture, and gender, especially on women's role in family life. However, the stress on women's activities as caregivers and mothers (already implied in the conference's subtitle associating women and youth as vulnerable populations)[6] elicited expressions of discomfort by some anti-smoking advocates and participants from women's groups, who felt that it defined women's lives too narrowly in terms of traditional obligations to the unborn and offspring and that it tended to relegate women to the status of victims and therefore passive objects of policy (Zeitlin 1999).[7]

Moreover, at the close of the gathering, in a late session in which conference organizers called upon women's NGOs to spell out their commitment to global tobacco control, the Swedish president of the International Network of Women against Tobacco (INWAT, founded in 1990), one of the rare NGOs of its kind, let it be known that anti-smoking activists had been waiting fifteen years for women's groups to express interest in tobacco control. As if in reply, when many invited participants actually came forth and did express their commitment to the struggle against smoking, they also remonstrated WHO organizers that the conference's structure demonstrated little real reciprocity, based as it was solely on the tobacco control community's agenda; and that women's NGOs expected the WHO to lend support to their own ongoing activities as well. Later, after the conference, the late director of a major Asian women's NGO—the Asia-Japan Women's Resource Center, based in Japan—privately claimed that to the best of her knowledge in the years prior to the conference INWAT and anti-smoking activists had done little in the way of contacting major women's NGOs she worked with in the region and inexplicably did not invite NGOs such as the International Women's Health Coalition (IWHC). In her view, matters weren't helped, either, by the very academic and technical emphasis of tobacco control activities over mass-based actions (Matsui 2000). In this regard, in the eyes of this writer, the nature of the gulf separating the two groups was underscored by the fact that existing women's anti-tobacco groups received little formal recognition at the conference. Thus, the unique activism of Women's Action on Smoking, Japan (founded in 1987), which at that time had no equivalent in other leading industrialized countries such as France, the U.K., the U.S., Italy, Russia, or Germany, was unacknowledged throughout most of the proceedings beyond the gesture of including its director on the roster of many speakers.

These points of tension were compounded by other difficulties that participants contended with, such as the dominance of Northern European and Anglo-American participants, especially among tobacco control advocates (few hispanophone representatives were present and even fewer francophone ones from Southern Europe, Africa, Latin America, and Asia). It was quite striking that during the gathering, organizers and Anglo-American anti-smoking advocates felt quite free to comment on Japan's relatively unregulated tobacco market and to display open expressions of boredom during presentations by female activists from the host country. It would be hard to imagine participants from countries such as France, Germany, the U.K., Canada, or the U.S. being treated in a similar manner in their capacity as conference hosts, and this led me to wonder whether I wasn't witnessing a replay of old colonial prerogatives in a new guise—one-upmanship in human rights and health issues—and the casting of Japan and the Japanese as a "hard to reach" nation and people and as ethically failed citizens in emerging global "communities of responsibility." Finally, other contentious issues occupied participants' attention, such as the difficulty of extrapolating data from highly industrialized countries to industrializing ones and differences between "North" and "South" policy agendas.

However, in the course of the conference, NGOs and organizers were able to overcome some of these obstacles, and in the final documents there was a notable shift in the relationship between anti-smoking advocates and women's NGOs. The conference concluded with the release of the Kobe Declaration, which had been hammered out by the organizers and those in attendance and incorporated suggestions and criticisms that emanated from the floor. As a joint statement by "women and youth leaders, non-governmental organization representatives, government delegates, media professionals, academics, health professionals, scientists, and policy-makers," the declaration decried the targeting of women and children by tobacco industry marketing campaigns and called for addressing through gender-specific strategies the tobacco epidemic among women and girls that puts them at risk in gender-specific ways. It resolved to integrate gender equality into tobacco control strategies and respect for diversity of women and girls in different cultural contexts and to incorporate recommendations to combat the negative impact of tobacco in sections dealing with "women and health" and "the girl-child" in the United Nations (UN) General Assembly Special Session on Women 2000 ("Kobe Declaration" 1999). In the final document, participants presented themselves as women who were already mobilized on behalf of tobacco control but who also were laying out demands that the WHO's TFI must meet regarding women's groups' own agendas (gender-specific policies in the general frame of gender equality). The declaration affirmed an aggregate identity based on gender, to be sure, but one that remained loosely defined (no single focus on caregiving and motherhood here) and respectful of its own internal cultural and geographic differences.[8]

The Question of Japan: The State and Japan Tobacco

If the WHO had as its ambition to recruit transnational women's NGOs into the Tobacco Free Initiative and in a sense to bypass the structures of the nation-state in favor of what Australian-based historian Tessa Morris-Suzuki calls the "sub-regimes" of global governance (Morris-Suzuki 1998, 177), it did so nevertheless in conjunction with a direct appeal to heads of national government and high state functionaries. During their stay in Japan, Brundtland and Yach met privately with Japanese Prime Minister Keizo Obuchi and officials from other ministries to press their case for higher cigarette taxes and the necessity of an international protocol regulating tobacco marketing and consumption. But although the MHW provided substantial funding for the conference, no high official from that ministry or any other showed up in Kobe. This did not surprise observers. The Japanese government's reticence in such matters stemmed from its controlling interest in the stock of Japan Tobacco Inc. (JT), the leading world manufacturer of cigarettes in terms of total profits (and fourth-largest in sales) and whose Mild Seven brand is second after Marlboro in global sales. The government's ties to Japan Tobacco had over the years been the object of local and international commentary in public health circles and the media, who had declared Japan a "smoker's paradise" in which weak health warnings ("Please remember to follow good smoking manners. As smoking might injure your health, please be careful not to smoke too much"), low cigarette taxes, and lack of state-sponsored health campaigns were the rule. This was a situation, so it was claimed, no longer matched anywhere else among leading industrialized nations.

The scandal of overt state sponsorship of a smoking culture has been uppermost in domestic and foreign commentators' minds going back to the 1980s. In 1984 the Conservative government, led by the Liberal Democratic Party

(LDP), passed the Tobacco Industry Law on the eve of the privatization of the tobacco monopoly (allotting 100 percent ownership to the finance ministry), despite opposition from major national newspapers and a large segment of public opinion. The stated purpose of the law was "to promote the sound development of the Japanese tobacco industry, thereby securing stable national revenues" (cited in TOPIC 2000, 1; Levin 1997, 4–8). The local and international press was unforgiving. In 1987 a *Washington Post* article titled, "In Japan, More Power to the Puffers" (Burgess 1987), and six years later the *New York Times* was even more blunt in a lead article on page one of the Sunday business section: "When Smoking Is a Patriotic Duty" (Sterngold 1993). That same year the English-language Tokyo newspaper *Japan Times* flatly claimed that little had changed in an editorial titled, "Still a Smoker's Paradise" ("Still a Smoker's" 1993), while one year later the French daily *Le Monde* ran an article, "Japan: Smokers' Last Paradise?" ("Le Japon" 1994). And in 1998 the *Japan Times* ran an op-ed piece by Kiroku Hanai, a well-known journalist and anti-smoking advocate, that reprised the same theme: "End the Smoker's Paradise" (Hanai 1998). Alongside his article also ran a large cartoon by Ryuji featuring a man smoking beneath the national flag in which the sun is portrayed as an ashtray full of lit and crushed cigarettes.[9]

Indeed, prior to the Kobe conference, the Japanese government's sponsorship of the conference on women and tobacco control worried some international female anti-smoking advocates, who thought that the gathering would be more theater than substance and that few women would actually be invited (Slama 1999). The fact that the official Japanese delegation to the preparatory work session for the WHO conference was composed solely of men only encouraged such a view. However, fears concerning attendance were not borne out by the conference. Still, in February following the conference, the most active state MHW functionary involved in laying the groundwork for the Kobe conference and in organizing important policy commissions and workshops on tobacco control in Japan lost her position. The precipitating cause was not the Kobe conference but rather the release the preceding August of an MHW report that outlined health goals for the next ten years ("Health Ministry" 1999). Titled *Health Japan 21*, it called for a dramatic reduction of smoking in Japan (50 percent by 2010). The new policy document provoked consternation among Japan Tobacco's powerful political allies in the ruling LDP and the MOF that detains JT's stock, and they successfully lobbied to have the official removed from her post.

The government's heavy-handed action offered the raw image of the state's casual disregard for the consequences of its own policies on populations with whose welfare it has been entrusted. This image was a local and global coproduction of Japanese and foreign anti-smoking advocates, the Japanese print press, international media, and the Japanese government's own actions and is one fa-

National symbol.
Courtesy of Ryuji/*Japan Times.*

miliar to students of discourses on Japan's political system, that of a presumably unresponsive government and culture, authoritarian and illiberal, almost pre-modern in practice—as Dan Rather put it, a "land that time forgot" (Morris-Suzuki 1998).[10] A milder version of this can be found in an otherwise careful assessment of the future of tobacco control by California-based researchers in an article comparing the political cultures of France, Canada, the U.S., and Japan. According to its authors, Japan is like France but even more bound by a hierarchical political culture, yet by the same token may be better poised to take action, for a government so effective in stymieing regulation of smoking could be equally effective in promoting it, if officials chose to do so:

> Like France, Japan has a highly centralized, elite-dominated political system with strong traditions of governmental control over business. Moreover, Japan is culturally a much more hierarchical society, in which citizens more readily defer to authority. Strongly entrenched norms of politeness govern personal and business relations. Thus if the Japanese government ever decides to make a priority of reducing smoking rates or protecting non-smokers, it will probably have little difficulty in achieving its goals. (Vogel, Kagan, and Kessler 1993, 322)

This image of centralized government and obedient population confirms a conventional trope of Japanese deference to authority. Conventionally, the figure that symbolically has embodied this view of Japanese culture and governance is the emperor, a transnational image that comforts the perspective of many critics of Japan, who either are from "the West" or share Western definitions of modernity. Thus, according to one revisionist scholar of the emperor system, Takashi Fujitani, for many observers "politically, Japan and its people have never been modern enough." He argues that Japan often serves as a screen upon which Western commentators project their own misgivings about modernity but by relocating them in Japan's pre-modern past:

> Such a view, though critical of contemporary politics in Japan, would continue to allow the displacement of our discontent with modernity into a Japanese past called "feudal"; furthermore it would preclude the possibility of turning the gaze emanating from the modern West back onto itself. Most important, it would keep us from recognizing that the subject-citizen produced by the Japanese emperor-system regime no longer appears so different from the hero of modern bourgeois society, the supposedly autonomous subject that people like Maruyama Masao [a postwar Japanese philosopher and cultural critic] have so long sought. (Fujitani 1996, 26–27)

In various constructions of Japan and things "Japanese," the questions of what counts as modern and who owns (political) "modernity" are constantly at issue and are often inflected in terms of cultural difference.[11] This is no less true in public health debates around smoking, where these questions serve as at once rhetorical resource and perceptual frame.

The Anti-smoking Movement in Japan: The Interplay of Domestic and International Forces

Faced with state sponsorship of tobacco production and consumption (through direct subsidies, low excise taxes—a practice largely shared by all tobacco-producing countries) but also the open promotion of smoking culture by the Tobacco Industry Law, in the 1990s in Japan there stood a panoply of anti-smoking organizations, many of which belonged to the umbrella All-Japan Anti-smoking Liaison Council (Zenkoku kin'en bun'en suishin kyōgikai), made up of more than fifty groups.[12] They included health education and information organizations that targeted the general public, journalists, and politicians, such as the Tobacco Problem Information Center (Tabako mondai jōhō sentā), or TOPIC; Women's Action on Smoking (Kitsuen to kenkō josei kyōkai, or, literally, Women's Smoking and Health Association); Japan Action on Smoking and Health (Nihon kin'en kyōkai, or Japan Stop Smoking Association, renamed in English in 1998 the Japan Association against Tobacco); and the Smoke-Free Environment Campaign for Kids in Japan (Kodomo ni hi'en kankyō o suishin kyōgikai, or the Council to Promote Smoke-Free Environments for Kids). Other groups promoted cessation, such as the Osaka Cancer Prevention and Detection Center (Osaka gan yobō kenshin sentā), Can Do Harajuku (Seventh-Day Adventist), the Stop-Smoking Friendship Association (Nippon kin'en yūai kai), and the Japan Anti-smoking Doctors' League (Nihon kin'en isshi rensei). Some epidemiological research centers participated, such as the Aichi Prefecture Cancer Center (Aichiken gan sentā), and still other groups focused on advocacy and legal actions, such as the Japan Anti-Tuberculosis Association (Zaidanhōjin kekkaku yobō kai, or literally, Tuberculosis Prevention Foundation Society), Japan Action for Nonsmokers' Rights (Ken'en ken kakuritsu o mezasu hitobito no kai, or the People's Organization for the Establishment of the Right to Hate Smoke), the Lawyers' Organization for Nonsmokers' Rights (Ken'en ken kakuritsu o mezasu hōritsuka no kai, or, the Lawyers' Organization for the Establishment of the Right to Hate Smoking), the Kinki Council on Smoking and Health (Kinki kin'en kyōgikai), the Aichi Prefecture Lung Cancer Countermeasure Association (Aichiken haigan taisaku kyōkai), the Hiroshima Anti-smoking Council (Hiroshima kin'en kyōgikai), the Kyushu Anti-smoking Association (Kyūshū kin'en kyōkai), and the Nonsmokers' Defense Association (Hikitsuenka o mamoru kai) in Sapporo. Voluntary health organizations such as the Japan Anti-Tuberculosis Association, Japan Cancer Society, Japan Heart Foundation, and Japan Health Promotion and Fitness Foundation also served as cosponsors of anti-smoking activities.

The groups were composed of activists, behavioral scientists, epidemiologists, health professionals, Christians, journalists, and lawyers, who often coordinated actions around public protests, policy statements, lawsuits, and the annual World No Tobacco Day sponsored by the WHO on May 31 each year. Research or work experience outside of Japan played a crucial role in professional biographies, especially for public health practitioners, since Japan had no schools of public health and public health was not part of the required curriculum in medical schools.[13] The earliest organized activities that were successful date back to the late 1970s, when three groups came into being: Women's Group to Eliminate Smoking (1977), an association based in Nagoya but now defunct; Japan Action for Nonsmokers' Rights (1978), located in Tokyo; and the Nonsmokers' Defense Association (1978) in Sapporo. Their efforts focused on cessation for women and "passive smoking" (in Japanese, "*judō-teki kitsuen*" but also "*kansetsuen*," or "indirect or secondhand smoke or smoking," a term that became common parlance in the early 1980s; more recently another term has come into use: "*fuku'en*," or "inhaling [others'] smoke or smoking"). The first two groups were founded by coworkers in a small environmentalist publishing house. The creation of the women's organization was related to the release of a special issue of the feminist publication *Women's Revolt* (*Onna hanran chaku*), edited by Ayako Kuno, who fought to get other feminists to quit smoking—an effort in which she enjoyed some success (Matsui 2000). Japan Action for Nonsmokers' Rights was created in 1978 at a public meeting that included in attendance Bungaku Watanabe (founder and director of TOPIC), Yoshio Isayama (who shortly afterward helped start the Lawyers' Organization for Nonsmokers' Rights), and Kenji Makino (editor of the science page at *Mainichi shinbun*). Their founding slogan was "the right to hate smoking" ("*ken'en ken*"), and their strategy was one that consciously refused to single out smokers and make them into the "enemy"; rather, it targeted public services such as transportation and hospitals. This approach was partly motivated by the fact that among the most committed smokers at that time were prominent actors, writers, politicians, and celebrities (Watanabe 1999). The group's early symbol featured the silhouette of a tulip framing the worried face of a child with the words, "Smoking. I'm sensitive to tobacco smoke" ("*Kitsuen. Tabako no kemuri ga nigate desu*").[14] This was in keeping with the frequent deployment of nature, children, and women as symbols by environmental, atomic bomb survivor, and anti-nuclear movements in Japan. However, in the end, unlike in the U.S., few environmental activists actually crossed over to the anti-smoking movement, and one tobacco control advocate even claimed that most ecologists have always been hostile to it (Nakano 1999). As if in confirmation of this, in the late 1990s the Japanese Environmental Protection Agency's primary focus still remained on other sources of pollution (Tominaga 2000).

It would appear that what prompted the formation of the groups was the circulation of new scientific information from overseas, especially the U.S. Already in 1964, when senior MHW officials read the U.S. surgeon general's first report on smoking, they sent out circulars to the forty-seven prefectural governments from the Public Health Bureau and the Children's Bureau drawing attention to the report; they also decided that they wanted confirmation by Japanese data, and in 1965 they commissioned Takeshi Hirayama, chief of epidemiology at Japan's National Cancer Institute, to conduct a major study on smoking based on a vast pool, using patients' records available through the extensive system of health centers. Results began to come in several years later. Capitalizing on accumulated data from abroad (including a Japanese translation of the 1964 U.S. surgeon general's report) and reports on anti-smoking struggles in Sweden, the U.S., France, the Soviet Union, and elsewhere, in 1977 science editor Kenji Makino began publishing an extended weekly series in *Mainichi shinbun*'s science page on smoking—sixty-eight articles in all—which he then rewrote and published in a single volume, which appeared the following year under the title, *Tobaccology* (*Tabakorojia* [or *The Science of Tobacco/Cigarettes*], 1978). The book contained fourteen chapters that covered everything from air pollution, toxins in cigarettes, and tobacco-related diseases to women, youth, addiction, and cessation to anti-smoking movements and the establishment of nonsmoking areas. The articles and subsequent book helped galvanize activists but had only a limited impact on the general public. Makino himself admitted that the book was before its time (Makino 1998). In the early 1980s Japan Action for Nonsmokers' Rights staged joint actions with the lawyers' group that involved letter-writing campaigns targeting hospitals and lawsuits against Japan National Railways in order to force the state-run rail system to create more nonsmoking cars on its shinkansen long-distance bullet trains. Although the lawsuits were ultimately dismissed in 1987 by a judiciary operating without a jury system and generally respectful of state prerogatives, pressure generated by the accompanying publicity did result in the creation of many more nonsmoking cars and spaces (Hozumi 1999; Matsumoto 1999). However, by then other events added momentum to the anti-smoking movement: the passage of the Tobacco Industry Law in 1984, the privatization of the tobacco monopoly in 1985, the highly publicized successful effort by U.S. Senator Jesse Helms (R-NC) and the U.S. Trade Representative to force open Japan's domestic market to American cigarette brands in 1986, and the hosting of the Sixth World Conference on Smoking and Health in Tokyo the following year. But before we turn to these events, it will be useful to review the intertwining histories of smoking, public health, and liberal governance in Japan leading up to that period.

The Culture of Smoking: Colonialism, the Nation-State, and Cinema

As in the case of France and the U.S., the production and consumption of tobacco in Japan goes back to the earliest period of European colonialism in the sixteenth century and was shaped by the rise of the nation-state and its growing need for revenue. The tobacco monopoly itself owes its birth in the late nineteenth century to the Japanese government's desire to protect itself against encroaching Western expansion and to subsidize its own rival colonial projects. The interplay of internal and external forces was particularly strong. Most historians agree that whereas the Spanish brought the tobacco habit to most of Asia by way of the Philippines, it was Portuguese traders who first introduced tobacco and pipe smoking to Japan around the end of the sixteenth century. The Portuguese were later expelled along with other foreign traders and missionaries; only Chinese and Dutch merchants were allowed by the Tokugawa shogunate to remain, whereupon the Dutch supplied Japan with the weed (J. Goodman 1993, 86). In 1612 the shogunate banned the cultivation, sale, and use of tobacco until 1624, when the government, like its European counterparts, faced with the failure of an ineffective policy, decided to turn the new consumption fashion to account by taxing it. In the early eighteenth century the government shifted its position once again and began to encourage tobacco cultivation in 1716, to protect rice fields. At that time smoking by men and women alike, in the form of long *kiseru* pipes, had rapidly become a much-sought-after luxury consumption among all ranks of society. Popular representations featured men of all occupations and women from courtesans and entertainers to shopkeepers' wives and aristocrats sporting tobacco pouches and holding long, graceful smoking pipes. However, new voices were raised against tobacco, most famously the neo-Confucianist scholar Ekken Kaibara, who in his treatise, *Guidelines for Maintaining Health* (*Yōjōkun*, 1713), warned against the deleterious health effects of smoking.

After the fall of the Tokugawa shogunate and with the advent of the Meiji era (1868–1912) and its ambition to create a modern Japanese nation-state, tobacco cultivation and production fell under the system of national taxation, and tobacco figured in government-sponsored publicity as one of nation's main agricultural products. At the end of the nineteenth century, large, private firms arose, such as Murai Bros. Co. (Kyoto), N. Kimura and Co. (Tokyo), Yezoye and Co. (Tokyo), Osaka Tobacco Co., and Iwaya and Co. (Tokyo), and introduced cigarettes to Japan (using either domestic or American leaf), while foreign concerns began to make inroads, as when James B. Duke's American Tobacco Co. (ATC), in response to steep tariffs on imported cigarettes, made direct investment in Japan. It entered a joint venture with Murai Bros. to build a factory to manufacture its cigarettes. Duke then promptly took a controlling interest in his Japanese partner's company.[15]

When war with China (1894) and imperial expansion created a significant drain on the national treasury, in order to increase revenues in 1898 the government began to reverse unequal treaties concerning tariffs struck with Western powers. It monopolized tobacco sales of domestic leaf and increased import tariffs on tobacco leaf and cigarettes. As the largest foreign direct investment in Japan, Duke's controlling interest in Murai Bros. provoked a nationalist backlash fomented in part by his domestic competitors. Then in 1904 the Diet effectively eliminated all production and marketing by private domestic and foreign firms by passing the Tobacco Monopoly Law, which nationalized the entire industry and formed the Imperial Tobacco Monopoly (Cochran 1986, 183–185; Cox 2000, 39–40, 106–111). With regard to tobacco consumption, colonial military policy shaped public health policy as well. Offering an early example of contradictory state policies that both promote and restrict tobacco consumption by citizens, military authorities acted to limit smoking by adolescent boys, for data had come to their attention indicating that early smoking by adolescents posed a hazard to their health (in particular, carbon monoxide's effect on height and weight) and thus to the fitness of potential military recruits. Thus, in 1900 the government implemented a law that prohibited smoking by minors and set the legal age at twenty—the highest in the industrialized world (Tominaga 2000). It is still in effect today.

As elsewhere in the industrialized and industrializing world, throughout the twentieth century the film industry was a major purveyor of smoking culture in Japan. One of the major technologies of modern culture, films articulated to a mass audience the tensions of a society undergoing rapid change. They highlighted, for dramatic and aesthetic effect, those moments of the day or night that smoking a cigarette served to mark off: from a pause in the workday; a moment's conversation; a visit to friend, neighbor, or lover; or a private negotiation to the entry into a different social milieu, as when men and women smoke together in the underworld or intellectual and artistic bohemia. The indexing of character, sexuality, and milieu was especially strong, with cigarettes given their association with (Western) modernity, urban life, and women's new public independence. The spread of cigarette smoking was steady through the 1920s and 1930s, especially among men. War accelerated propagation of the habit in Japan as it did throughout East Asia, Europe, and North America, for even as tobacco became rarer as a commodity on the open market during the Asia Pacific War (1931–45), packets of ten cigarettes remained a daily ration to officers, and military service constituted a privileged moment of initiation into smoking among Japanese men (Goto 1999). Following the end of the war and the U.S. occupation (1945–52), with the revival of uncensored domestic film production, Japanese screens were awash with commercial films that depicted the upheavals of postwar reconstruction, economic recovery, and emerging consumer society. Scenes abounded of smoky bars and nightclubs pouring out the sounds of American-style jazz, which were intercut with depictions of evolving family and workplace relationships, government and private-sector corruption, criminal activity, and, in the late 1950s, Japanese youth running amok.

In regard to the latter, one of the most remarkable films was New Wave director Yasuzo Masamura's *Giants and Toys* (*Kyojin to gangu*, 1958), which satirized the rise of postwar Japanese corporations, their ruthlessly competitive culture, and their obsession with emulating the methods of U.S. competitors. The film follows the adventures of three very young marketing employees—two men and one woman—from rival firms as they compete to promote their companies' products through absurd advertising campaigns. Their loose lifestyles, driven by blind ambition and irrepressible desires, are marked by fast convertibles, lovemaking, double-crossing, and endless smoking, and the overall visual and aural motif for the brave new world of U.S.-dominated mass-produced items and consumer marketing is the mechanical clicking of a gilded cigarette lighter shown in extreme close-up superimposed over images and sounds of the manufacturing and distribution process. A darker and more disturbing contemporary take on emerging consumer society was Ko Nakahira's *Juvenile Passion* (*Kurutta kajitsu*, or *Crazed Fruit*, 1956). It was based on a screenplay by Shintaro Ishihara, a young writer whose novels helped inaugurate the explosion of media interest in rebellious youth culture in Japan and spawned a wave of films based on the amoral, reckless lives of wealthy Japanese youth or "sun tribes" (*taiyōzoku*). Nakahira's film depicts a summer of sailing, wild parties, and outings to smoke-filled nightclubs catering to Americans through the rivalry of two brothers vying for the affections of a young Japanese woman married to a much older American officer. In *Juvenile Passion* cigarettes serve as key dramatic devices as the brothers' relationship deteriorates, culminating in the gentler brother's murder of his older, violent sibling in a fit of jealous rage. Both films are adolescent coming-of-age stories in which the young male protagonist plays the role of the younger observer of a world of power and pleasure that exists in the shadow of the American economic and military presence. His marginal status is signaled by both his naïveté and the fact that he doesn't smoke. In *Juvenile Passion* the themes of sexual maturity, national identity, and U.S. dominance dating from the occupation are even literalized in the figure of the Japanese woman to whom access is denied the two brothers by the presence of her American military husband. The tantalizing new modernity is perhaps within the grasp of rebellious Japanese youth, but it is not one entirely of their making—hence the films' ambiguity concerning the emerging consumer society before which they register both ironic dissent and obsessive fascination.

Outdoor, print, and broadcast advertising were also important purveyors of smoking culture throughout the twentieth century. In the absence of competitors, under state tobacco monopoly in Japan there was little need for cigarette advertising on a large scale. Adopting different visual styles—everything from the sleek lines of modern graphic design beginning in the 1930s to the rawness of photo journalism in the 1950s and 1960s—print ads depicted men either at leisure or taking a break from work in "contemporary" occupations far from the corporate office; for example, in the late 1950s they are portrayed as jazz musicians ("A smoke for every man. Midori: the cigarette with mint inside" ["*Kakujin kaku ketsuen. Hakka iri tabako*"]), news photographers ("You bet, that's it! Today, the cigarette that refreshes also tastes good" ["*Yoshi, are da! Kyō mo genki da tabako ga umai*"]), or construction engineers ("Work. Break. Work" ["*Shigoto. Koi. Shigoto*"]). As in contemporary tobacco advertising today in Japan, images of women smoking were largely absent. However, at the end of the 1950s the tobacco monopoly began to advertise in women's magazines and produced a series of print ads that featured composed studio shots of elegantly attired models and actresses, in Western dress and with fashionably short hairstyles, at leisure holding lit cigarettes. One set of ads for the "Peace" brand featured the slogan, "Cigarettes, an accessory that's on the go/that gets around" ("*Tabako wa ugoku akusesarii*"), followed by the tag, "The taste of genuine American leaf. Peace" ("*Honba Amerika ha no aji. Piisu*"). Another for "Three A" cigarettes promoted them as affording "Gentle taste. Three A" ("*Sofuto na aji. Suriiē*"). Cigarette smoking by women did increase among the cohort that came of age after the end of the Asia Pacific War (up to 15 percent) but would drop off in succeeding generations before rising again in the 1990s. Ads in the 1970s and 1980s would include depictions of artists and elderly successful businessmen who've reaped the rewards of economic prosperity, and throughout the 1990s ads for JT, U.S., and U.K. brands repeatedly tied cigarettes to the lifestyles of an utterly mature consumer society focused on suburban living, sports, leisure, and travel and made frequent use of well-known Japanese and Western male actors (Ken Ogata and Jean Reno) to catch the eye of consumers.

The 1980s: Neoliberal Backlash?

The passage of the Tobacco Industry Law (1984), the dismantlement of the tobacco monopoly and the opening of the cigarette market to foreign brands (1986), and the Sixth World Conference on Smoking and Health held in Tokyo

(1987) suggest that Japan, like other leading industrial nations, was undergoing a wave of neoliberal initiatives that were sweeping through government circles and policy elites at that time and that favored privatization of state enterprises, dismantlement of state-sponsored social services, budget reductions, tax cuts, rescinding trade barriers, and private initiatives of citizens concerning quality-of-life issues. Already the onset of the oil crisis in 1973–74, which hit energy-poor Japan particularly hard, had prompted the launch of the anti-welfare Administrative Reform Movement (Rinchō undō) by government ministries, the media, academics, and business and union leaders, which called for stemming the growing costs of health care and public pensions (Hiwatari 1993, 25–26; Carlile 1998).

However, even in the wake of reforms introduced by the U.S. occupation and of movements for expanded welfare benefits of the 1960s and 1970s, actual welfare practices in Japan never quite matched those in Europe or North America or, perhaps more accurately, were masked in part by arrangements with the private sector. In fact, Japan's mix of public and private welfare dating from the Meiji era, along with structures in large companies that already stressed employee flexibility and initiative, attracted great interest among Western proponents of neoliberal policies. Thus, according to some observers, a populist anti-welfare rhetoric never took root as deeply as it did elsewhere, and Japan emerged relatively untouched by the crisis of the welfare state (Vij 2001, 3, 19; Hiwatari 1993, 25–26).[16]

In the end, what the advocates of neoliberal measures proposed was more of a maintenance of the status quo than a fundamentally new policy. Moreover, proponents successfully turned back new demands for welfare entitlements by coming up with a countermodel of limited welfare dubbed the "Japanese welfare society" ("*Nihongata fukushi shakai*"), which they defended in the name of the "Japanese way of doing things," based on a picture of Japan rooted in traditions of cooperative, caring communities; stable family life; and neo-Confucian values (R. Goodman 1998, 150; Garon 1997, 222–230). This echoed revisionist thinking among modernization theorists with respect to Japan and was seconded by them in turn.[17] In 1982 the office of Prime Minister Yasuhiro Nakasone issued the Report of the Commission on Administrative Reform. It emphasized privatization over policies of deregulation that were popular in Anglo-American countries (Carlile 1998, 78–79). The culturalist argument that cited Japanese neo-Confucian values attracted the attention of many Western commentators, but it is one that perhaps is better understood as a descendent of the quite recent "invented traditions" created in the Meiji era that went into the making of Japanese modernity, which included not only neo-Confucianism but also the "timeless" emperor system itself (R. Goodman 1998, 150; Fujitani 1996).[18]

The State, Public Health, and Transnational Influences

Like welfare practices, medical and public health practices since the Meiji Restoration have blurred the line between the state and civil society. The initial Regulations on Medical Practice (*Isei*) promulgated by the government in 1874 drew on contemporary German concepts of health care. According to U.S. historian William Johnston, they enunciated for the first time in Japanese the concept of "public health" ("*kōshū eisei*"), which combined the Chinese character for police control over an area (*ei*) with the character for life (*sei*), to designate the policing of community health through state authority.[19] However, the importation of German Cameralist ideas of welfare underlying this concept was at variance with the views of physicians and the Japanese population. For the latter, *eisei* meant the individual's responsibility for monitoring his or her own health (Johnston 1995, 179–180); and until the mid-twentieth century, for most Japanese, health and hygiene were considered private matters, and they did not call for community health protection (Johnston 1995, 169–174). In actual practice, there functioned a mixed regime in which the Japanese state acted mainly through the sponsorship of private organizations, such as the Japan Health Society (Dai Nihon shiritsu eiseikai [literally, the Greater Japan Nongovernmental Health Society], 1883), the Welfare Association (Sansaikai, 1911), and the Japan Anti-Tuberculosis League (Nihon kekkaku yobō kyōkai, 1913; replaced by the Japan Anti-Tuberculosis Association [Nihon kekkaku yobō kai] in 1939), which focused on research and prevention. In 1931 the Japan Anti-Tuberculosis League actually set up health consultation centers (*Kenkō sōdanjo*) that were the direct forerunners of later government-funded health centers (*Hokenjo*). These private organizations along with individuals were forced to bear the costs of fighting diseases such as tuberculosis (Johnston 1995, 233–288; Maruchi and Matsuda 1991; Onodera 1991a; Tatara 1991). Much of their activity consisted of participating in "moral suasion" campaigns (*kyōka*) organized by the government. These campaigns were an array of mobilization efforts that recruited the help of middle-class citizens in the struggle to encourage everything from "self-management" among peasants in the nineteenth century to modern child-rearing methods, better hygiene and sanitation, higher savings rates, and life improvement ("New Life") among fellow citizens up through the postwar period (Garon 1997, 6–7).

The Mobilization of the Anti-smoking Movement

The state tobacco monopoly and its subsequent privatization in 1985 that remitted 100 percent ownership of Japan Tobacco stock to the powerful MOF essentially preempted any government-sponsored anti-smoking campaigns resembling the scale of earlier "moral suasion" efforts and restricted anti-smoking messages diffused by underfunded voluntary health organizations to media advocacy events (lawsuits, petitions, conferences, etc.) and the print media (posters, billboards, postcards, etc.), while tobacco industry marketing was able to take advantage of the full spectrum of the public media sphere. The 1984 Tobacco Industry Law retained the MOF's responsibility under the previous law for determining the content of health warning labels, which to this day remains weak.[20] The new law's brazen declaration that a robust tobacco industry was in the national interest, together with U.S. trade pressure, did much to galvanize public opinion and anti-tobacco advocates, and it provides a powerful example of how tobacco control is increasingly articulated though the interplay of local and global events. Compounding the sense of menace was the strong perception of an international threat concerning which the national government not only failed to offer protection but even conducted itself as an active collaborator. Most notable were the actions of the U.S. Trade Representative and Sen. Jesse Helms, chairman of the Senate Foreign Relations Committee, to force open East Asian tobacco markets (as well as markets elsewhere) dominated by state monopolies. Helms notoriously employed thinly veiled threats of economic retaliation to secure from Prime Minister Nakasone guarantees of 20 percent of the Japanese market for American cigarettes.[21] These threats were widely reported in the Japanese media ("Tobacco Negotiations" 1986), prompting the All-Japan Anti-smoking Liaison Council to write a letter of protest to President Ronald Reagan. Later in November 1987, at the Conference on Smoking and Health in Tokyo, copies of Helms' letter circulated among participants. As predicted, within two months of Japan's elimination of tariffs on foreign cigarettes 1 April 1987, U.S. and other cigarette imports' market share of domestic sales doubled from 4 percent to 8 percent, and by 1996 they enjoyed 22.3 percent of market share (Shibata 1987). In turn, JT's exports to East Asia grew considerably during the same period.

Even before the new trade agreement went into effect, anti-smoking advocates warned of its consequences. The director of the Japan Action for Nonsmokers' Rights published an opinion piece on the editorial page of the English-language newspaper *Japan Times* with the title, "Lung Cancer: The New Import?" He joined other nonsmokers' rights organizations in Japan, South Korea, and Taiwan in protest of the removal of tariffs, which would lead to an increase in the number of smokers (Watanabe 1986). The entry of foreign manufacturers into East Asian markets threatened not only to intensify competition for customers but also to increase the volume of advertising on radio and television (virtually banned in South Korea and Taiwan). The article also raised the specter of targeting women and youth as in the U.S. and the U.K., where smoking among males had been declining for some time. In fact, in Japan, JT itself was preparing to follow suit by bringing out a new brand called Misty for young women.

It was such a prospect—along with an awareness that lung cancer rates among Japanese women had doubled between 1961 and 1984 (from 5.1 percent to 11.8 percent)—that prompted the creation of Women's Action on Smoking. One of its founders declared,

> With the growing competition of the [tobacco] industry, cigarette companies will be placing their sales target on women. . . . There are things that men simply don't recognize, but we women can. . . . There is no develop[ed] country where men smoke as much as they do in Japan and the number of young women who smoke is increasing. . . . [Women's Action on Smoking is] the only women's group that focuses on the effects of smoking by women and that of the so-called "passive smoking" of which women and children are often the victims. (Cited in *Women's Action* 1987, viii; Fukami 1987)

The salience of the issue of secondhand smoke for the group was underscored by the fact that among its members (half of whom were men) was Takeshi Hirayama, author of a pioneering epidemiological study in 1981 on the risks of nonsmoking wives who live with husbands who smoke. Other issues that were of concern were smoking by users of birth control pills (once they became available in Japan) and by pregnant women, women who breastfeed and smoke, and the ¥1.627 trillion that smoking costs Japanese society annually. Also powering the drive for the creation of Women's Action on Smoking was doubtless the LDP-led government's long history of indifference and hostility to other women's issues, such as reproductive rights (expressed in public charges that women's desire to have fewer or no children jeopardized the nation's well-being and the repeated refusal to authorize the sale of the birth control pill) and equal employment opportunities outside the home (equal pay, job security). With respect to poor career prospects for women, leaders would later interpret the increase in young women's smoking as the substitution of the freedom to participate in the satisfactions of consumption for meaningful career choices denied to them in the workplace (Nakano 1999). Finally, if the entry of U.S.- and U.K.-based cigarette manufacturers into the Japanese market helped create Women's Action on Smoking, new domestic anti-smoking policies by the U.S. government were cited as an example by the group, which had its founding date, 6 February 1987, coincide with the day the U.S. government restricted smoking in all federal buildings.

The formation of Women's Action on Smoking emerged out of earlier women's activism. In the 1970s women already had taken the lead in consumer, anti-nuclear, and environmental movements, and many did so in the name of their roles as mothers and caretakers of children and guardians of public life. This stemmed in part from the women's and feminist movements and paralleled the growing numbers of women serving in public life as unpaid welfare commissioners and volunteers in social work. In a sense, women's public leadership in new social movements was also in keeping with older "moral suasion" or education campaigns, such as the "New Life" and savings campaigns, which the government conducted in collaboration with private organizations including women's groups up through the postwar era (Garon 1997, 117–118, 179–195). But state backing in combating the hazards of smoking was not forthcoming.

Cigarette manufacturers' successful breakthrough into protected domestic markets in Asia even caused some anti-smoking advocates in East Asia, North America, and Europe to claim that what was at hand was not only an "epidemic" but, more forcefully, a new "opium war," "global war," or "third world war" (Chen and Winder 1990). And one Massachusetts public health official at the Sixth World Conference on Smoking and Health held in Tokyo in November 1987 sarcastically titled his paper, "The American Liberation of the Japanese Cigarette Market," which, when published in the official conference proceedings, bore a less inflammatory title (Connolly 1988). The depiction of transnational corporate practices in terms of Western neocolonialism and world war was especially strong in Japan and East Asia and indicates how politically charged the issues of smoking and global tobacco control had become.

The local and global articulation of these questions and the framing of populations at risk (women and children) were destined to have a long transnational life in the ensuing years. For example, during negotiations of the proposed Master Settlement Agreement (1996–98) between U.S. state attorneys general and cigarette manufacturers, requiring reimbursement of states for medical tobacco-related medical expenses and prohibiting marketing to youth, one Japanese public health official privately expressed her disappointment to a leading American epidemiologist during the Tenth World Conference on Tobacco or Health in Beijing (1997). She termed the document the U.S. equivalent of the Tobacco Industry Law, the implication being that it gave tacit, unwarranted guarantees to the tobacco industry and contained no clause forbidding companies from recouping their declining U.S. revenues through increased exports abroad. In a word, it brought relief to depleted state treasuries and U.S. youth at the expense of peoples overseas (Mochizuki-Kobayashi 2000). Here, the promise of a smoke-free generation, bearer perhaps of a new American modernity, was based on transferring to other cultures and nation-states the practices of an earlier industrial society for which cigarettes were a privileged icon. Public discourse decrying the tobacco settlement also invoked the language of war and colonialism. The *Japan Times* published a sharply worded opinion piece that denounced the accords with the prediction that "Japan is likely to become a dumping ground for U.S. tobacco, which can be compared to industrial waste" (Hanai 1997). Complementing the metaphors of environmental pollution and implicit comparisons to high-handed neocolonial policy was an adjoining cartoon that lit-

U.S. planes bombing Japan with cigarettes.
Courtesy of Ryuji/*Japan Times*.

eralized the language of war: it pictures American aircraft (resembling the four-engine B-29s that leveled Japanese cities in 1945) bombing Japan with cigarettes.

It would hard to imagine a more powerful metaphor of the tobacco menace to present to a Japanese and international readership. Several months later the *Japan Times* reprinted a *New York Times* editorial but with a new title, "Poisoning Asia's Youth," that merges an image of pollution and violated boundaries with notions of criminal abuse and prepares for the comparison with the Opium Wars, which, in the Japanese version of the editorial, has been moved up to the opening paragraph:

> Hong Kong is one of the battlefronts of the modern-day Opium War. While Britain went to war last century to keep its Indian-grown opium streaming into Chinese ports, today American tobacco companies win profits and build addiction throughout Asia, where tobacco consumption is growing at the fastest rate in the world. ("Poisoning" 1997)[22]

The Sixth World Conference and the MHW White Paper

At the invitation of Japanese anti-smoking groups, the Sixth World Conference on Smoking and Health took place in Tokyo in November 1987, six months after the Tobacco Industry Law went into effect. Sponsored by four private Japanese foundations (Japan Anti-Tuberculosis Association, Japan Cancer Society, Japan Heart Foundation, and Japan Health Promotion and Fitness Foundation) and the American Cancer Society and with participants from over fifty countries, the conference was meant to give a boost to the local anti-smoking movement and, like the Kobe conference twelve years later, to expose Japanese government policies in the international court of public opinion. Although government agencies lent only nominal support, the MHW felt pressured to do something in order to avoid international embarrassment, and at the suggestion of a leading cancer epidemiologist (who was also the secretary-general of the upcoming meeting), ministry officials charged a committee to write a White Paper on Smoking and Health. Published the month before the opening of the conference, it was a review of the scientific literature on the health consequences of smoking, including sterility in women, the risk of giving birth prematurely, substantially higher risk for throat cancer and heart ailments, and higher lung cancer rates among nonsmoking wives of smokers (Tominaga 2000; "Smoking Linked" 1987; "Government Releases" 1987; "Japan Tobacco Reacts" 1987). There was little mention of economic groups, let alone any allusion to ethnic or regional differences. (Differences of regional origin [for example, the poorer regions of Tohoku, Hokkaido, Kyushu, and Okinawa] or ethnic and national origin [Korean, Filipino, Brazilian], while evoked in Japan in everything from casual conversation to bureaucratic record keeping, do not figure in epidemiological studies of tobacco in Japan.)[23] The report, like all official government White Papers, would carry potential weight within government circles and would serve as a reference in interministerial debates and negotiations.

The prospect of the White Paper's imminent publication alarmed American cigarette manufacturers. The U.S. embassy in Tokyo sent a cable to the State Department that reported on the committee's members, speculated on possible recommendations hostile to foreign cigarette brands, and promised to keep tabs on their activities ("Tobacco 301" 1986). Japan Tobacco and tobacco wholesalers also responded by creating the Tobacco Institute of Japan to establish industry codes and conduct campaigns against smoking by minors, organizing public relations campaigns by its members who decried the anti-smoking movement as "fanatical" and "anti-liberal," and establishing a foundation with an annual budget of ¥350 million (US$3.5 million) that funded distinguished researchers whose papers questioned scientific findings on the hazards of smoking.

Interministerial struggles followed the release of the White Paper and resulted in some changes in subsequent years. For example, the White Paper allowed the MHW to argue successfully for the inclusion of smoking regulations in the workplace in the amended Labor and Safety Law in 1992. And in 1993, according to one health official, when the MHW's annual White Paper (a review of all health and welfare issues) contained for the first time six pages on tobacco control, the MOF, which enjoys the power of review of other ministries' budgets, objected and demanded that the MHW remove them. The health and welfare minister refused. He successfully resisted MOF pressure by citing the government's responsibility for Japanese people's health[24] and charged that the finance ministry risked failing to fulfill its responsibilities concerning the health hazards of tobacco in the same way as the MHW did when it failed to protect citizens from HIV-contaminated blood supplies. He declared that, like the MHW, the MOF now possessed crucial information and must not fail to act on it.

Outside of specific events like the publication of the White Paper, anti-smoking advocates' influence over government proceedings remained limited to stimulating debates in the Diet and to participating in ad hoc governmental advisory panels (policy deliberation councils), which, however, included only officially sanctioned nonprofits, NGOs, and foundations (such as the Japan Anti-Tuberculosis Association, Japan Cancer Society, Japan Heart Foundation, Japan Health Promotion and Fitness Foundation, and Japanese Medical Society). But these paid the price of dependency for long-standing state sponsorship or funding and were reluctant to antagonize state bureaucrats or the ruling coalition. As the head of one leading foundation put it in an interview, organizations like TOPIC, which don't receive any government funding, enjoy real freedom of action (Shimao 1999). Anti-smoking activists thus found most of their activities relegated to spheres outside the halls of power. They formed their own groups such as the Japan Anti-smoking Doctors' League; pooled scientific and legal information (TOPIC); gathered petitions, which they submitted to national and municipal governments; filed lawsuits before an unsympathetic judiciary in order to generate publicity and pressure government officials; engaged in small-scale education campaigns; staged press conferences; and with the help of the WHO started an East Asian tobacco control organization, the Asia Pacific Association for Control of Tobacco (APACT), in 1989 in Taipei, to hold regular regional conferences.

Anti-smoking Poster Campaigns: Women and Children

Even as at the close of the millennium its membership had declined because of the difficulties recruiting members among the young,[25] the concerns of Women's Action on Smoking with risks posed to women and children by not only industry marketing campaigns but also men's secondhand smoke dominated anti-smoking discourse in late-twentieth-century Japan. The Japanese articulation of secondhand smoke as a male threat stemmed not only from the discrepancy in smoking rates between men and women in the late 1980s but also from the publication of the earliest major study on secondhand smoke of international importance by Takeshi Hirayama, the epidemiologist who had confirmed the findings of the 1964 U.S. surgeon general's report on smoking in a study performed at the request of the MHW. His new study, based on the 142,000 medical records between 1966 and 1981, discovered that wives of smokers were twice as likely to have lung cancer as wives of husbands who did not smoke (Hirayama 1981). In short, in Japan this study extended the strongly gendered and generational frame of the traditional dangers of smoking in general (older men as heavy smokers) to the hazards of secondhand smoke for everyone else. Finally, the focus on women and children as risk groups received an outside boost from the WHO, which launched two initiatives in 1989 and 1990 reprising many of these issues: "Women and Tobacco" and "Growing Up without Tobacco." Thus, at the time that women, along with youth, emerged as the audience of the new international marketing tactics of the tobacco industry, women and children also began to stand out as the primary victims of environmental tobacco smoke.

We have already seen one example of the articulation of the threat of tobacco in terms of risks posed to women and children in the 1978 emblem of one of the more active umbrella groups, Japan Action for Nonsmokers' Rights, which features the face of a small child superimposed on a tulip with the caption, "Smoking. I'm sensitive to tobacco smoke." Often messages were cast in terms of bonds with others as lived out in family life. For example, in anticipation of the upcoming international conference in 1987, Japan Action for Nonsmokers' Rights issued posters and postcards. One features a cartoon of smiling faces of family members around a table. Included are the mother and children, grandparents, an aunt and uncle, and household pets. Positioned between them are four drawings of a reclining husband and father in a bubble wearing a hat while smoking an enormous cigarette or cigar. His fumes literally enter the family circle and start their way around the table. The title reads, "What is the meaning of passive smoke?" ("*Passhibu sumōkingu [judōteki kitsuen] to wa?*"). Then follows a list of health consequences:

> It means being made to breathe others' smoke even against one's will. Toxic substances contained in tobacco smoke such as nicotine, tar, carbon monoxide and benzopyrene irritate mucous membranes, cause blood vessels to contract, increase the heart rate, and moreover cause blood pressure to rise, blood sugar to increase, etc.

Below is the group's name and address followed by the tag, "Because of the harm caused by passive smoke, we appeal for the protection of nonsmokers" ("*Judōteki kitsuen no gai kara hikitsuenka o mamoru koto o uttaemasu*").[26]

The same group also issued a second, more aggressive cartoon poster, but it concerns the more traditional hazards that smokers pose to themselves. It pictures two men and one woman, wearing the telltale smokers' hats, belonging to a single body whose bleeding heart is slowly being sawed in two. The blade's cutting motion is linked to each smoker's lungs and mouth by rods and cams that are powered in turn by the smoke, which fills the air around them. Below, an enigmatic caption reads in bold, "If you smoke one cigarette, 14 minutes 30 seconds . . ." Then follows an explanation of the health consequences of smoking:

> It is said that every time those who smoke more than twenty cigarettes a day smoke one cigarette they shorten their lifespan by fourteen minutes and thirty seconds (professor Sasamoto Hiroshi, Tokai University). Furthermore, their rate of heart disease and heart attack is more than twice than that of nonsmokers, and in the case of sudden death it jumps nearly five times higher.[27]

Here smokers are framed as alone, with no reference to their possible roles as parents or spouses or to them as a source of passive smoke to their loved ones. Taking the two posters together, it would appear that in this vein of anti-smoking discourse, on the one hand, women operate either as victims of male smokers or as smokers whose actions have consequences only for themselves, while, on the other, men stand as perpetuators of secondhand smoke and never its victims. It suggests that smoking ethically relegates those male subjects who indulge in the habit to the margins of family household life as threats to its well-being while female smokers receive no familial representation at all and reside outside the frame of the family circle altogether—marking them by virtue of their absence as single and nondomestic.

The protection of children and mothers against the hazards of environmental tobacco smoke and industry marketing organized much anti-smoking discourse in Japan through the 1990s. On the heels of the publication of new studies on secondhand smoke by the U.S. Environmental Protection Agency (1992) and by Hirayama (1992), when APACT held its Third Conference on Tobacco and Health in Omiya (Niigata), Japan, 6–8 June 1993, it adopted the theme, "Smoke Free World for Children." The conference poster itself combines the issues of secondhand smoke and industry marketing. It features a large bubble expressing the worries of four adults (a physician, a nurse, and two lay people) in the form of a fairy-tale nightmare of an industry salesman pictured as a wolf dressed in grandmother's clothing tempting two children with cigarettes drawn from a wicker basket. The caption in Japanese reads, "Because of the harm caused by cigarettes, let's protect Asia's children!!" ("*Tabako no gai kara Ajia no kodomotachi o mamorō!!*"). From the late 1980s through the 1990s in Japan it was the Smoke-Free Environment Campaign for Kids that most consistently exploited the hazard of secondhand smoke. In this group's publicity efforts, the perpetrators of cigarette smoke continue to be men and the victims women and children, who are almost always situated within the frame of the family circle. Founded in 1988 and based in Osaka, the organization runs annual anti-smoking poster competitions in schools and distributes up to 180,000 copies of the winning entries to schools, hospitals, public areas, and train stations.

One poster announces the Seventh National Smoke-Free Environment Campaign for Kids (1994–95) featuring a large drawing of a father in coat and glasses enveloping his wife and two children in his cigarette smoke. They react by coughing, sneezing, and, in the case of the toddler, crying. A title runs above and below the characters, exclaiming, "Cigarette smoke is dangerous for both those who exhale it and those who inhale it!" ("*Tabako no kemuri haku hito sū hito dōchira mo kiken*!") Two years later (1996–97), the winning campaign poster pictures simply the absent father's ashtray with a lit cigarette, whose spiraling smoke overcomes the mother, three children, and cat under the title, "Cigarette smoke bothers everybody" ("*Minna no meiwaku tabako no kemuri*").

Finally, in 1997–98 the poster selected to kick off the tenth national campaign expands the circle of danger to the public sphere but a sphere in which adult citizens are still held ethically accountable for the consequences of their behavior for those under their care directly or indirectly—in this case, children and teenagers as a group. It depicts a man with a cigarette descending a public staircase as cigarette smoke trails behind him, utterly oblivious to the discomfort of the school-age kids walking behind him, their mouths covered with their hands. A bubble that floats above the head of the child closest to him carries the thought, "Cigarette smoke is disgusting, isn't it?" ("*Tabako no kemuri iya da yo ne*"). What is interesting is that the national campaigns for a smoke-free environment for kids remain almost exclusively focused on adult men (in keeping with the fact that they constituted the vast majority of smokers in Japan even as their rates had been declining for fifteen years), whereas smoking mothers remain relatively rare in visual and verbal representations as a source of danger to others; indeed, they constitute one of the unspoken phenomena that anti-smoking activists and government officials are reluctant to acknowledge in public (Nakano 1999).[28] Similarly, while the rise of smoking rates of teenage girls is a subject of much commentary in the Japanese press and among health officials, in what little anti-smoking education material that does exist they are not present as sources of danger either to themselves or to others (siblings, elderly relatives, classmates, what have you). They are no more present than they are in the smoking prevalence statistics gathered by Japan Tobacco, which, like the monopoly before it, has been the main purveyor of studies of cigarette consumption practices in Japan but has never officially conducted studies of smoking by minors, female or male.

I can only speculate why smoking mothers and teenagers are absent from

"Cigarette smoke is dangerous for both those who exhale it and those who inhale it."

Courtesy of Smoke-Free Environment Campaign for Kids.

anti-smoking messages. It may have something to do with the women's movements of the 1970s and with an older ideology of victimhood dating from the atomic bomb survivors' and anti-nuclear movements of the 1950s and environmental protests of the 1970s in which mothers and children were symbols of choice for citizens' suffering (Yoneyama 1999).

JT's Response: The "Good Manners" Ads

In the late 1990s, Japan Tobacco didn't rest idle. Bending to public pressure and international trends running in favor of restrictions on smoking, it eventually agreed to a voluntary withdrawal of cigarette advertising on television starting

1 April 1998 while staving off major increases in taxes and changes in the health warning on cigarettes. In advance of the deadline it launched a series of public relations campaigns focusing on "good manners" ("*manā*"), in print and broadcast/cable media, destined to preempt any dramatic shifts in public opinion against smoking: "Smoking clean is JT" and "Such delight, JT." The first set of ads featured older, well-known Japanese male actors, like the macho Ken Ogata, who, speaking directly to the camera, declared their love of smoking ("I love to smoke"; "*watashi wa aikitsuenka desu*") while asking fellow (male) smokers to dispose of their cigarette butts properly (in ashtrays or hand containers) instead of littering public areas and thoroughfares. He appealed to male smokers with a phrase (*Watashi wa sutenai*") that has the double meaning of "I don't throw away (butts)" but also "I won't quit" or I won't give up (my habit)." The tag reads, "Smoking Clean Is JT" ("*Smoking Clean, JT desu*").

In the other ads stressing courtesy, this discourse that mixes mild defiance with the affirmation of conscientious public conduct gives way to playful vignettes played by white (Western) actors. They follow the encounters of a somewhat eccentric young train passenger in a bowler hat who refrains from smoking in the presence of others out of simple courtesy. Retro sets, clothing, and (in the case of TV spots) sound track evoke early-twentieth-century Europe or the U.S., an earlier moment of modernity in which good manners were practiced and smoking tolerated, and construct an arresting but charming frame for the drama of smoking in public space. In one TV ad, just as the young man is about to pull out a lighter, a striking Mediterranean-looking woman sits down next to him with a young baby, whereupon he excuses himself and moves to another bench, where a smiling old man moves an ashtray stand closer to him. As he silently signals thanks to the old man, birds alight on his hat, and he puts the lighter away and smiles. As the two men look at each other, a title flashes, "When you're happy, I'm happy," followed by the tag, "Ah, such delight, JT" ("*Anata ga ureshi to, watashi wa tanoshii. Ā, diraito, JT*"). Another TV spot follows the same passenger as he walks with his travel bags through the street. Just as he sits down on a bench and pulls out a lighter, a huge bubble floats by followed by laughing children, one of whom hands him a bubble blower, and he then joins in the fun. Identical title and tag follow. In neither ad are cigarettes to be seen.

These "Delight" campaigns cleverly rearticulate an anti-smoking sensibility that threatens to disrupt the delicate balance that has underlain smoking in advertising, cinema, television, and the general culture between the affirmation of individual pleasure and the expression of a social bond. To preempt the reframing of smoking in the presence of others by hostile public opinion into the simple expression of selfish, anti-social behavior, in the small vignettes the JT ads turn anti-smoking discourse on its head: here, the wry presentation of the young man's eccentric behavior renders smokers as both innocuous, childlike citizens and the very embodiment of charming, considerate behavior, especially with respect to women and children, who are deemed to be the victims of others' smoke and targets of tobacco advertising by anti-smoking activists, epidemiologists, and public health officials. But the finesse comes at a cost: the only way to reconcile the smoker's individual pleasure with obligations of the social bond is to substitute the pleasure of courtesy and social consideration for that of smoking and suppress momentarily the personal satisfactions afforded by the habit. Curiously, the only spaces at issue in the JT *manā* ads are the streets and public areas in and around railways. The ethical relation with others is a public one, not a private one involving family members as in many anti-smoking posters. The smoker's capacity for reflexive choice operates as the touchstone of social conduct, which in turn obviates any need for overt detailed regulation of normal citizens' behavior by the state through laws and health codes or through the vigilance of fellow citizens. However, just how JT would articulate its message in the case of family household relations is not clear and suggests that the domestic sphere may escape articulation by pro-smoking discourse.

World No Tobacco Day, 1998–1999

For the media event that kicked off the annual "World No Tobacco Day" organized by the WHO every 31 May, in 1998 the MHW and the Japan Public Health Association issued posters mocking the *manā* ads. Included as nonprofit cosponsors were the Japan Cancer Society, Japan Anti-Tuberculosis Association, Japan Health Promotion and Fitness Foundation, and Mother/Child Health Research Association. Cartoon figures in the form of anthropomorphic filtered and unfiltered cigarettes coughing and wheezing communicated the health message visually. In one poster, two medical attendants bear away to the hospital a dying patient on a litter with a lit cigarette stuck between his lips.

The title reads vertically, "Has he finished his last smoke?" ("*Saigo no ippuku sumasemashita ka*"). On the right-hand side, a vertical text announces: "Cigarettes contain a great many toxic substances, beginning with carcinogens. Now, cigarettes are not a 'manners problem.' They are a serious health problem for each and every one of us. Starting today, no smoking" ("'*Tabako' ni wa 'hatsuganbusshitsu' o hajime, kazuōku no yūgaibusshitsu ga fukumarete imasu. Ima ya, tabako wa 'manā no mondai' de wa arimasen. Watashitachi hitori hitori no taisetsu na 'kenkō mondai' desu. Kyō kara, nō sumōkingu*"). Below the characters stands the tag, "Eleventh World No Tobacco Day 1998/*Sekai kin'en dē* 5/31." In the lower right-hand corner, the poster exhorts viewers, "Let's cultivate a smoke-free generation! No Smoking Week 05/31– 06/06" ("*Muen sedai sodateyō.*

World No Tobacco Day, 1998.

Courtesy of Ministry of Health and Welfare, Japan,
and the Japan Public Health Association.

Kin'en shūkan 05/31–06/06"). Neatly and discreetly, the posters' convulsed cigarettes satirize the self-possession of JT's macho smokers in the "Smoking Clean" ads and appeal to the public to cultivate not "good manners" but a healthier generation.

The posters were reprised the following year, and they accompanied the official release of a free educational CD-ROM titled, "No Smoking: A Cigarette Device for Teaching about 'Tobacco and Health' with a CD-ROM" ("*Nō Sumōkingu tabako no shikumi. CD-ROM de oshieru,* [*tabako to kenkō*"]), which contained a wide range of tobacco control information from the history of the production, marketing, and consumption of tobacco to the latest scientific findings regarding prevalence rates and health effects to examples of anti-smoking spots and posters from around the world. Staged in a large auditorium located in central Tokyo to which were invited journalists, anti-smoking groups, public health experts and officials, teachers, and schoolchildren, the event was downplayed by the media, who claimed that the day's events had little echo outside the invited groups ("World No Tobacco Day" 1999).

As a small, portable medium, the CD-ROM can be easily circulated among various professional constituencies for the development of their own tobacco control "capacity" (anti-smoking groups, teachers, biomedical personnel, etc.) and distributed outside of the JT-dominated media sphere of TV and radio and of officially sanctioned circuits (Ministry of Education, health centers) in public and private gatherings after school and work. The authors took advantage of the CD-ROM format to narrativize and spatialize information. The disc's opening introduces viewers to an ultra-modern virtual space, an elegantly designed lobby of stone, steel, and glass that you might come across in the entry to a municipal government building, corporate headquarters, or museum in Japan. A middle-aged male host in a stylish but discreet jacket welcomes viewers and briefly outlines what awaits them. Different subjects offer different trajectories through various spaces: a "Timeline" containing a historical narrative divided into chapters; a "Laboratory" displaying recent data in the form of pie charts; a "Tobacco and Health Conference" featuring a dozen or more health experts and educators who speak briefly to the camera on a range of topics; and a "Quit Gallery" replete with video spots from the U.S. (California, Massachusetts, American Cancer Society), Australia, and Japan but for which no Japanese dubs or subtitles were provided.

While the interactive character of the CD-ROM remains limited to negotiating different topics and clicking on different video clips of ads and lectures, some materials stand out: a historical narrative linking the rise of cigarette smoking to war, industrialization, and colonialism; lectures by anti-smoking advocates belonging to groups usually excluded from official governmental ad hoc advisory boards (Can Do Harajuku, Women's Action on Smoking); foreign anti-smoking TV spots never seen in Japan; and the Japanese anti-smoking TV spot, previously mentioned, which aired in May 1994 during the month leading up to World No Tobacco Day (May 31), whose theme that year was media and tobacco.

The spot is the only Japanese anti-smoking ad ever broadcast in Japan and was produced by Yumiko Mochizuki-Kobayshi and the National Cancer Center, a WHO Collaborating Center. The ad features an elderly man, one of the two members of the 1960s fast-talking comedy team Top/Light (Toppu Raito), admonishing viewers in a raspy, inaudible voice to stop smoking. It reprises the testimonial style of spots first developed by the CNCT in France and then the State of Massachusetts featuring the model Janet Sachman, industry researcher Victor DeNoble, and the brother of the Marlboro Man, who bear witness against the tobacco industry (the latter two spots are also on the CD-ROM). Aired on CNN and CATV, it was instant success and was picked by most other stations, including the state-run NHK; the spot ended up being broadcast more than three hundred times.

The ad opens with a clip of a sketch in which a man lords it over a subordinate, demanding loudly, "Ashtray! Ashtray!" ("*Haizara! Haizara!*") while tapping ashes from his cigarette, as the other haplessly extends his cupped hands to catch the hot ashes. A vertical title appears—"Columbia Light"—followed by a horizontal title: "Working in a duo with Columbia Top, [Light] achieved national prominence as one of our most beloved comedians." The next visual is of the comedian today; visibly diminished, he stands in a dark suit against a gray backdrop and speaks with difficulty to the audience: "Since losing my vocal cords, I feel the wonder of words. And there is one thing I have to ask of you: Let's stop smoking. Just quit." His final gesture is to break a cigarette in two and toss it away. As he speaks, titles appear at the bottom of the screen: "In 1991, cancer of the larynx meant the removal of [Light's] vocal cords. Light is active today as a volunteer working to cheer up the spirits of cancer patients."

Then follows a voice-over set against a bright musical theme, "A Day the Whole World Gives Up Smoking," with a title in white lettering against a gray background: "Today, cancer is the leading cause of death for Japanese. And, it has been proven that tobacco is to blame for more than 95% of all larynx cancers." The spot concludes with a final title in English ("World No Tobacco Day") and Japanese ("World No Smoking Day"), with the date 5/31. The last image in the guise of a tag is of the English logo of the WHO against a blue background.[29]

Here the anti-smoking ad delivers a standard message about smoking cessation. It addresses an audience of smokers and their colleagues, friends, and loved ones who are arguably more differentiated by age (middle-age viewers who can remember the comedy duo) than by sex. It mixes personal testimony, biogra-

Ill member of comedy team Top/Light.

phical information, and public health discourse about the hazards of tobacco. Like the Janet Sachman spots, it depends for its effectiveness on the contrast between shared popular memories of the celebrity and his altered health. The melodramatic realization that it is already "too late" for Columbia Light perhaps spurs viewers to rethink their own life trajectories and to take action in order to preserve fond memories from an earlier time in life intact. However, unlike in the French and American testimonial spots, the narrative that suffuses the speaking subject with regret does not tie sentiments of needless loss to the machinations of the tobacco industry and its marketing practices. Rather, the forestalling of future trauma is left up to the prudential actions of individual citizen-subjects. Within the CD-ROM's collection of anti-smoking ads, that other, more political, task belongs to anti-industry spots produced by local governments and voluntary health organizations from overseas.

Other spots included in the Quit Gallery are the famous American Cancer Society ad of Yul Brynner warning, from beyond the grave, viewers not to smoke and the California spots depicting industry targeting of children in a rainfall of cigarettes and the wide availability of cigarette vending machines (a burning issue in Japan, where they are responsible for upward of 30 percent of cigarette sales). The sleek visual layout and the international origins of the materials attempt to reposition the MHW and its anti-smoking goals within an alluring hypermodern future that is global in scope and style. However, the cleverness of the materials could not hide the stark communication situation: day in and day out, JT's ads flooded the airwaves, newspapers, magazines, and billboards promoting cigarettes and smokers' "civil conduct," while the MHW posters grabbed public and media attention primarily during No Smoking Week.

Meanwhile, JT continued to stymie the MHW's bureaucratic initiatives to formulate future tobacco control policy. In 1998, when the MHW convened the Twenty-First-Century Tobacco Policy Deliberation Committee, its ambition was to draw on a wide array of expert opinion, and in a gesture of openness the MHW included four representatives of the MOF and JT. The president of the Japan Anti-Tuberculosis Association chaired the meetings, and transparency was the rule: all gatherings were open to the public, and minutes were published on the Internet. However, the committee was caught unprepared for the tactics adopted by industry sympathizers, who stalled the committee's work by disputing long-standing scientific claims about the dangers of smoking and in the national press openly questioned the legitimacy of the meetings. Their tactics succeeded in blocking consensus on any meaningful recommendations (Corliss 1998). And there is the previously cited case of another advisory panel, Health Japan 21, which in August 1999 recommended the reduction of adult prevalence rates by 50 percent by the year 2010. Upon the release of its proposals, LDP members of the ruling Conservative coalition revolted and conspired to have the MHW functionary in charge fired ("Health Ministry Sets Targets" 1999). Even so, a subtle generational shift may have started in the workplace and at home. In Tokyo meeting rooms, the obligatory ashtrays began to disappear, and the president of the Japan Anti-Tuberculosis Association recounted that in recent meetings with JT executives few senior officers smoked and none of the younger male executives did (Shimao 1999). And in 1999–2000 conversations were full of stories of young husbands and fathers, home from work, being forced by an alliance of wives and children to retreat to the porch or balcony outside to light up, which earned them the colloquial designation of "veranda" or "garden" smokers (*berandā sumōkā, gāden sumōkā*).

Several years later there was another shift, this time in governmental circles, when the MHW was able to sponsor in July 2003 a Health Promotion Law whose Article 25 addresses the prevention of passive smoking (Matsubara 2003; Oshima 2003). At the same time, the WHO's strategy of pursuing an international protocol finally began to bear fruit: the most recalcitrant states—Japan, the U.S., and Germany—in March 2004 signed the Framework Convention on Tobacco Control, which had been passed unanimously by the WHO General Assembly the previous May. The convention stipulated stricter health warnings, tax increases, advertising bans, and restrictions on smoking in public. On 19 May 2004 the Japanese Diet actually ratified the treaty but without enforcement provisions, and the MOF proposed a new health warning that anti-smoking advocates considered derisory: "Smoking could be one of the causes of lung cancer for you. Epidemiological studies show that smokers have a two to four times higher risk of dying from lung cancer than do nonsmokers" (Hanai 2004). And although the conservative Tokyo District Court ruled against plaintiffs in a major lawsuit against tobacco companies for their inadequate health warnings in October 2003, the same court awarded small damages to a municipal employee who sued the local government for not providing protection against secondhand smoke (Hanai 2003; "Worker Wins" 2004).

The struggle to reduce the hazards of smoking in late-twentieth-century Japan throws new light on the two contexts that have shaped tobacco control: the tradition of liberal government in the throes of free-market-based policy initiatives promoted by neoliberals and the latest wave of economic and cultural globalization. As in France and the U.S., long-standing practices of the liberal government remitted primary responsibility for care of the sick, the destitute, and the unemployed to local authorities and the private sector and respected the prerogatives of liberal medicine (fee for service, professional self-regulation). What is striking about Japan is that this mix of public and private assistance continued to dominate welfare and public health policy up through the 1980s to the point of preempting any neoliberal revolution from taking firm root in Japan. At the same time, the persistence of the state tobacco monopoly in privatized form (the MOF's control of Japan Tobacco) blocked any substantial state sponsorship of "moral suasion" campaigns by private groups, as had been done in the past, and left anti-smoking organizations without effective access to bureaucratic arenas (such as policy deliberation councils) and the public media sphere anywhere near matching that of JT itself.

In turn, this gave a peculiar twist to the interplay of local and global factors unmatched in California and France. Perhaps as a result of Japan's history of resisting and rivaling Western imperial powers, anti-smoking groups saw themselves as engaged on two fronts—against an irresponsible nation-state that had repeatedly chosen economic expediency over its obligation enjoined by the Constitution to secure the welfare of its people, and against the invasion of U.S. and U.K. tobacco companies and their aggressive marketing practices that were putting at risk its most vulnerable citizens, women and children. This may help explain why, when faced with similar incursions by the same cigarette manufacturers in France, French anti-smoking advocates rarely cast their struggle as one against foreign firms as imperial intruders.[30] At the same time, Japanese anti-smoking advocates also drew on data and strategies developed in the U.S. and played the example of U.S. government anti-smoking policies off the comparative indifference of the conservative LDP-led government. Commentators from outside Japan, including representatives of NGOs and voluntary health organizations, were more than happy to assist in this and subjected the Japanese government, Japan Tobacco, and daily life in Japan to critical scrutiny in the press and, as we saw previously, in international conferences such as the one in Kobe.

When viewed together, the forcing open of the Japanese domestic cigarette market by Sen. Jesse Helms and the U.S. Trade Representative and the pressuring of the Japanese government to change its domestic policies and support the WHO's Tobacco Free Initiative by Gro Harlem Brundtland underscore Japan's status as a routine object of external pressure and highlight the contradictory mode of governance of consumer societies and transnational accords that incite citizens at once to smoke and not to smoke. My point is that the postcolonial context is crucial for understanding the rhetoric and strategies of anti-smoking groups both inside and outside of Japan that target Japan for practices that other highly industrialized cigarette-producing nations share, but do so in a more disguised fashion. As one Japanese commentator put it, the U.S. now exports both tobacco and anti-smoking policies. By the same token, the case of Japan does ask the question of how in an era of bio-power—the bureaucratic management of "war and welfare"—twentieth-century nation-states' pursuit of turning populations into healthy, disciplined, and productive citizens (Foucault 1978) also encouraged people to smoke in the first place, especially in Japan, when military authorities early on in 1900 acted to prevent adolescent boys from consuming tobacco until the late age of twenty and then later issued cigarettes as part of soldiers' rations.

The answer lies, I think, in the notion of *management*—that is to say, in a calculus of risk by authorities and experts that differentiates according to age, sex, and so forth. Thus, what was considered dangerous for developing adolescents was not deemed sufficiently a threat to mature adults in their productive years to warrant governmental action. It was only with the rise of heavy cigarette smoking some fifty or sixty years later and the dramatic increase in citizens' life spans in industrialized countries did tobacco-related diseases such as lung cancer, emphysema, and coronary heart disease begin to incapacitate and kill citizens in large numbers, especially in middle age, and to show up in studies. Perhaps the great exception to this was the Nazi government in Germany, which between 1933 and 1945 went to great lengths to investigate the health risks of smoking and to discourage tobacco consumption for reasons that had to do with Hitler's own personal preoccupations and with the ascension of an ideology of health, which like earlier movements in the nineteenth-century U.S. and Europe stressed pure, natural living untrammeled by industrial civilization (Proctor 1999). As the example of the Japanese state and Japan Tobacco makes clear, the slow response of governments in liberal democracies to the health threat of tobacco may have much to do with the cynical calculus of government that protects citizens during their vulnerable years and permits unhealthy practices by adults that don't threaten the population during its most productive period (20–45 years), as the government profits from and yields to private interests with which the liberal state has long worked in tandem (the state's dependency on lucrative tax revenues, the power of lobbies) or, in the case of tobacco monopolies, with which the nation-state identifies, as it does with embedded cultural and social practices of smoking.

Conclusion

When I first became interested in anti-smoking campaigns, I was already working in southern California as a professor of French studies at the University of California, San Diego. Proposition 99 passed several months after my arrival from Vermont in 1988, and the ads that began to pop up around town and on television in the early 1990s were startling by virtue of their introduction of an aggressive tone in public discourse about smoking, an addictive habit, which, although tolerated by friends and family, had long fallen into disfavor as a dangerous practice if indulged in over time. I still remember the day in August 1964 when my four sisters and I loudly protested as our mother, driving an un-air-conditioned station wagon on a sweltering trip from Delaware to Ottawa, Ontario, to meet up with our father for a two-week vacation, took the opportunity of the long drive without her husband to resume smoking for the first time in almost twenty years. To her chagrin, we had already become the children of Surgeon General Luther Terry and his report released the previous winter. None of us smoked or would smoke other than socially in our teenage and adult years, so the new campaign in California offered to change something that had remained more or less in place throughout our early adult lives: smoking as an unfortunate but still acceptable part of social intercourse except in enclosed tight quarters.

The campaign sharpened what already was emerging at the time—a new sensibility regarding conduct in public and private space concerning the hazards of tobacco smoke and ethical and social distinctions based on what you did with your body in the presence of others. These echoed recent public debates over the transmission of HIV/AIDS and public policies meant to combat the pandemic and take care of the stricken. Traveling back and forth to the East Coast and Europe (and later on, living in Japan) made me aware of other regional and national distinctions that in the 1980s and 1990s became the subject of casual conversations and eventually of commentary in the media, which quickly reached for explanatory arguments based on either fundamental "cultural" differences or a theory of modernization that has all cultures and nations converging along a single timeline.

This book has followed the construction of these distinctions in public health practices and popular discourse and sought to challenge oversimplifications by looking at their points of emergence in the interplay of actions by the tobacco industry; social health movements; state policy bureaucracies; practices of population segmentation by epidemiologists, marketers, and tobacco control advocates; input by community representatives; and modes of collaboration between actors involved in anti-smoking campaigns. In particular, the diffusion of scientific findings through public health communication in the context of economic and cultural globalization and the ascension of neoliberal policy agendas has provided a window onto the tensions that shape anti-smoking efforts in the late twentieth century to spread nonsmoking as a social norm among citizen-consumers. Tensions between radical democratic impulses and traditional public health paternalism; aggregating and singularizing citizens and residents so as to constitute them as population groups amenable to intervention by public authorities; inclusive and exclusionist tendencies in public health policies; health movements' social base and the wider population they wish to address; and competing versions of liberal government have been the focus of this study. I have not attempted to delve into the subjective experience of participants in anti-smoking efforts, nor have I attempted to gauge the personal experience of smokers, ex-smokers, and nonsmokers in California, France, and Japan as shaped by the cultures of smoking and nonsmoking undergoing rapid change. Rather, I have concentrated on policies, strategies, public statements, and health promotion materials and the ethical and social subjects they tend to project or assume. So the example I just cited of my mother was less to raise issues concerning the status of new health information and how it is received, say, across generations or what complex reasons may have led someone to resume smoking the very year the surgeon general's report was released, than to signal a shift in attitudes toward smoking among college-bound members of what is roughly called the baby-boom generation in the U.S. (born between 1946 and 1964), who in turn witnessed and participated in a second revolution in their adult years concerning secondhand smoke that their changing attitudes earlier helped prepare.

As I made progress in my study, expanded its scope to France and Japan, and began to interview widely participants in anti-smoking efforts, my sense of surprise at the sharp rhetoric and tone of the California campaign gave way to a sentiment of being a witness to inevitable changes in the way we lived the life of the body and our relations with others. I had to agree with U.S. historian Allan Brandt, who claimed, even before the California campaign began to make headlines, that one of the tokens of twentieth-century U.S. modernity for a certain middle class was on its way out, and with its departure vanished perhaps an indefinable part of itself and of the ambient culture (Brandt 1990). Thus, it seemed to me that the very sensorium of public and private life in California was changing: in 1997, walking down Parnassus Avenue from the library of the University of California, San Francisco, I was struck for the first time by the mere smell of cigarette smoke on the street—not the presence of irritating fumes but simply the odor of cigarettes. Before, smoke outdoors causing physical discomfort used to catch my attention but not the simple smell of cigarettes as something unusual, as it did that day. (Smoking indoors had become of course another matter entirely.) My guess is that public smoking even in nonenclosed areas had already declined enough to change my own routine perception and underscored the degree to which what we consciously experience is shaped by the practices of everyday life.

The focus on the emergence of "global" nonsmoking has been limited to three countries all belonging to the G8 club of leading economic powers. Though restricted, the scope of my inquiry has had the advantage of bringing into focus the two-sided discourse of exceptionalism and universalism, whose subject always seems to be powerful nation-states (who else but the strong can claim cultural particularity of universal interest?) In tobacco control, depending on the nature of the discussion and participants' perspective, one's own exceptionalism is considered (or not) of universal interest, while that of other countries or regions is condemned (or not) as too particularist and narrow. The latter was the case of the Japanese group, Women's Action on Smoking, whose unique example and experience was for all intents and purposes unacknowledged during the WHO tobacco control conference on women and health in November 1999 in Kobe, Japan. I have argued that the functioning of this discourse has roots that go back to the practice and rhetoric of the respective empires of France, Japan, and the U.S. and their rivalry as competing models of modernity. Given the sheer dominance of global tobacco control discourse by Anglo-American countries (Commonwealth nations and the U.S.) and the claims of wide applicability made for approaches favored in those nations, the tendency of my analysis in this book on anti-smoking efforts and liberal practices of government has been to highlight the exceptionalism of the California and the U.S. when compared to non-Anglophone countries: the U.S.'s unusual history of cigarette smoking unbroken by the devastation of the Second World War; the ascendancy of social marketing methods and the weight given to media campaigns; the targeting of underserved populations by ethnicity and race in California and the reliance on U.S. epidemiologists' and marketers' respective segmentation practices; and the unexpectedly limited circulation of Californian anti-smoking ads. Conversely, this book has tended to attenuate the exceptionalism of France and Japan, in relation to which particularizing arguments are frequently made, by highlighting liberal governmental practices such as the long-standing reliance on the private sector and local authorities in matters of public health and welfare, the adoption of cigarette smoking in the 1920s and its promotion by the film industry, early French media campaigns, or Japanese state policies both favoring and discouraging consumption of tobacco.

The focus on California, France, and Japan has also foregrounded the advantages of adopting a posture of "studying across," for it avoids some of the pitfalls of the older field narrative of studying "up" or "down" that replays many of the assumptions of both modernization theory and discourses of exceptionalism by assigning modernity (good or bad) to the powerful and a nonmodern status (good or bad) to the disenfranchised. Yet for a scholar, even to try to work with or modify the rhetoric of modernization and exceptionalism can be a trap, for lurking within it is the logic of a disqualifying otherness or exoticism on the one hand and that of point-by-point comparisons between a select group of nations on the other, which exert a pull on any analysis. Ultimately, this book seeks to move beyond this logic toward one based on juxtaposing different sites in the context of globalization and liberal governance in order to draw out better the effective material situations and struggles of each in the arenas of tobacco control and the management of populations. My hope is that through my account of the clash and interpenetration of transnational and local constraints, the antismoking efforts in California, France, and Japan emerge as neither the simple expression of (forward-looking or backward) national or regional histories nor the product of global circumstance but stand somewhere in between—what I have termed "global singularities." Indeed, even as the World Health Organization wins ratification of its new international protocol, the Framework Convention on Tobacco Control, the future of effective tobacco control may also lie in regional agreements between governments such as the European Union or ASEAN countries in the Western Pacific and East Asia. This constitutes a limited transnationalism, if you will, in which capital, labor, policies, expertise, data, and personnel circulate more freely than in what passes today as a the global landscape and that moves toward decentering assumptions of orthodox theories of globalization that hark back to old modernization narrative according to which the one-way arrows of progress radiate out from the usual cosmopolitan centers whose exceptional exceptionalism strive to define and manage at a distance the collective life of the body.

Notes

Introduction

1. The Tobacco Free Initiative was launched by the new director-general, Gro Harlem Brundtland, when she took office on 1 May 1998.

2. Another example: one participant from an international NGO argued against the standard scholarly view, voiced during one session, that women have more trouble quitting because of psychological and social reasons, whereas a rereading of data, at least in industrialized countries, suggested that women quit proportionately at the same rates as men but that in the late twentieth century, women, in taking up heavy smoking twenty years later than men, therefore did so at higher rates than current men and quit in fewer absolute numbers.

3. Japanese studies scholar Naoki Sakai has even argued that the classic era of industrialization in the nineteenth century, dominated by the example of Western colonial powers, actually nourished awareness of national differences among those who adopted "Western" techniques and science in their own countries such as Japan.

4. For the most recent critical review of the status of "culture" within Western-based modernization theory as that which stands in the way of "development," see Chatterjee 1993, 226, 235–237. Similarly, according to Tessa Morris-Suzuki, culture is what falls outside of modernity (1998, 196). Yet historian Takashi Fujitani drew my attention to the fact that proponents of modernization theory had to make an exception in the case of Japan, whose spectacular economic success some of them attributed to the legacy of Tokugawa-era neo-Confucianism and its stress on work, discipline, and worldly success (Bellah 1985).

5. In a sense, Balibar's thesis concerning the status of "universal" culture finds confirmation in the realm of cigarette consumption, for the "global" circulation of the Marlboro, Gauloises, Mild Seven, Dunhill, Camel, Fine, and Rothmans brands roughly follows the borders of former colonial empires and spheres of influence and the flow of population movements from those regions.

6. Foucault even polemically weights this process toward a potentially open-ended dynamic of individuation and particularization by virtue of the infinite gradations of the normal and the pathological relative to which each person or group can be defined and recorded in bureaucratic files (Foucault 1978).

7. However, the ethnographic aspects of my study concern only participants in the conceptualization and implementation of tobacco control information campaigns and do not give an account of actual reception by audiences. That would be the subject of another study.

8. For an analysis of the complexities of fieldwork's directional metaphors, see Reid 2000. It would seem that the recourse to the metaphor of studying "up" to name fieldwork in institutional and bureaucratic settings stems from several factors: traditions of committed research often undertook research in direct antagonism to patriarchal, racist, and homophobic practices—that is to say, in contexts of asymmetrical power relations; forms of graduate student training in fieldwork; and an investment in maintaining a fieldwork epistemology based on radical otherness.

1. Global and Local Strategies

1. During and after the meeting, Yach not only spoke about global tobacco control; he also practiced it: the Lake Tahoe conference was followed by a lengthy three-hour press conference at the California DHS in Sacramento, at which Yach unveiled the WHO's "Don't Be Duped" media campaign that was broadcast via satellite hookup to hundreds of stations around the world. The WHO was scheduled to launch the campaign simultaneously in fourteen countries, and incorporated some of California's antismoking advertising, including the famous parody of Marlboro ads featuring two cowboys, one confessing to the other, "Bob, I've got emphysema." The billboard was adopted as the poster for World No Tobacco Day's official poster, whose theme was "entertainment," for the year 2000. For the purposes of the campaign, the poster was later altered to read, "Bob, I've got cancer," to broaden its international appeal. Interestingly, the adoption of the California billboard met with dissent by the international representative of the consortium of pharmaceutical conglomerates for tobacco control and by French advertising executives, who weren't convinced that it would work in British and French contexts.

2. In California the principal alternative political venue was the state referendum, a reform mechanism introduced in 1910 by California Republican Progressives and ratified by voters the following year. It allowed voters to bypass special interests dominating the legislature (Traynor and Glantz 1996, 544). Not only progressives but also conservative populists have availed themselves of this mechanism. This was amply demonstrated by the passage of Proposition 13 in 1978, which rolled back annual local residential real estate taxes to 1 percent, costing local treasuries $51 billion over the next five years and thus effectively defunding primary and secondary education in California (Stocker 1991; Sears and Citrin 1985). In many eyes, the passage of Proposition 13 marked the beginning of the neoliberal revolution in the U.S.

3. A study one year later showed continuing strong public support for the Berkeley ordinance and speculated that the presence of a large, professional middle class was decisive (Edwards 1978). The use of local ordinances proved to be spectacularly successful, and throughout the 1980s in California and elsewhere local ordinances spread rapidly (as of 1986, 117 ordinances had been passed in California, 246 nationally; Glantz and Balbach 2000, 39). The popularity of these ordinances would soon constitute a "California" phenomenon whose roots in the Minnesota and Arizona legislation were forgotten by the larger public.

4. In 1985 Minnesota was the first state in the nation to devote cigarette tax revenue to smoking control—7¢ per pack, of which 1¢ was allotted to a tobacco education campaign that included paid advertising (Minnesota health official 1997).

5. For a detailed account of the uneven commitment of voluntary health organizations and nonsmokers' rights groups to Proposition 99 and the ensuing legislation, see Haile et al. 1987; Breslow and Johnson 1993, 590; Traynor and Glantz 1996, 557–558. In 1992 the California Medical Association and the California Association of Public Hospitals and Health Systems would lend support to the diversion of funds earmarked for tobacco control by Gov. Pete Wilson (backed by the tobacco industry) and declared illegal by a state judge (Balbach and Glantz 1998, 399–400).

6. The neoliberal emphasis on the fiscal costs of smoking would have a future life in the Tobacco Control Program's reports to the California legislature and in local ordi-

nance campaigns but not in the statewide health promotion campaigns proper. However, other neoliberal issues of governance and recruitment of populations as responsible subjects would figure prominently.

7. Benjamin Hooks Jr., executive director of the NAACP in the late 1980s, remarked on industry funding of African American organizations: "If black leaders ask for funding from the tobacco industry, we are accused of 'selling out.' Whites get billions of dollars from these same companies and for some unknown reason they're not viewed as a sellout. So why do they single out black groups? To me, it's another form of racism" (Williams 1987, 8). For African Americans, complexity lies in the fact that on the one hand the cultivation of tobacco was one of the driving forces behind their enslavement in the New World, but on the other it became also a financial means to escape slavery and, later, to achieve economic sustenance for as farmers, laborers, and production workers and, later still in the twentieth century, as managers (even as small farmers suffered greatly as the industry reorganized). Moreover, the industry was the first in the U.S. to use middle-class African Americans in its ads and at the end of the century remained the largest national employer of African Americans (Robinson and Headen 1999).

8. The unspoken thought in some quarters was that, at least in the case of women, they would be served through the networks (Health promotion researcher B 2000). Still, the fate of largest group of female smokers—white women—was left apparently unaddressed.

9. See also U.S. Dept. of Health and Human Services 1980; American Cancer Society n.d.; American Lung Association 1988. When Proposition 99 was filed, smoking data, which were submitted with a letter to legislative analysts and the California Department of Finance, claimed even greater disparities: males, 32 percent; females, 29 percent; whites, 29 percent; Hispanics, 30 percent; blacks, 44 percent; and Asians, blank ("Health Consequences" 1987). The ties between matters of social class, gender, and race and these health concerns are deep and pervasive in the U.S. tobacco control discourse. What tends to be forgotten is that these ties have also shifted over time. Early on, the very first U.S. surgeon general's report on smoking noted that among men in the U.S., "white collar, professional, managerial, and technical professions contain fewer smokers than craftsmen, salespersons, laborers," while the correlation between smoking and education was not definitive, and smoking rates between blacks and whites were about the same, with more heavy smokers among whites (U.S. Dept. of Health, Education, and Welfare 1964, 363–364). By the 1980s a changed picture emerged in the research literature. The 1979 surgeon general's study reported that among women, white-collar workers had a higher incidence of smoking than blue-collar workers (U.S. Dept. of Health, Education, and Welfare 1979, 18–16); ten years later, the surgeon general presented a now-familiar epidemiological picture in which smoking "remains higher among blacks, blue-collar workers and less educated persons than in the overall population" and newer smokers are being recruited among minority groups, the less educated, the economically disadvantaged, and women (U.S. Dept. of Health and Human Services 1989, i–vii).

10. A private poll conducted by the coalition seemed to confirm the director's fears: only 39 percent of blacks interviewed declared themselves in favor of Proposition 99, compared with 56 percent of Hispanics and 63 percent of Asians (Holm and Rund 1988).

11. Reformers of the Progressive Era (1890–1920) were generally less concerned with improving overall social conditions than with promoting interventionist, expert-based public health programs with a strong moralizing tone that began "targeting" whole classes of citizens—namely, the poor and newly arrived immigrants—as vectors of con-

tagious diseases. This orientation was embodied in the many schools of public health created in the U.S. at that time, often with Rockefeller Foundation money (Brandt 1987; Fee 1994).

12. According to most accounts, the tobacco industry as well as the California Medical Association sponsored efforts to reallocate tobacco control education funding to medical services but failed (Traynor and Glantz 1996, 573–579; Glantz and Balbach 2000, 76–119).

13. Community empowerment's origins extend still further back in time to the nineteenth century, with the proliferation of private philanthropy's programs of self-help for the urban poor in Europe, Japan, and the U.S., and Progressive Era public health's preoccupation with the influx of immigrants into the U.S.

14. These five categories were first established by federal Office of Management and Budget Directive no. 15 in 1977.

15. In particular, the CDC was forced to drop its singular preoccupation with epidemiological surveillance and focus on problems of achieving behavioral change (Hornik 1997).

16. Knowles headed a research group in the 1970s on health care, many of whose papers were later published in a special issue of *Daedalus*, a publication of the U.S. National Academy of Sciences ("Doing Better and Feeling Worse" 1977).

17. As noted previously, it is also from the Alma Ata Declaration in 1978 that Richard Manoff, one of the earliest proponents of social marketing, dates the first steps taken in public health toward integrating social marketing approaches into its practices by virtue of the emphasis the document placed on community participation and on abandoning the old paternalism endemic to older programs (Manoff 1985, 3– 4).

18. The earliest definition of *social marketing*, published in 1971, is more spare and reads: "Social marketing is the design, implementation, and control of programs calculated to influence the acceptability of social ideas and involving considerations of product planning, pricing, communication, distribution, and marketing research" (Kotler and Zaltman 1971, 5).

19. Translated into public health terms, key components of social marketing are "price" (what the consumer must give up), "product" (the behavioral change desired by the program), "promotion" (appeals used), and "place" (channels of communication from mass media to interpersonal contacts) (National Cancer Institute 1989, 1; Kotler and Zaltman 1971, 4, 7; Altman and Piotrow 1980, 393–395; see also Wallack et al. 1993, 21–25).

20. The populist element is particularly strong. Early U.S. exponents of social marketing reveal a persistent self-promotional element, as in the title of an article, "The Outlandish Idea: How a Marketing Man Would Save India," published in 1968 (N. A. Martin 1968).

21. The promotion of a tangible product like condoms through social marketing techniques stood a much better chance of success than persuading citizens to abandon ingrained, unhealthy practices (such as smoking) for less palpable benefits. Marketers themselves had long expressed among themselves that marketing even of commercial products is a highly uncertain enterprise and that changing consumers' behavior is a difficult task (Schudson 1984, 86). And as one overview of social marketing put it, "Public health doesn't have the flexibility to adjust products and services to clients' interests and preferences. Commercial companies often drop a product when products prove unpopular" (Ling et al. 1992, 356).

22. Drawing on the international experience of the authors, the manual combined social marketing with health education models (stressing aspects of audiences' willingness to act) and mass communication studies of message transmission (Romano 2000). The authors emphasized accelerating shifts in social norms, setting public agendas, and reinforcing existing trends in healthy behavior. In other words, the goal was to intervene in already existent currents and build upon them. In most of the literature, social marketing media campaigns were understood to be only one of several elements of an integral campaign involving everything from advocacy and ordinances to policy measures.

Retrospective assessments of social marketing programs of the 1970s and 1980s acknowledged that full-fledged social marketing campaigns (including consumer research and evaluation) didn't make their appearance much before the mid-1980s and that their relative paucity hampered any accurate understanding of social marketing's potential in the realm of public health at that time (Altman and Piotrow 1980; Manoff 1985, 221; Novelli 1990).

23. Students of advertising like Michael Schudson drawing on the advertising world's own studies and tacit understandings are skeptical of the effectiveness of advertising in getting customers to purchase particular items. However, concerning social norms, lifestyles, and ways of living with respect to an American middle-class ideal, they do concede that advertising—and by extension, social marketing—can be influential but remain unconvinced whether public health social marketing can go so far as to set agendas unless they are supported by politicians (Schudson 1984, 1997).

24. U.S. researchers Lawrence Wallack and Russell Sciandra argue that media can indeed affect community norms and discussion in three ways: "First, media is often able to set the public agenda. It is often said that the media may not tell people what to think but do tell people what to think about. Second, the media confer status and legitimacy on various topics and points of view which are selected out for attention. The presence of a topic in the media helps to establish its importance. Third, the mass media can activate and stimulate public discussion. It also provides the framework for the discussion by roughly establishing the boundaries of legitimate discourse" (208).

25. Early social marketing programs had been carried out in several dozen countries by 1980, including India, Sri Lanka, Colombia, Bangladesh, Nepal, Thailand, Ghana, Indonesia, Philippines, and Mexico.

26. Manoff writes: "Problems may differ from country to country and within countries. But the social marketing of public health is a systems approach with universal application regardless of problem or local situation" (7). As one consultant in health communication and social marketing from the National Cancer Institute to the California DHS media campaign put it, "The model for health communication works anywhere. . . . You know, the whole process of developing solid communication works in any country" (Romano 2000).

27. Moreover, some worried that public health advocates ran the risk of throwing away large sums of money that could be better spent on "media advocacy," the tactical use of punctual advertising, press conferences, and "media events" that addressed power inequities and targeted specific law- and policymakers over a particular piece of legislation or administrative decision (Wallack and Sciandra 1991; Wallack et al. 1993; Health promotion researcher A 1997).

28. Even as committed a proponent of social marketing in the 1980s as Manoff openly worried that poorly conceived campaigns that don't take into account marketing's cus-

tomary bias toward better-off consumers may inadvertently widen the health gap between rich and poor (Manoff 1985, 92).

29. While underscoring these reservations, Wallack argued nonetheless for the usefulness of mass media in public health campaigns: "Although it may be the case that most in need are not helped by mass media campaigns, it is important to continue to disseminate information through well-designed campaigns based on mass communication theories, social marketing principles, formative research methods, and community organization techniques. These campaigns will help some, probably those who are relatively better off, but will not be sufficient to stimulate significant change. Nonetheless, public communication campaigns are an important part of a comprehensive strategy of health promotion" (49). William Novelli, president of Porter Novelli Associates (an international social marketing firm based in New York) and later director of the Campaign for Tobacco-Free Kids, a major U.S. anti-smoking NGO, strongly contested these reservations, claiming that ongoing social marketing campaigns intervened in social environments including government policy, public relations, and politics (Novelli, 1990, 3).

30. Goldstein notes the tautological nature of studies that purport to explain participation in the health movement and "healthy" activities by different social groups by basing their analysis on psychologizing, middle-class constructs such as "internal locus of control" and "self-sufficiency" (127).

31. The major intervention studies are the Stanford Three Community Study (1972–75; chronic heart disease [CHD] risk); the North Karelia Project, Finland (1972–82, CHD risk); the Australia North Coast Project (1978–80, smoking); the Stanford Five-City Study (1980–86, CHD risk); the Minnesota Heart Study (1980–86, CHD risk, smoking); the Pawtucket Heart Health Program (1981–93, cardiovascular disease risk); and the National Cancer Institute's Community Intervention Trials for Smoking Cessation (COMMIT, 1988–92, 11 paired communities).

32. This model has come in for criticism for overlooking the complexity of what epidemiologist David Burns has called the "sociocarcinogenesis" of smoking. In 1990 the NCI launched its follow-up study, the Americans Stop Smoking Intervention Trial (ASSIST), which included such policy measures and more extensive use of media. Both ASSIST and the California campaign were quite close to one another in terms of design and method, and each contributed to the refinement of the other (Pierce 1997; Burns 1997; Hornik 1997).

33. One of its authors, Rose-Mary Romano, served as an important consultant to the California DHS at the beginning of its campaign and was a member of its Media Advisory Committee (Romano 2000). The California media campaign would be part of a comprehensive approach based on the "multichannel" health promotion planning model that had three axes: target groups, channels (health care providers, work sites, schools, community networks, the community environment), and interventions (media, policy, and program services). Policy could include taxes and restricting smoking and access, while program services included cessation and prevention services (TEOC 1991, 10–11).

34. Thus, for example, Nielsen Market Research conducted its U.S. TV market research entirely in English addressing primarily white audiences until 1997, when it added Spanish-language polling (see Cerone 1994; Weinstein 1995).

35. For the purposes of generating meaningful data for the anti-smoking program, Native Americans, Pacific Islanders, and Asian Americans are often lumped together under various rubrics or simply absent altogether in studies.

36. Indicative of the problem in California may be the fact that in a major collection of articles on California tobacco control in diverse communities, there are no contributions by African American network members (Forst 1999).

37. Other early members included Jenny Cook (American Cancer Society); Perry Dyke, M.D. (Department of Education); Kenneth Kizer, M.D. (Director, DHS); David E. Hays-Bautista, Ph.D. (University of California, Los Angeles); Leroy S. Naman (businessman); Paul Torrens, M.D., M.P.H. (University of California, Berkeley); Kristina Sermersheim (Health Care Employees, SEIC); and Curtis Weidner, M.D. (local health department). See Bal 1990.

38. The persistent problems of oversight and evaluation mechanisms eventually led some anti-smoking advocates to recommend the independent funding for TEROC and evaluation and their full autonomy from the TCS and the DHS (Kleinschmidt 2000).

39. Other members included Lester Breslow, M.D. (University of California, Los Angeles); David Burns, M.D. (University of California, San Diego, and former general science editor of the surgeon general's 1986 report on passive smoking); Dorothy Rice, Sc.D. (University of California, San Francisco); Richard Lainard (Tobacco Products Liability Group); Michael Cummings, M.D. (Roswald Park Cancer Institute); Thomas Houston, M.D. (American Medical Association); David Sweeny (Canada); Lawrence Wallack, Ph.D. (University of California, Berkeley); Gregory Connolly, M.D. (Massachusetts Department of Public Health); Larry Gruden, Ph.D. (Tobacco Control Research Fund, Office of the President, University of California); and the health director of Alameda County, California.

40. Original members were Lester Breslow, M.D. (University of California, Los Angeles); Sharon Muraoka (American Cancer Society); Peggy Toy (Black Leadership Initiative on Cancer); Curtis Weichmer, M.D. (El Dorado County Health Department); Rose-Mary Romano (CDC); Kirsty Flynn (California Chamber of Commerce); Kathleen Harty, M.Ed. (Minnesota Health Department); J. W. Farquhar, M.D. (Stanford University, former director of the Stanford Five City Study); and Robert W. Denningston (substance abuse expert). See Bal 1989. Later they were joined by Robert Robinson, Dr.P.H., who had helped lead the successful African American protest in 1990 against the planned marketing of R. J. Reynolds' Uptown brand, the first cigarette brand to target a specific racial or ethnic community in the U.S.

2. The Dynamics of Collaboration and Community Input in the Media Campaign

1. The most notorious example was the cosponsorship by Philip Morris with the National Archives of a national tour of an exhibit of the 200–year-old Bill of Rights in 1990–91, during which the cigarette manufacturer gave away two million copies of the document.

2. These reformulated rights are the explicit theme of a series of radio ads released in the early 1990s by the DHS.

3. In point of fact, just who the executives are modeled after is not altogether clear. At first they seem to be a synthesis of Southern autocrats and Mafiosi. But close inspection of the spot reveals that they do not sport the black suits, mustaches, and cigars of Hollywood Mafiosi, nor are they identifiably Italian American. Yet they are clearly parasitical

on larger society, to which they owe no allegiance and to which they are not accountable. One colleague, a specialist on U.S. popular culture, who viewed the ad remarked that such a social position comes uncomfortably close to one traditionally assigned Jewish businessmen by standard anti-Semitic discourse. On the association between tobacco and Jews in Western cultures, see Gilman 2003.

4. This process of identification of California tobacco control with the media campaign even extended to nonsmokers' rights activists, who advocated local over statewide action, and it put them in the unusual position of defending a program that, while complementary to their own efforts, was centralized and top-down.

5. For one evaluator, the lack of observance of professional boundaries was evidenced by the fact that at least one published report was cosigned by the evaluators and five officials of the DHS responsible for the program (see Popham et al. 1993).

6. The original plan was for the University of California, San Diego, to conduct an eighteen-region survey every three years and to assist the DHS, which would conduct a smaller survey every year for statewide estimates only.

7. This view of the limits of academic evaluation is sometimes cited by tobacco control researchers themselves. One senior scientist remarked in an interview, "The problem that academics have is that we want to know exactly everything about it, and by the time we know exactly everything about it, it is too late" (Burns 1997).

8. It must be said that the report contained other results revealing weaknesses in the school program, ads targeting youth, as well as citizens' perceptions of the tobacco industry. Also, indicators of teenagers' likelihood of experimenting with cigarettes had gone up sharply. Moreover, since 1993, while the consumption of tobacco per smoker has continued to decline, the rate of decline in prevalence rates has been slightly lower than the national rate of decline. Finally, the report's results reflected an adjustment of the overall drop in prevalence rates to take into account the influx of immigrants into California during the 1990s whose smoking rates were generally lower than those of the resident population.

9. It would appear that an early recommendation to do research on industry targeting of minorities, women, and youth was not pursued initially by the Tobacco Control Section (Minnesota health official 1997).

10. The joint ethnic networks organized two conferences, "United against Tobacco Use," in 1994 and 1996, and in 1998 they convened a multi-ethnic tobacco education youth summit, "Tomorrow's Leaders Planning Today," on tobacco advertising and representations in films (Hong and Yu 1999, 57–58). And, as noted in my introduction, in November 1999 they organized a joint press conference denouncing Philip Morris' Virginia Slims "Find Your Own Voice" campaign aimed at their communities.

11. TEOC's first report in 1991 called attention to the scant prevalence data available on Hispanics/Latinos, Native Americans, and Asians/Pacific Islanders (TEOC 1991, 57–58). The initial 1990 California Tobacco Survey used in the report was conducted by John Pierce and David Burns at the University of California, San Diego, in two languages (English and Spanish). Although they did look at Alaskan Natives/Native Americans, Pacific Islanders, and different Asians (Chinese, Filipino, Vietnamese, Japanese, Korean, and Laotian communities), they had difficulty obtaining reasonable numbers, partly because of the digital dial method (about whose limits they had warned the DHS) and the prohibitively costly nature of such an undertaking. To arrive at meaningful data, they thus collapsed population categories together and listed Hispanic, white, black, Asian/Pacific Islander, and other in their detailed tables in the appendices and

Hispanic, white, black, and Asian/other in the body of their conclusions (Burns and Pierce 1992; Pierce 2002a, 2002b). Three years later, in 1994, the first overall evaluation of the Tobacco Control Program conducted by a team led by Pierce used the same dual classification in its appendices and conclusions (Pierce et al. 1994). Whereas the second overall evaluation released in 1998 by Pierce had adopted the even more simplified rubric "Asian" (with "other" dropping out) based on the California Tobacco Survey that itself continued to use the rubric "Asian/Pacific Islander" and "other" (Pierce et al. 1998, 2–16), TEROC employed the older category "Asian/other" in its 2000–03 report based on the DHS's in-house English-only California Adult Tobacco Survey (TEROC 2000, 14). As of the year 2000 there were still no distinct rubrics for Pacific Islanders or Native Americans. The state of the data sets reminds us that when it comes to aggregating and particularizing populations and communities, the necessity of articulating "diversity within diversity" outstrips dominant methods and current commitments of resources. For recent attempts to review ethnic/racial categories, see Root 2001; U.S. Dept. of Health and Human Services 1998, 7.

12. In fact, William Novelli, a leading social marketer, states that "micro" or "guerilla" marketing was already practiced in social-sector campaigns that used nontraditional media and nonmedia venues such as the Stanford Disease Prevention Program and the Pawtucket Heart Health Program (Novelli 1990, 4), a claim disputed by researchers in the Stanford study (Pierce 2002a).

13. This commitment to a statewide approach seemed to stem in part from the researchers' common assessment of the COMMIT intervention studies. They faulted COMMIT's research design for using what little media it did employ in its health campaign only at the local level, which effectively meant print media alone (Wallack and Sciandra 1991; Burns 1997).

14. As initial buys but not on a per capita basis compared with traditional TV markets, where the overall purchase price is quite high but presumably the audience is huge.

15. This may have seemed analogous to observers to the DHS's apparent reluctance to use Proposition 99 monies earmarked for indigent health care to start up new health services in the black community.

16. The original request for proposals for the media campaign stipulated that spots would be made with the ability to have "local" logos attached other than the standard one of the DHS. Apparently local logos were never used.

3. The Campaign against Secondhand Smoke

1. Several televised spots targeting the discomfort and hazard of secondhand smoke were actually featured in the earliest U.S. anti-smoking public service announcements in the late 1960s and early 1970s; they warned against the potentially lethal effects of tobacco smoke on loved ones and friends. Until the California campaign, few or none were produced since that time.

2. According to the 1998 evaluation report, in California children's exposure to ETS dropped from 29 percent in 1992 to 12 percent in 1996; workplace exposure of adults declined from 29 percent in 1990 to 13 percent in 1996 (Pierce 1998, 2/2–2/4).

3. One of the stipulations of the Master Settlement Agreement was for the tobacco companies to remove all outdoor advertising from billboards immediately and to replace the tobacco ads with anti-smoking billboards for the remaining duration of the rental

lease. Most of those spaces were filled by Californian counteradvertising (California health official A 1999).

4. For a consideration of current links between concepts of ecology and those of public health, see Young and Whitehead (1993).

5. On the rise of probabilistic thinking and risk analysis in the late nineteenth century and its applications in policy circles in the twentieth century, see Castel 1981, 1991; Nelkin 1985; Ewald 1986, 1991; Hacking 1990; O'Malley 1996, 1999.

6. On the concept of virtual witnessing as a tool for producing truth for and eliciting consent by absent (scientific) observers, see Shapin and Schaffer 1985, 60.

7. As Jan Zita Grover puts it in her article on "keywords" of the AIDS pandemic, in media, journalistic, and public health discourse "the 'general population' is virtuously going about its business, which is not pleasure seeking" (Grover 1989, 23).

8. For press reports on these developments, see Lee 1994; Verhovek 1996; Hicks 1994; Stolberg 1994b.

9. While the 1986 surgeon general's report and the 1992 Environmental Protection Agency report on ETS were crucial documents that generated wide publicity concerning the hazards of secondhand smoke (U.S. Dept. of Health and Human Services 1986; EPA 1992), it was the Glantz and Parmley study published in *Circulation* that spectacularly confirmed numbers derived several years earlier by A. Judson Wells (Wells 1988) and revolutionized the morbidity and mortality calculus for ETS in the public health community. After alcoholism and smoking, they declared passive smoke the third-highest cause of (premature) death in the U.S. In their review of eleven previous studies, the authors found that ETS accounted for ten times as many cases of heart disease as for lung cancer. Consequently, according to the study, nonsmokers' exposure to secondhand smoke increased their relative risk for death from heart disease by 30 percent, and ETS caused 37,000 annual deaths from heart disease in the U.S. and 15,700 deaths from all kinds of cancer, for a total of 52,700 deaths per year (Glantz and Parmley 1991). In 1995 Glantz and Parmley released a second study that detailed the biological mechanisms of ETS's effects on the coronary system (Roan 1995). The wide array of illnesses caused by ETS were recapitulated in the California EPA's study of all previously released ETS studies in 1997; it also listed spontaneous abortions/miscarriages as a probable effect of ETS (Morain 1997).

10. A later study published in 2000 tended to refute other claims concerning the incidence of breast cancer related to ETS; see Recer 2000.

11. It is interesting to note that sociologist Paul Starr, author of the standard history of U.S. medicine (Starr 1982) and one of the architects of the stillborn Clinton Health Plan (Starr 1994; White House Domestic Policy Council 1993), steers clear of any extended consideration of conventional public health discourse as a social practice that produces middle-class culture and bodies in opposition to unhealthy "others" (Starr 1982, 180–197). See also Conrad and Schneider 1980; Armstrong 1993; Crawford 1994; Stoler 1995.

12. Older internal threats to the stability of the (white) middle-class household have been, starting in the nineteenth century, hysterical women (mothers, daughters, sisters, aunts, and emancipated women, generally), masturbating children, servants, and effeminate men (homosexuals). In the 1990s in the U.S., one must add high school dropouts, pregnant runaways, lesbians, bisexual husbands (carriers of HIV), and deadbeat dads (divorced men who default on child support payments).

13. In fact, it would appear that one of the reasons creative departments in ad agencies are reluctant to incorporate extensive research in their work is the fear that the data would dictate the creation of family-based ads and little else (Schudson 1984, 57).

14. This ad was apparently a joint production of the DHS, the ad agency, the American Cancer Society, and the Pan-American Health Organization (PAHO) (California health official B 1998).

15. This does not mean that pregnant women, mothers, and other women who smoke aren't also targeted by DHS ads; they are, but not so much as vectors of secondhand smoke. Most often, they either threaten the fetus directly by abusing their bodies as smokers (in low-birthweight ads) or, as corrupting role models, pass the smoking habit on to their children. (There are similar ads targeting husbands and fathers as poor role models.) In the U.S. during the 1980s and 1990s pregnant women who smoked were not usually labeled "child abusers" but rather criminally negligent mothers on par with crack users; women who were accused of child abuse tended to be teachers, babysitters, or child day care workers—in other words, suspect outsiders and thus unworthy parental surrogates (see Nathan and Snedeker 1995).

16. For a provocative analysis of the intersection racial and gender stereotypes of Asian Americans with questions of citizenship and national "cultures" in legal arguments, see Volpp 1994.

17. Since its initial release in 1994, the billboard has reappeared in various venues, including as the cover photo for TEROC's 2000–03 report (TEROC 2000). Appearing after this ad and far fewer in number were billboards that symmetrically reverse the gender roles by depicting a black boy as a victim of his mother's smoking ("He has his daddy's eyes and his momma's lungs"). As such, they would attenuate the gender narrative that I have analyzed but preserve the ethnic and racial one.

Meanwhile, in 1996, radio versions of both billboard ads were broadcast in which the nonsmoking spouse assumes the mantle of public health admonition. For example, the radio equivalent of the billboard we have analyzed reads this way:

> ***Mother***: "When Nia was born, she was so beautiful, everybody told me she has her momma's eyes. But it's her daddy she really takes after. And, you know, she gets more and more like him every day. I had to take Nia to the doctor. She was always sniffling, wheezing, and coughing. But I thought that was normal for a child. Well, it's not. The doctor told me that Nia has asthma and was suffering from chronic bronchitis. He said that Nia was so sick, because she inhales smoke from her daddy's cigarettes. They call it secondhand smoke."
>
> ***Male voice-over***: "If you smoke around your kids, you're putting them in great danger. Secondhand smoke can cause serious respiratory problems in children, some of them fatal, as well as increased allergic reactions, ear infections, and flu symptoms."
>
> ***Mother***: "My daughter may have my eyes; now she's got her daddy's lungs."
>
> ***Male voice-over***: "Secondhand smoke kills. For your family's sake, don't smoke."

The following year, as if in response to criticisms of the billboards' punitive style, the DHS released a new TV spot in 1997 titled, "Better." It featured a young African American man holding his five-year-old daughter in his arms with the tag: "California. We know better."

18. A later Spanish-language ad released in 1997 and titled "Gravesite" stages a similar melodrama. It observes a widower with his two children at the grave of their mother, who died from breathing his cigarette smoke. However, the spot remains strictly a third-

person narrative: as the father breaks down and cries, a male voice-over recounts his story.

19. In the years that followed, other anti–secondhand smoke ads appeared featuring white men. A thirty-second TV spot features a cartoonish father obliviously smoking and laughing as he watches TV while his baby son on the floor spells out with his letter blocks the names of various illnesses that ETS can provoke in children: bronchitis, asthma, and, finally, SIDS. Most famous is the billboard of a man and a woman dressed in formal eveningwear in which the man queries the woman, "Mind if I smoke?" to which she replies, "Care if I die?" As aggressive as these may be, they don't approach the nasty tone of other ads featuring men of color.

20. Interestingly, this inverts the discourse of Philip Morris' "Find Your Own Voice" print ad campaign examined previously, which articulated questions of community identity, Americanness, and assimilation with, on the one hand, whiteness and smoking for viewers of color and, on the other, an exoticized nonwhiteness and smoking for viewers of European descent.

21. An early wave of DHS spots deglamorizing smoking among youth (the so-called Clifford ads) was apparently even less successful; they received uniformly low ratings by DHS officials; the contracted ad agency; and a committee of public health officials, U.S. media professionals, and health promotion researchers brought together by the health research organization Isosphere in 1997 to develop new approaches to underage smoking (Fievelson and Rabin 1997).

22. One widely cited example is the town of Sharon, which banned smoking completely in indoor and outdoor public spaces.

23. Other anti-ETS ads feature the same cross-cutting visuals with a well-known child actor ("Kids") and a twenty-five-year-old youth ("Warning") who each speak facing the camera; in both cases the flat, public health lecture tone dominates the less arresting mise-en-scène. A perhaps more powerful anti-ETS ad ("Baby Monitor") depicts a monitor that emits the cries of a baby as a voice-over explains: "Every year 300,000 babies get sick from secondhand smoke." As an unidentified finger switches off the monitor, the voice continues, "But the tobacco industry doesn't want to hear it." It garnered great interest on the part of representatives from other countries when it was presented in an international forum (Connolly 1997).

24. On the x-ray as symbol of twentieth-century modern medicine in the U.S., see Cartwright 1995.

25. Ultimately, I think that the "Unborn" ad still suffers from the colorless tone of public health discourse, while the California split-screen ad's authoritative voice-over is shot through with irony and sarcasm less typical of PSAs.

26. In interviews it was never suggested that these protests were simply orchestrated by the tobacco industry.

4. Revising Late Modernity

1. Here, modernity is understood as the perpetually evolving process of capitalist development and democratic government.

2. As far as I've been able to determine, only one early PSA aired between 1967 and 1970 made the analogy between industrial pollution and cigarette smoke.

3. More specifically, all but two of the smokers are middle-aged (and of those, all but one are white); the two youths are black and male.

4. As spokesman of the Canadian Cancer Society, he became the "Cancer Prevention Man."

5. The failure of the youth components of most anti-smoking campaigns was not widely discussed at that time, even if among anti-smoking advocates it has been acknowledged as the tobacco industry's preferred approach to smoking for precisely its weakness. See, for example, Stanton Glantz's article on the "youth trap" of youth prevention programs that the tobacco industry generally supported (Glantz 1996).

6. This section and the conclusion have benefited from exchanges with Robert Robinson, whose suggestions are here gratefully acknowledged.

7. Still, *Pulp Fiction*'s most stylized representation of smoking was restricted to the publicity poster featuring the crime boss's moll Mia Wallace (played by Uma Thurman) in her retro haircut and clothes, facing the camera, provocatively smoking a cigarette on a bed.

8. Not even the heinous neo-Nazis in Singleton's *Higher Learning* (1995) indulge in cigarettes.

5. France

1. Interestingly, Malraux's photo has its own dynamic history. According to Gisèle Freund, it was in response to Malraux's desire for a photo that would enhance a macho image of strength that she suggested that he place a cigarette between his lips ("Timbre Malraux censuré" 1996). The static icon of the famous writer photo turns out to be part and parcel of a conscious process of self-production. The controversy was also picked up by the U.S. press; see "Sanitizing History" 1996, which focused on the propaganda uses to which the U.S. government has put stamps.

2. Still, the French opponents of anti-smoking measures framed their stance as not only "French" but also implicitly part of a universal resistance against tobacco control compromised by American particularism (read puritanism).

3. Much the same could be said in particular for science, technology, and medicine at the end of the nineteenth century. According to Japan studies scholar Tessa Morris-Suzuki, citing the work of Naoki Sakai, as "Western" science and medicine circulated ever more widely in late-nineteenth-century Japan, so did its local practitioners become acutely aware of their "Japanese" ways of practicing them (Morris-Suzuki 1994).

4. The *loi Evin* was passed on 10 January 1991 and implemented in stages: the restrictions on smoking in public areas began 1 November 1992 and the ban on advertising and promotion on 1 January 1993.

5. Most recently, in the fall of 2002, U.S. comedian Robin Williams treated audiences in an HBO special to the portrait of a chain-smoking middle-aged French woman, stylishly but callously indulging in her habit and catty commentary ("Robin Williams" 2002).

6. Interestingly, the author of the design was a U.S. artist who was chosen after an international competition (Le Net 1997).

7. Simone Veil was an influential minister who had sponsored the final legislation legalizing abortion in France in 1974, which bears her name. The anti-smoking campaign

also deployed eighteen radio PSAs and 100 billboards as well as distributing 500,000 posters and 1.5 million brochures (Le Net 1981, 76–77).

8. Tubiana has had an interesting international career, which began as a student of the effects of nuclear radiation under Ernest O. Lawrence at the University of California Radiation Laboratory at Berkeley in the 1930s. Later, as an medical intern in France, he was presented by his supervisor with a rare tumor, which he was assured that he would never see again: lung cancer.

9. Goodman points out that it is inaccurate to claim that the adoption of cigarettes was directly related to economic development.

10. It has to be added that not only did French smokers smoke fewer cigarettes but more dark tobacco, they adopted filter cigarettes later than did U.S. smokers and, unlike Americans, tended to smoke each cigarette down to the butt.

11. On the crucial question of smoking rates of the unemployed (which were reportedly as high as 60 percent in the mid-1990s), the surveys could not provide details, for they lumped retirees together with those who were out of work as "inactive" ("*sans activité*").

12. In France under the Fifth Republic, the legislative process entails several steps. To begin with, it is initiated not by the Parliament but by the executive branch. There is the core legislation and then a second bill that lays out its implementation (the latter is vaguely similar in function to the supplemental U.S. legislation that authorizes funding for a newly passed bill).

13. Thus, the Medical Assistance Act (1893) made departments and communes responsible for indigent health care; the 1902 law made the reduction of disease a goal of French administrators but contained little in the way of funding; the national anti-tuberculosis and anti-cancer campaigns were joint initiatives with local committees; and when the national insurance law, which targeted low-income employees, was passed in 1930, the program's management was handed over to autonomous private regional and departmental-based organizations, and it preserved liberal medicine's guiding principles (fee for service, free choice of physicians) that are still respected today.

14. On the CNAM board sit representatives of trade unions, employers' associations, and insurance companies, along with health professionals and government officials.

15. The report called for the launching of a public information campaign urging citizens to use condoms; the creation of a National AIDS Research Agency and National AIDS Council; the revamping of the CFES by endowing it with a scientific advisory board and a long-term plan; suppression of all forms of discrimination against people with AIDS; and full access to health care by patients. The campaign would take as its inspiration campaigns conducted by other nations and would avoid standard public health language.

16. The authors were Gérard Dubois (professor of public health), Claude Got (heart specialist, professor of pathology), François Grémy (professor of medical statistics), Albert Hirsch (pulmonary specialist, professor of medicine), and Maurice Tubiana (professor of bioradiology and honorary director of the Gustave Roussy Cancer Research Institute, Paris). "*Tabagisme*" is a term difficult to translate into English. Derived from "*tabac*" (tobacco), it signifies at once excessive chronic tobacco use and health-related disorders stemming from abusive consumption of tobacco (*Le Petit Robert* 1978). In tobacco control literature, depending on the context, it is commonly translated as "tobacco," "tobacco abuse," or "tobacco-related illnesses." In French there is no equivalent of the English word "tobacco control"; the phrase "*anti-tabac*" ("anti-tobacco") comes closest.

17. See chapter 1 of the present book; also Beaglehole and Bonita 1997, 215–217.

18. The WHO definition reads: "Health is a state of complete mental, physical, and social well being and not merely the absence of disease or infirmity" (cited in Beaglehole and Bonita 1997, 3).

19. In the late 1980s France had only one school of public health, the École nationale de santé publique, located in Rennes (compared with more than forty in the U.S. and three in Belgium), and its primary function was the training of future health managers and bureaucrats (*L'Action politique* 1989, 33).

20. Tariffs were lifted progressively beginning in 1977, and the Council of State (Conseil d'État) struck down the government practice of fixing prices of imported cigarettes in 1992 ("Le Conseil d'État" 1992).

21. Gitanes, whose elegant motifs of Spain and a dancer recalled the early origins of cigarettes and Merimée and Bizet's myth of Carmen, the exotic Spanish bohemian. In fact, the two brands' logos together (Gitanes and Gauloises) neatly encapsulate the story of Carmen: the gypsy and her soldier lover. On contemporary French advertising and nationalism, see Duncan 2000; Kidd 2000.

22. The French anti-alcohol movement was never wedded to a prohibitionist platform. In France, a wine-growing country for thousands of years, the principal target was hard liquor, most famously absinthe (68 percent alcohol), which was banned in 1915.

23. Donzelot points out that social assistance to those citizens frequently in trouble always contained paternalistic and repressive aspects directed at particular populations deemed potentially disruptive. The social protection of citizens was based on the model case of workers facing temporary unemployment from time to time because of the evolution of the social division of labor. Services for both groups included everything from medical care and subsidized housing to retraining of workers and psychological services. The overarching practice was one of socializing of risk and of creating a life for citizens beyond the tyranny of the marketplace and economic necessity. According to Donzelot, the intractable long-term unemployment and jobless youth caused the difference between social assistance and social protection to blur. At the same time, the infrastructure of services and housing began to degrade, as "assistance" of permanently unemployed and never employed absorbed a larger share of resources. "Social solidarity" (whose roots go back to the "solidarism" of early-twentieth-century social reformer Léon Bourgeois) with its paternalist overtones of charity replaced the notion of "social protection" that embraced not simply the indigent but all citizens (Donzelot 1992, 19–21).

24. The relative freedom with which physical sex is evoked in the public media sphere by a state agency—almost unthinkable in countries like the U.S. and Japan—stems from the French HIV/AIDS campaigns, limited though they may have been.

25. An important but little-known fact is that by 1931 France had a higher proportion of foreigners than any other nation (Shain 1999, 220 n21).

26. The legacy of Nazi scientific experimentation on human subjects is fundamental also for understanding the continuing dominance in France of Freudian-based psychology as opposed to the methods and approaches of motivational and decision-science studies developed by experimental psychology in North America and elsewhere involving human subjects. Much early marketing practice and human resource literature and many important models of health promotion derive from this field. Most recently, the legacy of Nazi science has also underwritten French state reticence about experiments involving manipulation of the human genome (Rabinow 1999, 7–23, 89–111).

27. In France, statistical analysis enjoyed very low status as a research field until

the 1970s (which was reflected in the fact that women constituted the majority of medical statisticians).

28. The surveys were conducted by private marketing research organizations such as the BVA Institute and Démonscopie (now integrated into the international research company Synovate, part of the Aegis Group) in collaboration with INSERM and were published in its annual technical report *Baromètre santé* and in its occasional publication for general audiences, *La Santé de l'homme*.

29. Interestingly, they did not attribute these weaknesses per se to the reliance by a semi-public agency on private-sector marketing organizations.

30. In France the legal age for smoking is fifteen (which is also the age of consent for both sexes), whereas it is eighteen in the U.S. and twenty in Japan. So the phrase "children who smoke" evokes different images of young smokers in the three countries.

31. "*Je veux que les gens sachent ce qui m'est arrivé en fumant. La chose la plus importante que chacun peut faire est de ne pas fumer. Nous devons dénoncer les publicités qui encouragent les enfants à fumer. J'ai incité les gens à fumer et maintenant j'essaie de leur expliquer ce qui m'est arrivé.*"

32. Even more aggressively, but also more anonymously, the CNCT reprised in another poster repulsive pictures of decimated lungs of dead smokers, more common in Anglo-American anti-smoking campaigns, with the caption, "Tobacco-sickened lungs" ("*Poumons tabagiques*"). There was no tag identifying the either original sponsor or the CNCT.

33. For example, the CNCT's leading lawyer, Francis Caballero, holds a degree in internal law from Harvard University; one of the presidents of the governing board, Gérard Dubois, has a doctorate in public health from Johns Hopkins University; Maurice Tubiana trained in radiology at the University of California, Berkeley; and Pascal Mélihan-Chenein actively contemplated obtaining a doctorate in public health from Johns Hopkins University.

34. One advocate cited Stanton Glantz of the University of California, San Francisco, and former director of Americans for Nonsmokers' Rights (ANR) as an example of a someone who, in a typically "American" fashion, transcended these differences in his dual role as medical researcher and grassroots activist.

35. On the other hand, some felt that the CFES's focus on youth initiation was pointless as long as teenagers saw that smoking by adults remained a tolerated or legitimate activity, and targeting adolescents was a politically easy way out that ruffled few feathers as compared with campaigning for adult cessation and sharply raising taxes.

36. As an indication of the expansion of public-private collaboration in public health, by the late 1990s social marketing of health issues had become a large enough market in France and Europe that many large ad agencies had created entire departments devoted to it (Maruani 1999).

6. Japan

1. I thank Takashi Fujitani for this reference and for clarifying this debate for me.

2. The account that follows is based mainly on interviews conducted in English or Japanese (with a translator) and on English-language print sources. It necessarily reflects the research tools at my disposal and the fieldwork situation in which I found myself: namely, that my limited Japanese reduced my access to archival and print material and

to non-English and non-French speakers. As a result, in this chapter I exercise greater circumspection in my analysis and conclusions than in the preceding ones.

3. The conference was held 14–18 November 1999. Major funders of the conference besides the WHO, UNICEF, and the United Nations included the Ministry of Health and Welfare; the Hyogo Prefecture government; the Rockefeller Foundation; and Pharmacia Upjohn, the Swedish-based multinational pharmaceutical company that markets nicotine gum and patches, later bought out by U.S.-based Pfizer.

4. Since 1900, twenty has been the legal age of for consuming tobacco and alcohol. It is also the legal age of adulthood. Smoking rates for men in Japan hit their peak in the 1960s (in 1966, 82.7 percent, according to Japan Tobacco) and have declined slowly ever since, whereas smoking-related deaths have risen. Per capita consumption (including light, filtered, and mentholated tobacco) remained among the highest worldwide—peaking at 3,400 cigarettes around 1975–80 to drop to 2,400 cigarettes in 1998 ("33.63 Million Japanese Smoke" 1999; "Anti-Smoking Plan Comes under Fire" 2000; National Survey 1999; Corrao et al. 2000, 394; WHO 1997, 451).

5. Other organizations included the National Council on Women and Development (Ghana), Vietnam Women's Union, Women's Legal Action Watch (Mauritius), Third Wave Fund for Young Women (Sri Lanka), Consumers Association of Penang (Malaysia), Pan American Health Organization (PAHO), SIS International (Philippines), National Alliance of Women's Organizations (India), and the Conference of Non-Governmental Organizations.

6. Although announced as a topic, "youth" were the subject of no panel. No reasons were given by conference organizers.

7. Judith Mackay (Asian Tobacco Control Consultancy), the conference cochair, publicly conceded during the conference that putting women and children on the agenda together did in fact run the danger of infantilizing women.

8. There was also an identity rhetorically defined by age ("youth"), but it was never operational in the conference proceedings for reasons that were never explained. At best, "children" were invoked but most always as the offspring and traditional responsibility of women.

9. Other articles in the same vein periodically appeared thereafter. The month following the Eleventh World Conference on Tobacco or Health held in August 2000 in Chicago, the English edition of *Yomiuri shinbun*, one of Japan's largest dailies, featured a report on the WHO's TFI with the title, "Japan—A Smokers' Paradise under Siege" (Tamura 2000); and in 2001 the *Chicago Tribune* ran an article with the title that identified smoking with an entire country and nation-state, "The Land of the Rising Smoke" ("The Land of the Rising Smoke" 2001). Still other Japanese and American examples include the following: the English-language *Asahi Evening News* reprinted an Associated Press article with the title, "Big Tobacco Still Calls the Shots in Japan" ("Big Tobacco" 1997); also, the title of a *Los Angeles Times* article summarized the contradictory position of the Japanese government that promotes both health and smoking: "Burning the Cigarette at Both Ends" (Efron 1997). The intractable nature of the government's ties to tobacco interests is evoked in a piece published in the English edition of *Mainichi shinbun*, "Where There's Smoke, There's Bureaucracy" (Hadfield 1997). The theme persisted in reporting into the twenty-first century: a *New York Times* headline ran, "Japan and Tobacco Revenue: Leader Faces Difficult Choice" (Strom 2001).

10. For a useful overview of reciprocal images of U.S. and Japanese cultures in the advertising of both countries, see O'Barr 1994.

11. This was driven home for me within a week of my arrival in Japan in June 1998, when the International House of Japan in Tokyo (founded by the Rockefeller Foundation in 1952) hosted a colloquium chaired by Donald Ritchie on a bilingual book titled *Japan: Made in USA / Warawareru nihonjin* [literally, *Japanese who are made fun of*]. Edited by Japanese journalists working in the U.S., it contained interviews and articles by Japanese and U.S. journalists and historians sharply critical of reporting on Japanese daily life in the *New York Times* that exoticized the Japanese, especially women (*Japan: Made in USA* 1998).

12. A word on Japanese and English terms: Many Japanese organizations have adopted both Japanese and English names, but since there exist few exact linguistic equivalents in both languages, I have given the Japanese names followed by a literal English translation to give a sense of this gap. In Japanese as in French there is no word for the current English phrase, "tobacco control." "Tobacco" and "cigarettes" are both *tabako* in Japanese. "Public health" and the Japanese *kōshū eisei* aren't exact equivalents of each other either (see below). Also, the phrases *kin'en* and *ken'en* are usually rendered in English by the single phrase "anti-smoking," whereas the literal translation would be respectively, "stop smoking" or "nonsmoking" and "hate smoking." For the purposes of this book and following common practice in English-language international publications on public health, Japanese personal names are given in Western order.

13. Enjoying little in the way of official state sponsorship, advocates complained in interviews of having to work in isolation from professional colleagues and mainstream society.

14. In 1999 the design was altered to feature a smiling face of an indeterminate age with the tag, "Smoking Rights. (Clean) Air Belongs to Everyone" ("*Kitsuenken. Kūki wa minna no mono*").

15. British American Tobacco Co. was a major player after its founding in 1902 to serve as the overseas operations of both the American Tobacco Co. and the Imperial Tobacco Co. Although registered in London, it remained under American control as late as 1923 (Cochran 1986).

16. They cite several reasons for this that are relevant to public health and tobacco control. First, Japan had a welfare system in which families, communities, and local companies provided welfare services but over which the central government maintained control. Second, the mainstays of the late-twentieth-century provisions were the corps of unpaid local welfare commissioners (200,000 by the 1980s) drawn from community notables plus four million other volunteers and the system of national health insurance funded by employers and workers based on liberal medicine's principle of fee for service that kept levels of direct state assistance low (as in Germany). As a result, in the 1980s government social spending on family allowance and public assistance was among the lowest in the industrialized world (S. J. Anderson 1993, 109–113, 133). As one author put it, even as progressives and conservatives struggled over the issue, the welfare debate in Japan did not neatly pit advocates of self-reliance and private welfare against proponents of economic planning and state-sponsored welfare (Hiwatari 1993, 40–41).

17. For an interesting critique of assessments of Japan in terms of convergence theory, see S. J. Anderson 1993, 22–23. He argues that in this regard both Japan and the U.S. constitute exceptions to the claims of this theory.

18. Until the Meiji period, the much-cited Confucian values were largely restricted to the samurai class alone (6 percent of the population) and enjoyed little in the way of a following among the rest of the population.

19. In the following years, the government set up a Sanitary Bureau (Eisei kyoku) and later a Central Sanitary Board (Chūō eiseikai), and ordered prefectural governments to establish sanitary departments (*eiseika*); when it came to enforcement, that was literally a matter assigned to the police.

20. In conjunction with the power of the Ministry of Finance within the bureaucracy, the industry exerted considerable influence over politicians in the Diet, whose actions in the 1990s were limited to budgeting ¥2 million (US$20,000) annually for the WHO-sponsored World No Tobacco Day (Sato 1999, 585–586).

21. Anti-smoking activists in the U.S. later obtained copies of the letter through the U.S. Freedom of Information Act.

22. The original editorial bore the less dramatic title, "Selling Cigarettes in Asia" ("Selling Cigarettes" 1997), and it was reprinted in toto in the *International Herald Tribune* (of which the *New York Times* was part owner) under the title, "Cigarettes for Asia" ("Cigarettes for Asia" 1997).

23. In the post-1945 settlement, older policies dating from the colonial era encouraging immigration of colonial populations to Japan, the extension of Japanese citizenship to Koreans living in the empire, and intermarriage between Japanese and Koreans were rescinded in favor of strict citizenship laws based on blood ties and the myth of Japan as an ethnically homogeneous society (Fujitani 2004).

24. Unlike in the U.S., responsibility for public health is explicitly assigned to the State in the 1947 Constitution: "All people shall have the right to maintain the minimum standards of wholesome and cultured living. In all spheres of life, the State shall use its endeavors for the promotion and extension of social welfare and security, and of public health." (Constitution of Japan [1947], Art. 25).

25. This is a problem shared with many activist groups whose members are mostly middle-aged. What new members Women's Action on Smoking does get join after spending some time abroad, especially in the U.S. (Nakano 1999).

26. *Passhibu sumōkingu (judōteki kitsuen) to wa? Honnin no ishi to wa mukankei ni tanin no kitsuen ni yotte tabako no kemuri o suwasareru koto to iimasu. Tabako no kemuri ni fukumarete iru nikochin, tāru, issankatanso, bentsupiren nado no yūgai busshitsu wa, nenmaku o shigekishi, kekkan no shūshū, shinpakusu no zōka, sara ni ketsuatsu kōyō, kettō no zōka nado o hikiokoshimasu. Ken'en ken kakuritsu mezasu hitobito no kai. Jūdōteki kitsuen no gai kara hikitsuenka o mamoru koto o uttaemasu.*

27. *Ippon sueba jūyoppun sanjū byō ichinichi nijuppon ijō tabako o sū hito wa ippon sū goto nijūyoppun sanjū byō jumyō ga chijimu (Tōkai Daigaku. Sasamoto Hiroshi kyōju) to iwarete imasu. Mata, kyōshinshō, shinkinkōsoku nado no shinshikkan ni kakaru ritsu wa non sumōkā no nibai ijō, toku ni shunkanshi wa gobai chikaku ni mo hane agarimasu.*

28. One runner-up poster for the 1996–97 campaign features a mother who, while holding a crying baby, glances ruefully at the lit cigarette in her left hand as its smoke envelopes her child. The poster reads, "Don't envelop the baby in smoke, bundle it up in love" ("*Kemuri de kurunaide, aide kurunde*").

29. Transliterated into romaji:

> ***Right vertical title***: *Koromubia raito*
> ***Horizontal title***: *Koromubia toppu to tomo ni manzai konbi o kumi zenkokuteki ni ichijidai o fuubi shita. 1991 nen kōtōgan de seitai o setsujo. Raitosan wa genzai, mizukara gankanja o hagemasu tame no borantia katsudō o tsuzukete imasu.*

Speaker: *Seitai o ushinatte kara kotoba no subarashisa o kanjimasu. Sore de iitai koto ga hitosu aru. Sore wa tabako yamemashō. Dame.*

Voice-over: *Sekaijiu ga tabako o yameru hi.*

Full-screen title: *Genzai, Nihonjin no shiin nanba wan wa gan desu. Soshite, kōtōgan no 95% ijō ga tabako ni gen'in ga aru koto ga jisshō sarete imasu.*

30. But when U.S. companies began to market an entirely new product—fast food—in France, French advocates didn't hesitate to denounce it as an imperial invasion. However, regarding tobacco control measures examined previously, from the very outset opponents of the 1992 French *loi Evin* didn't hesitate to characterize the regulations as the inappropriate importation of U.S. methods into France.

Works Cited

"33.63 Million Japanese Smoke." 1999. *Japan Times*, 12 September.
L'Action politique dans le domaine de la santé publique et de la prevention. 1989. Paris: n.p.
Advertising executive A. 1997. Interview. 20 May.
Advertising executive B. 2000. Interview. 9 August.
Advocacy Institute. 1988. "Smoking Control Media Advocacy." Paper presented to Mass Communications and Public Health: Complexities and Conflict. Rancho Mirage, Calif.
Altman, Diana L., and Phyllis T. Piotrow. 1980. "Social Marketing: Does It Work?" *Population Reports* 8 (21): 393–434.
American Academy of Otolaryngology. n.d. "Poisoning Your Children: The Perils of Secondhand Smoke." Educational video.
American Cancer Society. n.d. *Report on Smoking and Health among Minorities*. Proposition 99 Campaign Files. "Minority Issues in Tobacco Use, 1988." Mss. 94–52, Carton 1, Folder 26. Tobacco Control Archives. Archives and Special Collections. University of California, San Francisco.
———. 1988. California Division. "Yes on 99: Promotional Guide." Proposition 99 Campaign Files. "American Cancer Society, 1987–88." Mss. 94–52, Carton 1, Folder 33. Tobacco Control Archives. Archives and Special Collections. University of California, San Francisco.
American Lung Association and Minority Outreach Initiative, Office of Minority Health, U.S. Public Health Service. 1988. "Challenge: The Impact of Respiratory Disease on Minority Populations." First Leadership Forum on Respiratory Health in Minority Populations. Los Angeles, Calif., 15–17 June. Proposition 99 Campaign Files. "Minority Issues in Tobacco Use, 1988." Mss. 94–52, Carton 1, Folder 26. Tobacco Control Archives. Archives and Special Collections. University of California, San Francisco.
Americans for Nonsmokers' Rights. 1996; rev. 1998. "Recipe for a Smokefree Society." 15 October.
Anderson, Benedict. 1991. *Imagined Communities: Reflections of the Origin and Spread of Nationalism*. London: Verso.
Anderson, Stephen J. 1993. *Welfare Policy and Politics in Japan: Beyond the Developmental State*. New York: Paragon.
"Anti-Smoke Torch Flickers." 1990. *Advertising Age*, 16 April. 1.
"Anti-Smoking Campaign Angers Blacks." 1990. *Sacramento Observer*, 12–18 April, G-3.
"Anti-Smoking Plan Comes under Fire." 2000. *Asahi shinbun*, 4 February.
APITEN (Asian/Pacific Islander Tobacco Education Network). 1999a. "Minority Communities Outraged at New Virginia Slims' Ad Campaign Targeting Minority Women." Press release. 18 October.

——. 1999b. "California's Ethnic Leaders Blast Philip Morris for Targeting Their Communities." Press release. 9 November.

Appadurai, Arjun. 1996. *Modernity at Large: Cultural Dimensions of Globalization.* Minneapolis: University of Minnesota Press.

Armstrong, David. 1993. "Public Health and the Fabrication of Identity." *Sociology* 27 (3): 393–410.

——. 1995. "The Rise of Surveillance Medicine," *Sociology of Health and Illness* 17 (3): 393–404.

"As the U.S. Clamps Down, Europe Is Smokers' Paradise." 1997. Reuters, 20 June.

Atkin, Charles, and Elaine Bratic Arkin. 1990. "Issues and Initiatives in Communicating Health Information." In *Mass Communication and Public Health: Conflicts and Complexities,* ed. Charles Atkin and Lawrence Wallack, 13–40. Newbury Park, Calif.: Sage.

Baezconde-Garbanati, Lourdes. 2000. Interview. 3 January.

Baezconde-Garbanati, Lourdes, James A. Garbanati, Cecilia Portugal, Radon Lopez Rodriguez, Aurora Flores, and Holly Sisneros. 1999. "Entering a New Era: Strategies of the Hispanic/Latino Tobacco Education Network for Organizing and Mobilizing Hispanic Communities." In *Planning and Implementing Effective Tobacco Education and Prevention Programs,* ed. Martin L. Forst, 112–131. Springfield, Ill.: Charles C. Thomas.

Bal, Dileep G. 1989. Letter to Members of Media Advisory Committee. 13 November. Lester Breslow Tobacco Control Papers. "Media Advisory Committee, 1989." Mss. 94–50, Carton 1, Folder 23. Tobacco Control Archives. Archives and Special Collections. University of California, San Francisco.

——. 1990. Letter to Lester Breslow. 6 September. Lester Breslow Tobacco Control Papers. "Tobacco Education Oversight Committee (TEOC), 1989–92." Mss. 94–50, Carton 1, Folder 28. Tobacco Control Archives. Archives and Special Collections. University of California, San Francisco.

Bal, D.[ileep] G., K. W. Kizer, P. G. Felten, H. N. Mozar, and D. Niemeyer. 1990. "Reducing Tobacco Consumption in California: Development of a Statewide Anti-Tobacco Use Campaign." *JAMA* 264 (12): 1570–1574.

Balbach, Edith, and Stanton A. Glantz. 1998. "Tobacco Control Advocates Must Demand High-Quality Media Campaign: The California Experience." *Tobacco Control* 7 (4): 397–408.

Balibar, Etienne. 2001. *Nous, citoyens d'Europe? Les frontières, l'état, le peuple.* Paris: La Découverte.

Barry, Andrew, Thomas Osborne, and Nikolas Rose, eds. 1993. "Liberalism, Neo-Liberalism, and Governmentality." Special Issue of *Economy and Society* 22 (3). Repr. 1996 as *Foucault and Political Reason: Liberalism, Neo-Liberalism, and Rationalities of Government.* London: UCL Press.

Baudier, François, Danielle Grizeau, Jacques Draussin, and Bernadette Roussille. n.d. "1976–1996: Vingt ans de prévention du tabagisme en France." Paris: Comité français d'éducation pour la santé.

Baudrillard, Jean. 1988. *America.* Trans. Chris Turner. London: Verso.

Beaglehole, Robert, and Ruth Bonita. 1997. *Public Health at the Crossroads: Achievements and Prospects.* Cambridge: Cambridge University Press.

Beattie, Alan. 1991. "Knowledge and Control in Health Promotion: A Test Case for Social Policy and Social Theory." In *The Sociology of the Health Service*, ed. Jonathan Gabe, Michael Calnan, and Michael Bury, 162–202. London and New York: Routledge.
Beck, Ulrich. 1992. *The Risk Society: Towards a New Modernity*. Trans. Mark Ritter. London: Sage.
Becker, Marshall H. 1986. "The Tyranny of Health Promotion." *Public Health Reviews* 14 (1): 15–23.
Bellah, Robert Neely. 1985. *Tokugawa Religion: The Cultural Roots of Modern Japan*. New York: Free Press.
Berkeley City Council. 1977. Berkeley Smoking Pollution Control Act of 1977 (Ordinance no. 4083). Peter Hanauer Tobacco Control Papers. Mss. 94–44, Carton 2. Tobacco Control Archives. Archives and Special Collections. University of California, San Francisco.
Biener, Lois, Jeffrey E. Harris, and William Hamilton. 2000. "Impact of the Massachusetts Tobacco Control Programme: Population Trend Analysis." *British Medical Journal* 321 (5): 351–354.
"Big Tobacco Still Calls the Shots in Japan." 1997. *Asahi Evening News*, 8 July.
Björkman, James Warner, and Christa Altenstetter. 1997. "Globalized Concepts and Local Practice." In *Health Policy Reform, National Variations, and Globalization*, ed. James Warner Björkman and Christa Altenstetter, 1–16. New York: Saint Martin's.
"Black Newspaper Publishers Meet." 1988. *Sacramento Observer*, 3–9 November, G-2.
"Blue Haze over California." 1997. *Los Angeles Times*, Home ed., 27 March, A1.
Blum, Alan. 2002. "Resistance to Identity Categorization in France." In *Census and Identity: The Politics of Race, Ethnicity, and Language in National Censuses*, ed. David I Kertzer and Dominique Arel, 121–147. Cambridge: Cambridge University Press.
Boltanski, Luc, and Laurent Thévenot. 1991. *De la justification: les économies de la grandeur*. Paris: Gallimard.
Boucher, Philippe. 1990. "Hygiènisme." *Le Monde*. 30 June.
Boucher, Philippe. 1996. Interview. 1 July.
———. 1997. Interview. 15 September.
Bowker, Geoffrey C., and Susan Leigh Star. 1999. *Sorting Things Out: Classification and Its Consequences*. Cambridge, Mass.: MIT Press.
Braithwaite, Ronald L., Cynthia Bianchi, and Sandra E. Taylor. 1994. "Ethnographic Approach to Community Organization and Health Empowerment." *Health Education Quarterly* (Fall): 407–416.
Braithwaite, Ronald L., Fredrick Murphy, Ngina Lythcott, and Daniel S. Blumenthal. 1989. "Community Organization and Development for Health Promotion within an Urban Black Community: A Conceptual Model." *Health Education* 20 (5): 56–60.
Brandt, Allan M. 1987. *No Magic Bullet: A Social History of Venereal Disease in the United States since 1880*. New York: Oxford.
———. 1990. "The Cigarette, Risk, and American Culture." *Daedalus* (Fall): 155–176.

Braxton, Greg. 1990. "Stations Smoke-Out Anti-Smoking Ads," *Los Angeles Times*, Home ed., 24 April, F2.

Breslow, Lester. 1990. "A Health Promotion Primer for the 1990s." *Health Affairs* 9 (2): 6–21.

Breslow, Lester, and Michael Johnson. 1993. "California's Proposition 99 on Tobacco, and Its Impact." *Annual Review of Public Health* 14: 585–604.

Breslow, Lester, and Thomas Tai-Seale. 1996. "An Experience with Health Promotion in the Inner City." *American Journal of Health Promotion* 10 (3): 185–188.

Brody, Jane E. 1996. "In Smoking, Study Sees Risk of Causes of Breast Cancer." *New York Times*, National ed., 5 May.

Brundtland, Gro Harlem. 1999. Keynote address. WHO International Conference on Tobacco and Health. Kobe, Japan. 15 November. Geneva: WHO.

Buell, Frederick. 1994. *National Culture and the New Global System*. Baltimore: Johns Hopkins University Press.

Burgess, John. 1987. "In Japan, More Power to the Puffers." *Washington Post*, 3 April, B1.

———. 1997. Interview. 7 April.

Burns, David M., and John P. Pierce. 1992. *Tobacco Use in California 1990–91*. Sacramento: California Department of Health Services.

Caballero, Francis. 1999. Interview. 27 September.

Calavita, Kitty. 1996. "The New Politics of Immigration: 'Balanced-Budget Conservatism' and the Symbolism of Proposition 187." *Social Problems* 43 (3): 284–305.

California health official A. 1996. Interview. 4 November.

———. 1999. Interview. 9 September.

California health official B. 1998. Interview. 9 June.

———. 1999. Interview. 26 August.

California Statutes 1989. Chapters 1328–1332 (California Assembly Bill 75).

Carlile, Lonny E. 1998. "The Politics of Administrative Reform." In *Is Japan Really Changing Its Ways? Regulatory Reform and the Japanese Economy*, ed. Lonny E. Carlile and Mark C. Tilton, 76–110. Washington: Brookings Press.

Carlyon, William H. 1984. "Reflections: Disease Prevention/Health Promotion—Bridging the Gap to Wellness," *Health Values: Achieving High-Level Wellness* 8 (3): 27–30.

Carol, Julia. 2000. Interview. 5 January.

Cartwright, Lisa. 1995. *Screening the Body: Tracing Medicine's Visual Culture*. Minneapolis: University of Minnesota Press.

Castel, Robert. 1981. *La gestion des risques: de l'anti-psychiatrie à l'après-psychanalyse*. Paris: Minuit.

———. 1991. "From Dangerousness to Risk." In *The Foucault Effect: Studies in Governmentality*, ed. Graham Burchell, Colin Gordon, and Peter Miller, 281–298. Chicago: University of Chicago Press.

CDC (Centers for Disease Control and Prevention). 1995–98. *Media Resource Book for Tobacco Control* (EL/1041). 2 vols. Washington: Government Printing Office.

———. 1998. National Health Interview Surveys, 1965–95.

Cerone, Daniel. 1994. "KMEX Finds Discrepancy in World Cup TV Ratings; Tele-

vision: The Spanish-Language Station Says Latinos Are Underrepresented in Nielsen's Audience Measurement." *Los Angeles Times,* 24 June, F-2.
Chatterjee, Partha. 1993. *The Nation and Its Fragments.* Princeton, N.J.: Princeton University Press.
Chavez, Lydia. 1998. *The Color Bind: California's Battle to End Affirmative Action.* Berkeley: University of California Press.
Chen, T. T., and A. E. Winder. 1990. "The Opium Wars Revisited as U.S. Forces Tobacco Exports in Asia." *American Journal of Public Health* 80 (6): 659–662.
Chilcote, Samuel D. Jr. 1990. Memo to members of the Executive Committee. Tobacco Institute. 11 April. Proposition 99 Campaign Advertising Documents. 3610.01. [Memo—California Ad Campaign and Tobacco Institute's efforts to thwart.] California/Minnesota Depository. Tobacco Control Archives. Archives and Special Collections. University of California, San Francisco.
"La Chine." 1996. France 3, 9 May.
"Cigarettes for Asia." 1997. *International Herald Tribune,* 11 September.
Cimons, Marlene. 2000. "State Programs Credited for Dip in Lung Cancer." *Los Angeles Times,* Home edition, 1 December, A1.
Coalition for a Healthy California. 1987. "Initiative Will Take On the Tobacco Goliath in a Fight for a Healthier California." Press release. 16 December.
———. 1988a. "The Tobacco Tax and Health Promotion Act of 1988." Proposition 99 Campaign Files. "American Lung Association of California, 1987–88." Mss. 94–52, Carton 1, Folder 36. Tobacco Control Archives. Archives and Special Collections. University of California, San Francisco.
———. 1988b. "The Tobacco Tax and Health Promotion Act of 1988. Questions and Answers." Proposition 99 Campaign Files. "American Lung Association of California, 1987–88." Mss. 94–52, Carton 1, Folder 36. Tobacco Control Archives. Archives and Special Collections. University of California, San Francisco.
———. 1988c. Proposition 99 Campaign Files. Audio and Video Tapes. "Coalition for a Healthy California. Five Radio Spots Tied. 8/25/88." Mss. 94–52, Carton 5, Video Tape no. 23. Tobacco Control Archives. Archives and Special Collections. University of California, San Francisco.
———. 1988d. Proposition 99 Campaign Files. Audio and Video Tapes. "Coalition for a Healthy California. Five Radio Spots Tied. 8/25/88." Mss. 94–52, Carton 5, Audio Tape no. 7. Tobacco Control Archives. Archives and Special Collections. University of California, San Francisco.
Cochran, Sherman. 1986. "Commercial Penetration and Economic Imperialism in China: An American Cigarette Company's Entrance into the Market." In *America's China Trade in Historical Perspective: The Chinese and American Performance,* ed. Ernest R. May and John K. Fairbank, 151–203. Cambridge, Mass.: Harvard University Press.
Coleman, Jennifer. 2000. "Anti-Tobacco Measures Lessen Cancer." AP Wire Report. 30 November.
Coleman, William. 1982. *Death Is a Social Disease: Public Health and Political Economy in Early Industrial France.* Madison: University of Wisconsin Press.
"Le Complot des marchands de fumée." 1999. *Libération,* 25–26 September.
Connolly, Gregory. 1988. "Tobacco and United States Trade Sanctions." In *Smok-*

ing and Health 1987: Proceedings of the Sixth World Conference on Smoking and Health, Tokyo, 9–12 November 1987, ed. Masakazu Aoki, Shigeru Hisamichi, and Suketami Tominaga, 351–354. Amsterdam: Elsevier.

———. 1997. Interview. 13 August.

"Le Conseil d'État interdit au gouvernement de fixer le prix des tabacs importés." 1992. *Le Monde,* 1–2 March, 7.

Coontz, Stephanie. 1988. *The Social Origins of Private Life*. London: Verso.

———. 1992. *The Way We Never Were: American Families and the Nostalgia Trap*. New York: Basic Books.

Cooper, Richard, and Brian E. Simmons. 1985. "Cigarette Smoking and Ill Health among Black Americans." *New York State Journal of Medicine* (July): 344–349.

Corbett, K., et al. 1991. "Process Evaluation in the Community Intervention Trial for Smoking Cessation (COMMIT)." *International Quarterly of Health Communication* 11 (3): 291–309.

Corliss, Mick. 1998. "Japanese Panel Fails to Come Up with New Recommendations." *Japan Times,* 1 September.

Corrao, Marla Ann, G. Emmanuel Guindon, Namita Sharma, and Dorna Fakhrabadi Shokoohi, eds. 2000. *Tobacco Control Country Profiles*. Atlanta: American Cancer Society.

Cox, Howard. 2000. *The Global Cigarette: Origins and Evolution of British-American Tobacco 1880–1945*. Oxford: Oxford University Press.

Crawford, Robert. 1980. "Healthism and the Medicalization of Everyday Life." *International Journal of Health Services* 10 (3): 365–388.

———. 1994. "The Boundaries of the Self and the Unhealthy Other: Reflections on Health, Culture, and AIDS." *Social Science and Medicine* 38 (10): 1347–1365.

Crimp, Douglas, ed. 1989. *AIDS: Cultural Analysis/Cultural Activism*. Cambridge, Mass.: MIT Press.

Cruikshank, Barbara. 1999. *The Will to Empower: Democratic Citizens and Other Subjects*. Ithaca, N.Y.: Cornell University Press.

Cunningham, Stuart, Gay Hawkins, Audrey Yue, Tina Nguyen, and John Sinclair. 2001. "Multicultural Broadcasting and Diasporic Video as Public Sphericules." In *Citizenship and Cultural Policy*, ed. Jeffrey Minson and Denise Meredyth. London: Sage.

Davis, R. M. 1987. "Current Trends in Cigarette Advertising and Marketing." *New England Journal of Medicine* 316 (12): 725–732.

Dean, Mitchell. 1999. *Governmentability: Power and Rule in Modern Society*. London: Sage.

———. 2002. "Liberal Government and Authoritarianism." *Economy and Society* 31 (1): 37–61.

"Declaration of Alma-Ata." 1978; 2000. Copenhagen: WHO Regional Office for Europe.

Deleuze, Gilles. 1995. "Postscript on Control Societies." In *Negotiations*, trans. Martin Joughin. New York: Columbia University Press.

Denoix, Pierre, Daniel Schwartz, et al. 1958. "Enquête française sur l'étiologie du cancer bronchite." *Bulletin de l'association française pour l'étude du cancer* 45: 3630–3643.

DHS (California Department of Health Services). 1995. "Toward a Tobacco Free California." l.
"Doing Better and Feeling Worse: Health in the United States." 1977. Special issue of *Daedalus* 106 (1).
"Doing Drugs in the Womb." 1993. Editorial. *Los Angeles Times*, 18 September, 37.
Donzelot, Jacques. 1979. *The Policing of Families*. Trans. Robert Hurley. New York: Pantheon.
——. 1984. *L'Invention du social: essai sur le déclin des passions politiques*. Paris: Fayard.
——. 1992. "Le Social du troisième type." In *Face à l'exclusion. Le modèle français*, ed. Jacques Donzelot, 15–39. Paris: Esprit.
Donzelot, Jacques, and Joël Roman. 1992. "Le Déplacement de la question sociale." In *Face à l'exclusion. Le modèle français*, ed. Jacques Donzelot, 5–11. Paris: Esprit.
Douglas, Mary. 1992. *Risk and Blame: Essays in Cultural Theory*. London and New York: Routledge.
Dubois, Gérard. 1999. Interview. 29 September.
Duncan, Alistair. 2000. "Advertising Culture in France: No Coca-Cola Please, We're French!" In *Contemporary French Cultural Studies*, ed. William Kidd and Sian Reynolds, 179–192. New York: Oxford University Press.
Durant la nuit 1992. TF1, 2 November.
Duyvendak, Jan Willem. 1995. *The Power of Politics: New Social Movements in France*. Boulder, Colo.: Westview.
Edelmann, Frédéric. 1993. "Histoire d'une chaîne de solidarité." *Le Monde*, 9 December, xviii–xix.
Edwards, Eleanor. 1978. "No-Smoking: Sweet Smell of Success." *Berkeley Independent-Gazette*, 22 July. Peter Hanauer Tobacco Control Papers. "Legislation." Mss. 94–44, Carton 2, Folder 48. Tobacco Control Archives. Archives and Special Collections. University of California, San Francisco.
Efron, Sonni. 1997. "Burning the Cigarette at Both Ends." *Los Angeles Times*, 17 May, A2.
Elder, John. 1997. Interview. 29 July.
Ellis, Virginia. 1993. "Whiff of State Health Chief's Past Makes Smoking Foes Gag." *Los Angeles Times*, Home ed., 23 November, A3.
Encyclopédie du tabac et des fumeurs. 1975. Paris: Le Temps.
"L'Enfer du tabac." 1994. *Envoyé spécial*. Antenne 2, 27 October.
Enquête sur la prevalence du tabagisme en décembre 2003. 2004. Paris: Institut national de prévention et d'éducation à la santé and IPSOS.
EPA (U.S. Environmental Protection Agency). 1992. *Respiratory Health Effects of Passive Smoking: Lung Cancer and Other Disorders*. Washington: Government Printing Office.
Epstein, Steven. 1996. *Impure Science: AIDS, Activism, and the Politics of Knowledge*. Berkeley and London: University of California Press.
Erickson, Allan C., Jeffrey W. McKenna, and Rose-Mary Romano. 1990. "Past Lessons and New Uses of the Mass Media in Reducing Tobacco Consumption." *Public Health Reports* 105 (3): 239–245.

"États-Unis: la fin du mégot." 1987. *Le Monde*. 21 March.
"États-Unis: les fumeurs en enfer." 1991. *Le Figaro*. 2 June.
Ethnic network director A. 2000. Interview. 5 January.
Ethnic network director B. 2000. Interview. 5 January.
Ethnic network director C. 2000. Interview. 5 January.
Ethnic network director D. 2000. Interview. 3 January.
Evaluation Advisory Committee. 1994. Minutes. 25 and 26 February. Lester Breslow Tobacco Control Papers. "Tobacco Education Oversight Committee (TEOC), 1989–92." Mss. 94–50. Carton 1, Folder 7. Tobacco Control Archives. Archives and Special Collections. University of California, San Francisco.
Ewald, François. 1986. *L'État providence*. Paris: Grasset.
——. 1991. "Insurance and Risk." In *The Foucault Effect: Studies in Governmentality*, ed. Graham Burchell, Colin Gordon, and Peter Miller, 107–210. Chicago: University of Chicago Press.
"Exemple USA." 1989. TFI, 7 October.
FAIR (Federation for American Immigration Reform). 2001. "California: Current Population Survey Data." www.fairus.org/html/042cacps.htm.
Fassin, Eric. 1997. "Du Multiculturalisme à la discrimination" *Le Débat* 97: 131–136.
——. 1998. "Homosexualité et mariage aux États-Unis." *Actes de la recherche en sciences sociales* 125: 63–73.
——. 1999a. "The Purloined Letter: American Feminism in the French Mirror." *French Historical Studies* 22 (1): 113–138.
——. 1999b. "'Good to Think': The American Reference in French Discourses of Immigration and Ethnicity." In *Multicultural Questions*, ed. Christian Joppke and Steven Lukes, 224–241. Oxford: Oxford University Press.
"Faut-il interdire le tabac?" 1995. *J'y crois, j'y crois pas*. TF1, 30 November.
Featherstone, Mike, ed. 1990. *Global Culture: Nationalism, Globalization, and Modernity*. London and Thousand Oaks, Calif.: Sage.
Federal Trade Commission Cigarette Report for 2000. 2002. www.ftc.gov/os/2002/05/2002cigrpt.pdf (17 January 2005).
Fee, Elizabeth. 1994. "Public Health and the State: The United States." In *The History of Public Health and the Modern State*, ed. Dorothy Porter, 224–275. Amsterdam and Atlanta: Rodopi B. V.
"Femmes et tabac." 1998. Paris: Comité français d'éducation pour la santé and Le Planning Familial.
Fertas, Samica. 1999. Interview. 24 September.
Fievelson, Diana, and Steven Rabin. 1997. Isosphere. Interview. 17 December.
Folléa, Laurence. 1994. "Nouvelle croisade." *Le Monde*, 24 October.
——. 1997. Interview. 17 September.
Ford, Gene. 1992. *The French Paradox and Drinking for Health*. San Francisco: Wine Appreciation Guild.
Forst, Martin L., ed. 1999. *Planning and Implementing Effective Tobacco Education and Prevention Programs*. Springfield, Ill.: Charles C. Thomas.
Forsythe, Diana. 1995. "Ethics and Politics of Studying Up." Paper presented to the American Anthropological Association, Washington, D.C. December.
Foster, Catherine. 1990. "'Uptown' Cancellation: A Turning Point for Tobacco Ad-

vertising? (R. J. Reynolds Tobacco Co. Ends New Product's Test Marketing to Blacks)." *Christian Science Monitor*, 23 January, 7.
Foster, Robert. 2002. *Materializing the Nation: Commodities, Consumption, and Media in Papua, New Guinea*. Bloomington and Indianapolis: Indiana University Press.
Foucault, Michel. 1978. *An Introduction*. Vol. 1 of *The History of Sexuality*. Trans. Robert Hurley. New York: Viking.
———. 1981. "Omnes et Singulatim: Towards a Criticism of Political Reason." In *The Tanner Lectures on Human Values 1981*, vol. 2, ed. Starling M. McMurrin, 224–254. New York: Cambridge University Press.
———. 1985. *The Uses of Pleasure*. Vol. 3 of *The History of Sexuality*. Trans. Robert Hurley. New York: Random House.
Fox, Richard Wightman, and T.J. Jackson Lears, eds. 1983. *The Culture of Consumption: Critical Essays in American History, 1880–1980*. New York: Pantheon Books.
French AIDS activist. 1994. Private conversation.
Fujitani, Takashi. 1996. *Splendid Monarchy: Power and Pageantry in Modern Japan*. Berkeley and London: University of California Press.
———. 2004. Private conversation. 29 June.
Fukami, Akiko. 1987. "New Group Attacks Tobacco's Effects on Women." *Japan Times*, 1 February.
"Fumer, c'est pas ma nature!" 1991. Special issue on tobacco. *La Santé de l'homme*. Paris: CFES.
Garfield, Bob. 1990. "California's Anti-Smoking Ad Fans Flames of Racial Paranoia." *Advertising Age*. 16 April.
Garon, Sheldon. 1997. *Molding Japanese Minds: The State in Everyday Life*. Princeton, N.J.: Princeton University Press.
George, Susan. 1996. "Comment se fabriquer une idéologie." *Le Monde diplomatique*. August.
Giddens, Anthony. 1990. *The Consequences of Modernity*. Stanford, Calif.: Stanford University Press.
Gilman, Sander L. 1995. *Health and Illness: Images of Identity and Difference*. London: Reaktion Books.
———. 2003. *Jewish Frontiers: Essays on Bodies, Histories, and Identities*. New York: Palgrave Macmillan.
Glantz, Stanton A. 1996. "Preventing Tobacco Use—The Youth Access Trap." *American Journal of Public Health* 86 (2): 156–158.
———. 1997. Interview. 20 August.
———. 2000. Interview. 4 January.
Glantz, Stanton A., and Edith Balbach. 2000. *Tobacco War: Inside the California Battles*. Berkeley and London: University of California Press.
Glantz, Stanton A., John Slade, Lisa Bero, Peter Hanauer, and Deborah Barnes, eds. 1996. *The Cigarette Papers*. Berkeley and London: University of California Press.
Glantz, Stanton A., and William W. Parmley. 1991. "Passive Smoking and Heart Disease: Epidemiology, Physiology, and Biochemistry." *Circulation* 83 (1): 1–12.

Goldfarb, Brian. 2002. *Visual Pedagogy: Media Cultures In and Beyond the Classroom.* Durham, N.C., and London: Duke University Press.
Goldstein, Michael S. 1992. *The Health Movement: Promoting Fitness in America.* New York: Twayne.
Goodman, Jordan. 1993. *Tobacco in History: The Cultures of Dependence.* London and New York: Routledge.
Goodman, Roger. 1998. "The 'Japanese-style Welfare State' and the Delivery of Personal Social Services." In *The East Asian Welfare Model: Welfare Orientalism and the State,* ed. Roger Goodman, Gordon White, and Hock-ju Kun. London: Routledge.
Gordon, Colin. 1991. "Government Rationality: An Introduction." In *The Foucault Effect: Studies in Governmentality,* ed. Graham Burchell, Colin Gordon, and Peter Miller, 1–52. Chicago: University of Chicago Press.
Gordon, Diana. 1994. *The Return of the Dangerous Classes: Drug Prohibition and Policy Studies.* New York: W.W. Norton.
Got, Claude. 1989. *Rapport sur le sida.* Paris: Flammarion.
——. 1992. *La Santé.* Paris: Flammarion.
——. 1997. Interview. 18 September.
Goto, Yuichiro. 1999. Interview. 5 December.
"Government Releases Study on Smoking." 1987. *Mainichi Daily News,* 18 October.
Grace, Victoria M. 1991. "The Marketing of Empowerment and the Construction of the Health Consumer: A Critique of Health Promotion." *International Journal of Health Services* 21 (2): 329–343.
Graphismes et créations SEITA: années 30 40 50. 1996. Paris: SEITA.
Greco, Monica. 1993. "Psychosomatic Subjects and the 'Duty to Be Well': Personal Agency within Medical Rationality." *Economy and Society* 22 (3): 357–372.
Grewal, Inderpal, and Caren Kaplan. 1994. *Scattered Hegemonies: Postmodernity and Transnational Feminist Practices.* Minneapolis: University of Minnesota Press.
Grizeau, Danielle. n.d. "Bilan descampagnes anti-tabac en France (1976–1991)." Paris: Comité français d'éducation pour la santé. Service études, stratégie et coordination du réseau.
——. 1993. "Tabac." In *Baromètre santé 1992,* 94–113. Paris: Editions CFES.
——. 1997. Interview. 16 September.
Grizeau, Danielle, and Pierre Arwidson. 1997. "Tabac: consommation et régementation." In *Baromètre santé 95/96,* 175–202. Paris: Editions CFES.
Grizeau, Danielle, and François Baudier. 1995. "Tabac." In *Baromètre santé 93/94,* 58–81. Vanves: Comité français d'éducation pour la santé.
Grossberg, Lawrence. 1993. "Can Cultural Studies Find True Happiness in Communication?" *Journal of Communication* 43 (4): 89–97.
Grover, Jan Zita. 1989. "AIDS: Keywords." In *AIDS: Cultural Analysis/Cultural Activism,* ed. Douglas Crimp. Cambridge, Mass.: MIT Press.
Gupta, Akhil, and James Ferguson. 1992. "Beyond 'Culture': Space, Identity, and the Politics of Difference." *Cultural Anthropology* 7 (1): 6–23.
——. 1997. *Anthropological Locations: Boundaries and Grounds of a Field Science.* Berkeley: University of California.
Gusfield, Joseph R. 1993. "The Social Symbolism of Smoking and Health." In *Smok-*

ing Policy: Law, Politics, and Culture, ed. Robert L. Rabin and Stephen D. Sugarman, 49–68. New York: Oxford University Press.
Guttman, Nurit. 1997. "Ethical Dilemmas in Health Campaigns." *Health Communication* 9 (2): 155–190.
Hacking, Ian. 1990. *The Taming of Chance*. Cambridge and New York: Cambridge University Press.
———. 1991. "The Making and Molding of Child Abuse." *Critical Inquiry* 17: 253–288.
Hadfield, Peter. 1997. "Where There's Smoke, There's Bureaucracy." *Mainichi shinbun*, 15 March.
Haile, Robert, E. Richard Brown, and William J. McCarthy. Los Angeles Coastal Cities Unit, California Division, American Cancer Society. 1987. Letter to Sally Westbrook, Public Issues Committee, California Division, American Cancer Society, 11 August. Proposition 99 Campaign Files. "American Cancer Society 1987–88." Mss. 94–52, Carton 1, Folder 33. Tobacco Control Archives. Archives and Special Collections. University of California, San Francisco.
Hall, Stuart. 1991. "Old and New Identities, Old and New Ethnicities." In *Culture, Globalization, and the World System*, ed. Anthony King. London: Macmillan.
"Halte au tabac." 1989. *C'est pas juste*. France 3, 6 January.
Hamamoto, Darrell Y. 1994. *Monitored Peril: Asian Americans and the Politics of TV Representation*. Minneapolis: University of Minnesota Press.
Hamilton, William, and Lynne Harrold. 1996. "Independent Evaluation of the Massachusetts Tobacco Control Program." Abt Associates.
Hamlin, Christopher. 1994. "State Medicine in Britain." In *The History of Public Health and the Modern State*, ed. Dorothy Porter, 132–164. Amsterdam and Atlanta: Rodopi B. V.
Hanai, Kiroku. 1997. "Smoke Out Double Standards." *Japan Times*, 21 July.
———. 1998. "End the Smoker's Paradise." *Japan Times*, 23 March.
———. 2003. "Stub Out the Smoking Habit." *Japan Times*, 24 November.
———. 2004 "Common Sense Up in Flames." *Japan Times*, 24 January.
———. 2004. "Blowing Smoke on Tobacco." *Japan Times*, 24 May.
Harvey, David. 1990. *The Condition of Postmodernity: An Enquiry into Cultural Change*. London: Blackwell.
Hastings, Gerard, and Amanda Haywood. 1991. "Social Marketing and Communication in Health Promotion." *Health Promotion International* 6 (2): 135–145.
Hazan, Anna Russo, Helene Levens Lipton, and Stanton A. Glantz. 1994. "Popular Films Do Not Reflect Current Tobacco Use." *American Journal of Public Health* 84 (6): 998–1000.
"Health Consequences and Economic Costs of Tobacco Use in California and the U.S. Data Compiled by Colton Seale (444–8726) for the Planning and Conservation League and the American Lung Association." 1987. Mss. 94–52, Carton 4, Folder "Proposition 99 Initiative Language." Tobacco Control Archives. Archives and Special Collections. University of California, San Francisco.
"Health Ministry Sets Targets." 1999. *Japan Times*, 13 August.
Health promotion researcher A. 1997. Interview. 6 May.
Health promotion researcher B. 2000. Interview. 9 August.

Hicks, J. 1994. "Council Votes to Eliminate Most Smoking in Public Sites." *New York Times*, National ed., 22 December, B3.

Hinsberg, Pat. 1990a. "Warning: Anti-Smoking Ads Could Burn Rest of Industry." *Adweek*, 16 April, 1.

——. 1990b. "Will K/D/P's New Campaign Fire Up the Agency's Stature?" *Adweek*, 16 April, 1.

Hirayama, Takeshi. 1981. "Non-Smoking Wives of Heavy Smokers Have a Higher Risk of Lung Cancer: A Study from Japan." *British Medical Journal* 282 (6259): 183–185.

Hirsch, Albert. 1995. "Enacting Tobacco Control Policy in France." In *Tobacco and Health*, ed. Karen Slama, 135–138. Proceedings of the Ninth World Conference on Tobacco or Health. New York: Plenum.

——. 1997. Interview. 17 September.

Hirsch, Albert, and Serge Karsenty. 1992. *Le Prix de la fumée*. Paris: Odile Jacob.

Hiwatari, Nobuhiro. 1993. "Sustaining the Welfare State and International Competitiveness in Japan: The Welfare Reforms of the 1980s and the Political Economy." Tokyo: Institute of Social Science, University of Tokyo.

Holm, Paul, and Chuck Rund. 1988. "September Survey Results on Prop 99." Memo to Jack Nicholl. Charlton Research Company. Proposition 99 Campaign Files. Mss. 2000–04. Carton 1, Folder 2, "Prop. 99 Initiative." Tobacco Control Archives. Archives and Special Collections. University of California, San Francisco.

Hong, Betty and Joon-Hu Yu. 1999. "Asian and Pacific Islander Tobacco Control Network: A Statewide Partnership for the Wellness of Asian and Pacific Islander Communities." In *Planning and Implementing Effective Tobacco Education and Prevention Programs*, ed. Martin L. Forst, 51–68. Springfield, Ill.: Charles C. Thomas.

Hornik, Robert C. 1997. Interview. 31 July.

Hozumi, Tadao. 1999. Interview. 10 March.

"Innocent Victims of Smoking." 1991. Editorial. *Los Angeles Times*, Home ed., 30 June, M4.

"Interdiction tabac." 1991. TF1, 2 January.

Iwabuchi, Koichi. 2002. *Recentering Globalization: Popular Culture and Japanese Transnationalism*. Durham, N.C.: Duke University Press.

Jackson, Derrick Z. 1997. "African Americans and Big Tobacco." *Boston Globe*, 9 July, A19.

Japan: Made in USA / Warawareru nihonjin. 1998. Tokyo: Jipangu.

"Japan Tobacco Reacts against Government White Paper." 1987. *Daily Yomiuri*, 21 October.

"Le Japon: le dernier paradis des fumeurs?" 1994. *Le Monde*, 8 November.

"Les Jeunes et le tabac." 1994. Special issue of *La Santé de l'homme* 313 (Sept.–Oct.).

Johnston, William. 1995. *The Modern Epidemic: A History of Tuberculosis in Japan*. Cambridge, Mass.: Council on East Asian Studies.

Joseph, Isaac, Philippe Fritsch, and Alain Battegay. 1977. *Disciplines à domicile: l'édification de la famille*. Fontenay-sous-Bois: Recherches.

Kahn, Joel S. 1995. *Culture, Multiculture, Postculture*. London and Thousand Oaks, Calif.: Sage.

Karsenty, Serge. 1997. Interview. 17 September.

Karsenty, Serge, and Albert Hirsch. 1992. "Une lutte contre le tabagisme fondée sur l'expertise scientifique." In *La Lutte contre le tabagisme est-elle efficace?* ed. Karen Slama, Serge Karsenty, and Albert Hirsh, 197–202. INSERM and CFES. Paris: La Documentation française.

Kaufman, Nancy. 1999. Interview. 23 December.

Kearney, Michael. 1995. "The Local and the Global: The Anthropology of Globalization and Transnationalism." *Annual Review of Anthropology* 24 (4): 547–565.

Keye, Paul. 1993. "What Don't We Know and When Haven't We Known It?" Paper presented at School of Hygiene and Public Health, Johns Hopkins University, Baltimore, Md. 13 October .

———. 1999. Interview. 18 December.

Kidd, William. 2000. "Frenchness: Constructed and Reconstructed." In *Contemporary French Cultural Studies,* ed. William Kidd and Sian Reynolds, 154–162. New York: Oxford.

Klaus, Alisa, 1993. *Every Child a Lion: The Origins of Maternal and Infant Health Policy in the United States and France, 1890–1920.* Ithaca, N.Y.: Cornell University Press.

Klein, Naomi. 2000. "Culture Jamming: Ads under Attack." In *No Logo,* 279–309. New York: Picador.

Klein, Richard. 1993. *Cigarettes Are Sublime.* Durham, N.C.: Duke University Press.

———. 1997. "After the Preaching, the Lure of the Taboo." *New York Times,* National ed., 24 August.

Kleinschmidt, Kirk. 2000. "Rendez-vous with Kirk Kleinschmidt. (Rendez-vous no. 61)." Interview with Philippe Boucher. Globalink.org. 14 April.

Kluger, Richard. 1996. *Ashes to Ashes: America's Hundred-Year Cigarette War, the Public Health, and the Unabashed Triumph of Philip Morris.* New York: Knopf.

Knight, Peter, ed. 2002. *Conspiracy Nation: The Politics of Paranoia in Postwar America.* New York: New York University Press.

Knowles, John H. 1977a. "The Responsibility of the Individual." In "Doing Better and Feeling Worse: Health in the United States." Special issue of *Daedalus* 106 (1): 57–80.

———. 1977b. "Responsibility for Health." *Science* 198 (4322): 1103.

"Kobe Declaration." 1999. WHO International Conference on Tobacco and Health, Kobe, Japan. "Making a Difference to Tobacco and Health: Avoiding the Tobacco Epidemic in Women and Youth." Press Release (WHO/71). 18 November.

Koh, Howard K. 1996. "An Analysis of the Successful 1992 Massachusetts Tobacco Tax Initiative." *Tobacco Control* 5: 220–225.

Kotler, Philip. 1975. *Marketing for Nonprofit Organizations.* Englewood Cliffs, N.J.: Prentice Hall.

Kotler, Philip, and Gerald Zaltman. 1971. Social Marketing: An Approach to Planned Social Change." *Journal of Marketing* 35: 3–12.

Lamont, Michèle. 1992. *Money, Morals, and Manners: The Culture of the French and American Upper-Middle Class.* Chicago: University of Chicago Press.

"The Land of the Rising Smoke: Liberal Laws, Cheap Cigarettes—Even Patriotism—Encourage the Japanese to Keep Puffing Away." 2001. *Chicago Tribune,* 21 June.

Larner, Wendy. 1997. "The Legacy of the Social: Market Governance and the Consumer." *Economy and Society* 26 (3): 373–399.

Lash, Scott, and John Ury. 1994. *Economies of Signs and Space*. London: Sage.

Leclerc, Annie. 1979. *Au feu du jour*. Paris: Grasset.

Lee, D. 1994. "No More Butts about It; Statewide No-Smoking Law Comes In with New Year." *Los Angeles Times*, Home ed., 29 December, D1.

Lemaire, Jean-François. 1999. *Le Tabagisme*. 5th ed. Paris: Presses universitaires de France.

Le Net, Michel. 1981. *L'État annonceur*. Paris: Editions de l'organisation.

——. 1997. Interview. 18 September.

Levin, Mark A. 1997. "Smoke around the Rising Sun: An American Look at Tobacco Regulation in Japan." *Stanford Law and Policy Review* 99: 1–37.

Ling, Jack C, Barbara A. K. Franklin, Janis F. Lindsteadt, and Susan A. N. Gearon. 1992. "Social Marketing: Its Place in Public Health." *Annual Review of Public Health* 13: 341–362.

Lipset, Seymour Martin. 1996. *American Exceptionalism: The Double-Edged Sword*. New York: Norton.

La Loi relative à la lutte contre le tabagisme et l'alcoolisme. Rapport d'évaluation. By Guy Berger, Marie Mauffret, Anne-Chantal Rousseau-Gival, Catherine Zaidman. 1999. (Premier ministre, Commissariat général du plan, Conseil national de l'évaluation) Paris: La Documentation française.

Lupton, Deborah. 1995. *The Imperative of Health: Public Health and the Regulated Body*. London and Thousand Oaks, Calif.: Sage.

Luthra, Rashmi. 1998. "Communication in the Social Marketing of Contraceptives. A Case Study of the Bangladesh Project." Ph.D. dissertation, University of Wisconsin.

Lutte contre le tabagisme et communication en éducation pour la santé. Un cadre pour les actions nationales. 1996. Paris: Comité français d'éducation pour la santé.

Lynch, Michael, and Steve Woolgar, eds. 1990. *Representation in Scientific Practice*. Cambridge, Mass.: MIT Press.

Makino, Kenji. 1998. Interview. 20 November.

Manoff, Richard K. 1985. *Social Marketing: New Imperative for Public Health*. New York: Praeger.

Marmor, Theodore R. 1997. "Global Health Policy Reform: Mythology or Learning Opportunity." In *Health Policy Reform, National Variations, and Globalization*, ed. James Warner Björkman and Christa Altenstetter, 348–364. New York: St. Martin's.

Marrot, Bernard. 1995. *L'Administration de la santé en France*. Paris: L'Harmattan.

Martin, Brian. 1991. *Scientific Knowledge in Controversy: The Social Dynamics of the Fluoridation Debate*. Albany: State University of New York Press.

Martin, Carolyn. Chair of TEOC. 1992. Memo to Molly Coye, Director, Department of Health Services. 16 October. Lester Breslow Tobacco Control Papers. "Tobacco Education Oversight Committee (TEOC), 1989–92." Mss. 94–50. Carton 1, Folder 28. Tobacco Control Archives. Archives and Special Collections. University of California, San Francisco.

Martin, Emily. 1994. *Flexible Bodies: Tracking Immunity in American Culture from the Days of Polio to the Age of AIDS*. Boston: Beacon.
Martin, Nathaniel A. 1968. "The Outlandish Idea: How a Marketing Man Would Save India." *Marketing Communications* 297 (March): 54–60.
Maruani, Benoît. 1999. Interview. 29 September.
Maruchi, N., and M. Matsuda. 1991. "Provision and Financing of Health Care Services in Japan." In *Oxford Textbook of Public Health,* 2d ed., vol. 1, ed. Walter W. Holland, Roger Detels, and George Knox. 334–346. Oxford and New York: Oxford University Press.
Massood, Paula J. 2003. *Black City Film: African American Experiences in Film.* Philadelphia: Temple University Press.
Mathy, Jean-Philippe. 1993. *Extreme-Occident: French Intellectuals and America.* Chicago: University of Chicago Press.
———. 2000. *French Resistance: The French-American Culture Wars.* Minneapolis: University of Minnesota Press.
Matsubara, Hiroshi. 2003. "Japan Slowly Pulls Head Out of Sand on Smoking Ills." *Japan Times,* 22 October.
Matsui, Yayori. 2000. Interview. 14 February.
Matsumoto, Tsuneo. 1999. Interview. 8 April.
Mattson, M., et al. 1991. "Evaluation Plan for the Community Intervention Trial for Smoking Cessation (COMMIT)." *International Quarterly of Health Communication* 11 (3): 271–290.
Maugh, Thomas H. II. 1995. "Crib Death Risk, Secondhand Smoke Linked." *Los Angeles Times,* Home ed., 8 March, A1.
Mélihan-Chenein, Pascal. 1997. Interview. 15 September.
———. 1999. Interview. 22 September.
Meurisse, Hélène. 1999. Interview. 30 September.
Miller, John. 1988. Memo to Tony Najera, American Lung Association, California Division. 9 September. Proposition 99 Campaign Files. "Minority Issues in Tobacco Use, 1988." Mss. 94–52, Carton 1, Folder 26. Tobacco Control Archives. Archives and Special Collections. University of California, San Francisco.
Miller, Peter, and Nikolas Rose. 1990. "Political Rationalities and Technologies of Government." In *Texts, Contexts, Concepts: Studies on Politics and Power in Language,* ed. Sakari Hänninen and Kari Palonen, 167–183. Helsinki: Finnish Political Science Association.
Miller, Toby. 1993. *The Well-Tempered Self: Citizenship, Culture, and the Postmodern Subject.* Baltimore: Johns Hopkins University Press.
———. 1998. *Technologies of Truth: Cultural Citizenship and the Popular Media.* Minneapolis: University of Minnesota Press.
Miller, Toby, Nitin Govil, John McMurria, and Richard Maxwell. 1998. *Global Hollywood.* London: BFI.
Minkler, Meredith. 1989. "Health Education, Promotion, and the Open Society: An Historical Perspective." *Health Education Quarterly* 16 (1): 17–30.
Minnesota health official. 1997. Interview. 7 August.
Mochizuki-Kobayashi, Yumiko. 2005. Personal communication. 27 February.
———. 2000. Interview. 13 January.

Montes, J. Henry, Eugenia Eng, and Ronald L. Braithwaite. 1995. "A Commentary on Minority Health as a Paradigm Shift in the United States." *American Journal of Health Promotion* 9 (4): 247–250.

Morain, Dan. 1997. "Controversy Flares over State Anti-Tobacco Efforts." *Los Angeles Times*, Home ed., 11 February, 3.

Morelle, Aquilino. 1996. *La Défaite de la santé publique*. Paris: Flammarion.

Morris-Suzuki, Tessa. 1994. *The Technological Transformation of Japan: From the Seventeenth to the Twenty-first Century*. Cambridge and New York: Cambridge University Press.

——1998. *Re-inventing Japan: Time, Space, Nation*. Armonk, N.Y.: Sharpe.

Murard, Lion, and Patrick Zylberman. 1976. *Le petit travailleur infatigable: ou, le prolétaire régénéré: villes-usines, habitat et intimités au XIXe siècle*. Fontenay-sous-Bois: Recherches.

——. 1996. *L'Hygiène dans la République. La santé publique en France, ou l'utopie contrariée 1870–1918*. Paris: Fayard.

Nader, Laura. 1972. "Up the Anthropologist: Perspectives Gaines from Studying Up." In *Reinventing Anthropology*, ed. Dell Hymes, 284–311. New York: Random House.

Nakano, Nobuko. 1999. Interview. 27 January.

Nathan, Debbie, and Michael Snedeker. 1995. *Satan's Silence: Ritual Abuse and the Making of an Modern American Witch Hunt*. New York: Basic Books.

Nathanson, Constance A. 1996. "Disease Prevention as Social Change: Toward a Theory of Public Health." *Population and Development Review* 22 (4): 609 – 637.

National Black Caucus of State Legislators. n.d. Untitled doc. Proposition 99 Campaign Files. "Minority Issues in Tobacco Use, 1988." Mss. 94–52, Carton 1, Folder 26. Tobacco Control Archives. Archives and Special Collections. University of California, San Francisco.

National Cancer Institute. Office of Cancer Communication. 1989. *Making Health Communication Programs Work: A Planner's Guide*. Publication no. 89–1493. Bethesda, Md.: National Institutes of Health.

National Cancer Institute. 1995. *Community-Based Interventions for Smokers. The COMMIT Field Experience*. Washington: Government Printing Office.

National Survey on Smoking and Health in Japan, 1999: Summary of Findings. 1999. Tokyo: Ministry of Health and Welfare.

Nau, Jean-Yves. 1989. "La lutte contre le tabagisme, l'alcoolisme, la surconsommation medicamenteuse. Cinq experts proposent un plan d'urgence contre l'abus des drogues licites." *Le Monde*, 14 November.

——. 1991. "Les associations de lutte contre le tabagisme vont multiplier les plaintes visant des fabricants." *Le Monde*, 22–23 September.

——. 1997. Interview. 17 September.

——. 1999. Interview. 30 September.

Nau, Jean-Yves, and Franck Nouchi. 1992. "La 'Retraite' du professeur Got. Défenseur acharné de la santé publique." *Le Monde*, 17 June.

Neighbors, Harold, R.L. Braithwaite, and E. Thompson, 1995. "Health Promotion and African Americans." *American Journal of Health Promotion* 9 (4): 281–287.

Nelkin, Dorothy. 1985. *The Language of Risk: Conflicting Perspectives in Occupational Health.* Beverly Hills, Calif.: Sage.

———. 1995. "Science Controversies: The Dynamics of Public Disputes in the United States." In *The Handbook of Science and Technology Studies,* ed. Sheila Jasanoff et al., 445–456. Thousand Oaks, Calif.: Sage.

Nicholl, Jack. 1988a. Coalition for a Healthy California. Memo to Susan Magazine, Chris Edwards, Barbara Perzigian, Susan Smith, Karl Ory, Betsy Hite, and Cecilia DeCuir. "Minority Involvement in Prop 99." 10 August. Proposition 99 Campaign Files. "Minority Issues in Tobacco Use 1988." Mss. 94–52, Carton 1, Folder 26. Tobacco Control Archives. Archives and Special Collections. University of California, San Francisco.

———. 1988b. Press release. Coalition for a Healthy California. 9 November. Proposition 99 Campaign Files. Mss. 94–52, Carton 1, Folder 26. Tobacco Control Archives. Archives and Special Collections. University of California, San Francisco.

Nourrisson, Didier. 1988. "Tabagisme et antitabagisme en France au XIXe siècle." *Histoire, économie, société* 7 (4): 540–547.

"Nous licensions les fumeurs." 1988. *Le Monde,* 4 March.

"La Nouvelle campagne anti-tabac." 1976. *Le Monde,* 29 September.

Novelli, William D. 1990. "Achieving Black Belt Excellence in Marketing Health Care." *Journal of Health Care Marketing* 10 (4): 2–7.

Novotny, Thomas, Kenneth Warner, Juliette E. Kendrick, and Patrick Remington. 1988. "Socioeconomic Factors and Racial Smoking Differences in the United States." Unpub. ms. Rockville, Md.: Office on Smoking and Health, Centers for Disease Control and Prevention. Proposition 99 Campaign Files. "Minority Issues in Tobacco Use 1988." Mss. 94–52, Carton 1, Folder 26. Tobacco Control Archives. Archives and Special Collections. University of California, San Francisco. Later published as "Smoking by Blacks and Whites: Socioeconomic and Demographic Differences." 1998. *American Journal of Public Health* 78 (9): 1187–1189.

O'Barr, William. 1994. "Unexpected Audiences: American and Japanese Representations of One Another." In *Culture and the Ad: Exploring Otherness in the World of Advertising,* 157–198. Boulder, Colo.: Westview.

"Oblivious to Laws, the French Puff in Public." 2000. *USA Today,* 11 August.

O'Malley, Pat. 1992. "Risk, Power, and Crime Prevention." *Economy and Society* 21 (3): 252–275.

———. 1996. "Risk and Responsibility." In *Foucault and Political Reason: Liberalism, Neo-Liberalism, and Rationalities of Government,* ed. Andrew Barry, Thomas Osborne, and Nikolas Rose, 189–208. London: UCL Press.

———. 1999. "Governmentality and Risk Society." *Economy and Society* 28 (1): 138–148.

Omi, Michael, and Howard Winant. 1994. *Racial Formation in the United States from the 1960s to the 1990s.* New York: Routledge.

Ong, Aihwa. 1999. *Flexible Citizenship: The Cultural Logics of Transnationality.* Durham, N.C.: Duke University Press.

Onodera, Nobuo. 1991. "Public Health Policies and Strategies in Japan." In *Oxford Textbook of Public Health*, 2d ed., vol. 1, ed. Walter W. Holland, Roger Detels, and George Knox. 253–260. Oxford and New York: Oxford University Press.

Osborne, Thomas. 1993. "On Liberalism, Neo-Liberalism, and the 'Liberal' Profession of Medicine." In "Liberalism, Neo-Liberalism, and Governmentality," ed. Andrew Barry, Thomas Osborne, and Nikolas Rose, Special Issue of *Economy and Society* 22 (3): 345–356.

Oshima, Akira. 2003. "National Tobacco Business Law Requires Drastic Revision to Promote Anti-Tobacco Measures." *Yomiuri shinbun*, 12 March. Trans. Yumiko Toyoda.

Overholser, Geneva, 1999. "Paris Would Still Be Lovely without the Curls of Smoke." *International Herald Tribune*, 9 April.

Padioleau, Jean G. 1982. *L'État au concret*. Paris: Presses universitaires de France.

"Passive Smoke Found Third Highest Cause of Death." 1991. *Los Angeles Times*, Home ed., 14 January, B3.

Patton, Cindy. 1990. *Inventing AIDS*. New York: Routledge.

Peletier, Martin. 1990. "Tabacs: les mineurs sans fumée." *Quotidien de Paris*, 24 June.

Peterson, Christopher, and Albert J. Stunkard. 1989. "Personal Control and Health Promotion." *Social Science and Medicine* 28 (8): 819–828.

Le Petit Robert. Dictionnaire alphabétique et analogique de la langue française. 1978. Paris: Société du nouveau Littré.

Peyrot, Maurice. 1991. "Les limites de la contre-publicité." *Le Monde*, 26 October.

Philip Morris Communications. 1990. "Is It Health, or Is It Racism?" *Sun Reporter*, 6 June. 39.

Pierce, John P. 1997. Interview. 17 March.

——. 2002a. Personal communication. 23 January.

——. 2002b. Interview. 5 February.

Pierce, J.[ohn] P., R. N. Aldrich, S. Hanratty, T. Dwyer, and D. Hill. 1987. "Uptake and Quitting Smoking Trends in Australia 1974–1984." *Preventive Medicine* 16 (2): 252–260.

Pierce, John P., Petra Macaskill, and David Hill. 1990. "Long Term Effectiveness of Mass Media Led Antismoking Campaigns in Australia." *American Journal of Public Health* 80 (5): 565–569.

Pierce, John P., Leigh Thurmond, and Bradley Rosbrook. 1992. "Projecting International Lung Cancer Mortality Rates: First Approximations with Tobacco-Consumption Data." *Journal of National Cancer Institute Monographs* 12: 45–49.

Pierce, John P., Nicola Evans, Arthur J. Farkas, Shirley W. Cavin, Charles Berry, Michael Kramer, Sheila Kealey, Bradley Rosbrook, Won Choi, and Robert M. Kaplan. 1994. *Tobacco Use in California: An Evaluation of the Tobacco Control Program, 1989–1993—A Report to the California Department of Health Services*. La Jolla: University of California, San Diego.

Pierce, J.[ohn] P., E. A. Gilpin, S. L. Emery, A. J. Frakas, S. H. Zhu, W. S. Choi, C. C. Berry, J. M. Distefan, M. M. White, S. Soroko, and A. Navarro. 1998. *Tobacco Control in California: Who's Winning the War? An Evaluation of the Tobacco Control Program, 1989–96*. La Jolla: University of California, San Diego.

Pinell, Patrice. 1992. *Naissance d'un fléau. Histoire de la lutte contre le cancer en France 1890–1940*. Paris: Métailé.
"Poisoning Asia's Youth." 1997. *Japan Times*, 12 September.
"Polémique autour d'un spot anti-tabac." 1991. *Le Monde*, 3–4 November.
Pollay, Richard W. 1989. "Campaigns, Change, and Culture: On the Polluting Potential of Persuasion." In *Information Campaigns: Balancing Social Values and Social Change*, ed. Charles T. Salmon, 185–196. Newbury Park, Calif.: Sage.
Popham, W. James. 1997. Interview. 1 August.
Popham, W. James, Lance D. Potter, Dileep G. Bal, Michael D. Johnson, Jacquolyn M. Duerr, and Valerie Quinn. 1993. "Do Anti-Smoking Media Campaigns Help Smokers Quit?" *Public Health Reports* 108 (4): 510–513.
Porter, Theodore. 1995. *Trust in Numbers: The Pursuit of Objectivity in Science and Public Life*. Princeton, N.J.: Princeton University Press.
Pracontal, Michel de. 1998. *La Guerre du tabac*. Paris: Fayard.
"Les Premiers résultats de la campagne anti-tabac." 1977. *Le Monde*, 11 May.
Proctor, Robert N. 1999. *The Nazi War on Cancer*. Princeton, N.J.: Princeton University Press.
"La Progression du tabagisme est ralentie." 1978. *Le Monde*, 27–28 August.
"Quand l'Amérique diabolise le tabac; la prohibitiou est de retour." 1991. *Quotidien de Paris*, 29 August.
Rabinow, Paul. 1996. *The Anthropology of Reason*. Princeton, N.J.: Princeton University Press.
——. 1999. *French DNA*. Chicago: University of Chicago Press.
Ramsay, Matthew. 1994. "Public Health in France." In *The History of Public Health and the Modern State*, ed. Dorothy Porter, 45–118. Amsterdam and Atlanta: Rodopi B. V.
Rapport de contrôle du Comité national contre le tabagisme. 1998. Report no. 97107. January. Paris: Inspection générale des affaires sociales.
Recer, Paul. 2000. "Smokers' Wives Analyzed in Study." 1996. *Los Angeles Times*, Home ed., 17 October.
Reeves, Jimmie L., and Richard Campbell. 1994. *Cracked Coverage: Television News, The Anti-Cocaine Crusade, and the Reagan Legacy*. Durham, N.C.: Duke University Press.
Reid, Roddey. 1993. *Families in Jeopardy: Regulating the Social Body in France, 1750–1910*. Stanford, Calif.: Stanford University Press.
——. 1995. "Death of the Family; or, Keeping Human Beings Human." In *Posthuman Bodies*, ed. Judith Halberstam and Ira Livingston, 177–199. Bloomington: Indiana University Press.
——. 2000. "Researcher or Smoker? or, When the Other Isn't Other Enough in Studying 'across' Tobacco Control." In *Doing Science + Culture*, ed. Roddey Reid and Sharon Traweek, 119–150. New York: Routledge.
——. 2002. "Tobacco Industry." *Dictionary of American History*. New York: Scribner's.
Resnicow, Ken, Tom Baranowski, Jasjit S. Ahluwalia, and Ronald L. Braithwaite. 1999. "Cultural Sensitivity in Public Health: Defined and Demystified." *Ethnicity and Disease* 9: 10–21.

"Robin Williams: Live on Broadway." 2002. HBO Special Broadcast. 14 July.
Robinson, Robert G. 2000. Interview. 7 August.
Robinson, Robert G., and Sandra W. Headen. 1999. "Tobacco Use and the African American Community: A Conceptual Framework for the Year 2000 and Beyond." In *Planning and Implementing Effective Tobacco Education and Prevention Programs*, ed. Martin L. Forst, 83–111. Springfield, Ill.: Charles C. Thomas.
Romano, Rose-Mary. 2000. Interview. 25 August.
Root, Michael. 2001. "The Problem of Race in Medicine." *Philosophy of the Social Sciences* 31 (1): 20–39.
Rose, Lisle Abbott. 1999. *The Cold War Comes to Main Street: America in 1950*. Lawrence: University Press of Kansas.
Rose, Nikolas. 1993. "Government, Authority, and Expertise in Advanced Liberalism." *Economy and Society* 22 (3): 283–299.
——. 1996. "The Death of the Social? Refiguring the Territory of Government." *Economy and Society* 25 (3): 327–356.
——. 1999. *Powers of Freedom: Reframing Political Thought*. Cambridge: Cambridge University Press.
Rotily, Michel, and Anne Bregeault. 1994. "Pas de campagne sans enquête: comment les jeunes Français fument en 1994." *La Santé de l'homme* 313 (Sept.–Oct.): 18–21.
Sakai, Naoki. 1997. "Modernity and Its Critique: The Problem of Universalism and Particularism." In *Translation and Subjectivity: On "Japan" and Cultural Nationalism*, 153–192. Minneapolis: University of Minnesota Press.
Salmon, Charles T., ed. 1989a. *Information Campaigns: Balancing Social Values and Social Change*. Newbury Park, Calif.: Sage.
——. 1989b. Preface. In *Information Campaigns: Balancing Social Values and Social Change*, ed. Charles T. Salmon, 7–13. Newbury Park, Calif.: Sage.
"Sanitizing History on a Postage Stamp." 1996. *New York Times*, 27 October, E5.
La Santé en France. Rapport remis au ministre des affaires sociales et de la solidarité nationale et au secrétaire d'état chargé de la santé. 1985. Paris: La Documentation française.
Sasco, Annie. 1992. "Evaluation des actions en milieu scolaire: problèmes méthodologiques." In *La Lutte contre le tabagisme est-elle efficace?* ed. Karen Slama, Serge Karsenty, and Albert Hirsh, 73–78. INSERM and CFES. Paris: La Documentation française.
Sato, Hajime. 1999. "Policy and Politics of Smoking Control in Japan." *Social Science and Medicine* 49: 581–600.
"Scapegoating the Black Family." 1989. Special issue, *Nation*, 24–31 July.
Schudson, Michael. 1984. *Advertising, the Uneasy Persuasion: Its Dubious Impact on American Society*. New York: Basic Books.
——. 1993. "Symbols and Smokers: Advertising, Health Messages, and Public Policy." In *Smoking Policy: Law, Politics, and Culture*, ed. Robert L. Rabin and Stephen D. Sugarman, 208–225. New York: Oxford University Press.
Sears, David O., and Jack Citrin. 1985. *Tax Revolt: Something for Nothing in California*. Cambridge, Mass.: Harvard University Press.
"Selling Cigarettes in Asia." 1997. *New York Times*, 10 September.

Setbon, Michel. 1993. *Pouvoirs contre SIDA: de la transfusion sanguine au dépistage: décisions et pratiques en France, Grande-Bretagne et Suede.* Paris: Seuil.
Shain, Martin. 1999. "Minorities and Immigrant Incorporation in France: The State and the Dynamics of Multiculturalism." In *Multicultural Questions,* ed. Christian Joppke and Steven Lukes, 199–223. Oxford: Oxford University Press.
Shapin, Steven, and Simon Schaffer. 1985. *Leviathan and the Air-Pump: Hobbes, Boyle, and the Experimental Life.* Princeton, N.J.: Princeton University Press.
Shibata, Yasuhiko. 1987. "Foreign Cigarette Sales Proving Critics Correct?" *Daily Yomiuri,* 28 August.
Shilts, Randy. 1987. *And the Band Played On: Politics, People, and the AIDS Epidemic.* New York: St. Martin's.
Shimao, Tadao. 1999. Interview. 17 March.
Shome, Roka, and Radha S. Hegde. 2002. "Cultural Communication and the Challenge of Globalization." *Critical Studies in Media Communication* 19 (2): 172–189.
Shorty, Lawrence A. 1999. "Native Population Tobacco Issues: A Native Tobacco Person's Perspective." In *Planning and Implementing Effective Tobacco Education and Prevention Programs,* ed. Martin L. Forst, 69–82. Springfield, Ill.: Charles C. Thomas.
"Si tu fumes, t'es viré." 1991. *Le Nouvel observateur,* 30 May–5 June.
Slama, Karen. 1997. Interview. 17 September.
———. 1999. Interview. 23 September.
"Smoking and Breast Cancer May Be Linked, Study Shows." 1996. National Public Radio. Morning Edition. 26 April.
"Smoking Linked to Sterility, Premature Births, Deformity." 1987. *Japan Times,* 17 October.
"Smoking No Smoking. Fumeurs victimes." 1996. ARTE, 20 March.
Soto, Tom. 1988. Memo to Jack Nicholl, Coalition for a Healthy California. 18 August. Proposition 99 Campaign Files. "Minority Issues in Tobacco Use, 1988." Mss. 94–52, Carton 1, Folder 26. Tobacco Control Archives. Archives and Special Collections. University of California, San Francisco.
Speisser, Béatrice. 1997. Interview. 16 September.
———. 1999. Interview. 24 September.
Spivak, Gayatri. 1985. "Can the Subaltern Speak? Speculations on Widow Sacrifice." *Wedge* 7–8: 120–130.
———. 1998. "Cultural Talks in the Hot Peace: Revisiting the 'Global Village.'" In *Cosmopolitics: Thinking and Feeling beyond the Nation,* ed. Pheng Cheah and Bruce Robbins, 329–348. Minneapolis: University of Minnesota Press.
Stacey, Judith. 1994. "Scents, Scholars, and Stigma: The Revisionist Campaign for Family Values." *Social Text* 40: 51–75.
Stall, B., and C. Decker. 1994. "Wilson and Prop. 187 Win." *Los Angeles Times,* 9 November, A1.
Starr, Paul. 1982. *The Social Transformation of American Medicine: The Rise of a Sovereign Profession and the Making of a Vast Industry.* New York: Basic Books.
———. 1994. *The Logic of Health Care Reform: Why and How the President's Plan Will Work.* New York: Penguin.

"State Set to Scrap Anti-Smoking TV Ads." 1992. *Los Angeles Times*, Home ed., 16 January, B8.
Steffen, Monica. 1996. *The Fight against AIDS: An International Policy Comparison between Four European Countries—France, Great Britain, Germany, and Italy.* Grenoble: Presses universitaires de Grenoble.
Stern, Lesley. 1999. *The Smoking Book.* Chicago: University of Chicago Press.
Sterngold, James. 1993. "When Smoking Is a Patriotic Duty." *New York Times*, 17 October, Sec. 3–1.
"Still a Smoker's Paradise." 1993. Editorial. *Japan Times*, 25 April.
Stocker, Frederick D., ed. 1991. *Proposition 13: A Ten-Year Retrospective.* Cambridge, Mass.: Lincoln Institute of Land Policy Study.
Stockwell, Theresa F., and Stanton A. Glantz. 1997. "Tobacco Use Is Increasing in Popular Films." *Tobacco Control* 6: 282–284.
Stolberg, Sheryl. 1994a. "Fetuses Affected by Secondhand Smoke." *Los Angeles Times*, Home ed., 23 April, A1.
——. 1994b. "Clearing the Air: How Dangerous Is Secondhand Smoke?" *Los Angeles Times*, Home ed., 26 May, A1.
Stoler, Ann L. 1995. *Race and the Education of Desire: Foucault's History of Sexuality and the Colonial Order of Things.* Durham, N.C.: Duke University Press.
Strom, Stephanie. 2001. "Japan and Tobacco Revenue: Leader Faces Difficult Choice." *New York Times*, 13 June, A1.
Stuyck, Stephen C. 1990. "Public Health and the Media: Unequal Partners?" In *Mass Communication and Public Health*, ed. Charles Atkin and Lawrence Wallack, 71–77. Newbury Park, Calif.: Sage.
Le Système de santé français. Réflexions et propositions. 1983. Paris: La Documentation française.
"Tabac: Clinton met le paquet." 1996. Antenne 2, 22 August.
"Tabac: la guerre du feu: les ayatollahs de l'oxygène." 1991. *Le Nouvel observateur*, 5 June.
Takaki, Ronald. 1993. *A Different Mirror: A Multicultural History of America.* Boston: Little, Brown.
Tamura, Yoshiko. 2000. "Japan—A Smokers' Paradise under Siege." *Yomiuri shinbun*, 1 September.
Tanaka, Stefan. 1993. *Japan's Orient: Rendering Pasts into History.* Berkeley: University of California Press.
Tatara, Kozo. 1991. "The Origins and Development of Public Health in Japan." In *Oxford Textbook of Public Health*, 2d ed., vol. 1, ed. Walter W. Holland, Roger Detels, and George Knox. 35– 48. Oxford and New York: Oxford University Press.
TEOC (Tobacco Education Oversight Committee). 1991. *Toward a Tobacco-Free California: A Master Plan to Reduce Californians' Use of Tobacco.* Submitted to the Legislature by the Tobacco Education Oversight Committee. Sacramento, Calif.: TEOC.
——. 1993. "Minutes of the TEOC Meeting, March 29, 1993." Lester Breslow Tobacco Control Papers. "TEOC Meeting 24 May 1993." Mss. 94–50. Carton 1, Folder 32. Tobacco Control Archives. Archives and Special Collections. University of California, San Francisco.

TEROC (Tobacco Education and Research Oversight Committee). 1995. *Toward a Tobacco-Free California: Mastering the Challenges, 1995–97*. Submitted to the Legislature by the Tobacco Education Oversight Committee. Sacramento, Calif.: TEOC.

——. 2000. *Toward a Tobacco-Free California: Strategies for the Twenty-first Century 2000–2003*. Submitted to the Legislature by the Tobacco Education Oversight Committee. Sacramento, Calif.: TEOC.

Terry, Jennifer. 1989. "The Body Invaded: Medical Surveillance of Women as Reproducers." *Socialist Review* 89: 13–43.

"Timbre Malraux censuré." 1996. *Téléjournal*. France 2, 20 October.

"Tobacco 301: MHW Will Study Smoking's Health Effects." 1986. Cable from United States Embassy, Tokyo, to the Secretary of State. 26 March.

Tobacco control advocate A. 1997. Interview. 25 August.

Tobacco control researcher A. 1997. Interview. 13 August.

Tobacco Education Media Campaign. Tobacco Control Section. 1989. *Request for Proposal* (No. 20–014). Chronic Diseases Branch. Preventive Medical Services Division. California Department Health Services. 1 December.

"Tobacco Negotiations: U.S. Informally Decides to Take Retaliatory Action. Dissatisfied with 'Market-Opening Measures.'" 1986. *Asahi shinbun*, 4 September.

"Tobacco under Fire, Part IV." 1997. *CBS Evening News*. 8 May.

Tominaga, Suketami. 2000. Interview. 17 January.

Tones, Keith. 1986. "Health Education and the Ideology of Health Promotion: A Review of Alternative Approaches." *Health Education Research* 1 (1): 3–12.

——. 1993. "Changing Theory and Practice: Trends in Methods, Strategies, and Settings in Health Education." *Health Education Journal* 52 (3): 125–139.

TOPIC (Tobacco Problems Information Center). 2000. "TOPICS in Japan: Japan's Tobacco Situation 2000 Profile." Tokyo: TOPIC.

Traynor, Michael P., and Stanton A. Glantz. 1996. "California's Tobacco Tax Initiative: The Development and Passage of Proposition 99." *Journal of Health Politics, Policy, and Law* 21 (3): 543–585.

Treichler, Paula A. 1989. "AIDS, Homophobia, and Biomedical Discourse: An Epidemic of Signification." In *AIDS: Cultural Analysis/Cultural Activism*, ed. Douglas Crimp, 31–70. Cambridge, Mass.: MIT Press.

Treichler, Paula A., Lisa Cartwright, and Constance Penley, eds. 1998. *The Visible Woman: Imaging Technologies, Gender, and Science*. New York: New York University Press.

Troyer, Ronald J., and Gerald E. Markle. 1983. *Cigarettes: The Battle over Smoking*. New Brunswick, N.J.: Rutgers University Press.

U.S. Dept. of Health and Human Services. Public Health Service. 1980. *The Health Consequences of Smoking for Women: A Report of the Surgeon General*. Washington: Government Printing Office.

——. 1986. *The Health Consequences of Involuntary Smoking: A Report of the Surgeon General*. Washington: Government Printing Office.

——. 1989. *Reducing the Health Consequences of Smoking: Twenty-five Years of Progress—A Report of the Surgeon General*. Washington: Government Printing Office.

———. 1998. *Tobacco Use among U.S. Racial/Ethnic Minority Groups: African Americans, American Indians and Native Alaskans, Asian Americans and Pacific Islanders, Hispanics—A Report of the Surgeon General.* Washington, D.C.: Government Printing Office.

U.S. Dept. of Health, Education, and Welfare. Public Health Service. 1964. *Smoking and Health: Report of the Advisory Committee to the Surgeon General.* Princeton, N.J.: Van Nostrand.

———. 1979. *Smoking and Health: A Report of the Surgeon General.* Washington: Government Printing Office.

Verhovek, Samuel H. 1996. "The Great Outdoors Is Latest Battlefield in War on Smoking." *New York Times,* National ed., 5 May, A1.

Vij, Ritu. 2001. "Self-Seeking and Welfare: Rethinking State and Civil Society in Modern Japan." Ph.D. dissertation, University of Denver.

Vogel, David, Robert A. Kagan, and Timothy Kessler. 1993. "Political Culture and Tobacco Control: An International Comparison." *Tobacco Control* 2: 317–326.

Volpp, Leti. 1994. "(Mis)identifying Culture: Asian Women and the 'Cultural Defense.'" *Harvard Women's Law Journal* 17: 57–101.

Wallack, Lawrence. 1984. "Social Marketing as Prevention: Uncovering Some Critical Assumptions." In *Advances in Consumer Research,* vol. 11, ed. T. C. Kinnear, 682–687. Provo, Utah: Association for Consumer Research.

———. 1990a. "Mass Media and Health Promotion." In *Mass Communication and Public Health: Conflicts and Complexities,* ed. Charles Atkin and Lawrence Wallack, 41–70. Newbury Park, Calif.: Sage.

———. 1990b. "Improving Health Promotion: Media Advocacy and Marketing Approaches." In *Mass Communication and Public Health: Conflicts and Complexities,* ed. Charles Atkin and Lawrence Wallack, 147–163. Newbury Park, Calif.: Sage.

Wallack, Lawrence, et al. 1993. *Media Advocacy and Public Health: Power for Prevention.* Thousand Oaks, Calif., and London: Sage.

Wallack, Lawrence, and Russell Sciandra. 1991. "Media Advocacy and Public Education in the Community Intervention Trial to Reduce Heavy Smoking (COMMIT)." *International Quarterly of Community Health Education* 11 (3): 205–222.

Warner, Kenneth E. 1981. "Cigarette Smoking in the 1970s: The Impact of the Antismoking Campaign on Consumption." *Science* 211 (4483): 729–731.

Watanabe, Bungaku. 1986. "Lung Cancer: The New Import?" *Japan Times,* 19 October.

———. 1999. Interview. 6 July.

Weinstein, Steve. 1995. "Nielsen Admits Discrepancy in Viewer Surveys; Ratings: The Media Monitoring Firm Concedes a Probable Undercount." *Los Angeles Times,* 6 April, D-2.

Wells, A. Judson. 1988. "An Estimate of Adult Mortality in the United States from Passive Smoking." *Environment International* 14: 249–285.

White House Domestic Policy Council. 1993. *The President's Health Security Plan.* New York: Times Books.

WHO (World Health Organization). 1997. *Tobacco or Health: A Global Status Report.* Geneva: WHO.

Williams, Lena. 1987. "Blacks in Debate on Tobacco Industry Influences," *New York Times,* National ed., 17 January, 1, 8.

Williamson, Judith. 1978. *Decoding Advertisements: Ideology and Meaning in Advertisements.* London: Boyars.

Women's Action on Smoking: Japanese Women Expressive and Active Now. 1987. Tokyo: Ahni Publishing House.

"Worker Wins 40,000 Yen over Passive Smoke." 2004. *Japan Times,* 13 July.

"World No Tobacco Day Widely Ignored by Japanese Smokers." 1999. *Japan Times,* 1 June.

"World without Fathers: The Struggle to Save the Black Family." 1993. *Newsweek,* 30 August, 16–29.

Yach, Derek. 1999a. "Nowhere to Run, Nowhere to Hide—World Community Joins WHO in Holding Up Mirrors to Big Tobacco." California Tobacco Control Project Director's Meeting. Lake Tahoe, California. 1 November.

———. 1999b. Interview. 17 November.

Yach, Derek, and Douglas Bettcher. 2000. "Globalisation of Tobacco Industry Influence and New Global Responses." *Tobacco Control* 9: 206–216.

Yoneyama, Lisa. 1999. *Hiroshima Traces: Time, Space, and the Dialectics of Memory.* Berkeley: University of California Press.

Young, Ian, and Margaret Whitehead. 1993. "Back to the Future: Our Social History and Its Impact on Health Education." *Health Education Journal* 52 (3): 114–119.

Zeitlin, June. 1999. Interview. 3 January.

Zivy, Pierre. 1965. *Le Tabac: son histoire et son bon usage.* Paris: Union générale d'éditions. Collection 10/18.

Tobacco Control Timeline

California, France, and Japan

1880s

—U.S.: manufacturers create nationally advertised brands of machine-made cigarettes; Duke and Sons Inc. seize 40% of domestic cigarette market

1890

—U.S.: 26 states have laws banning tobacco sales to minors

1895–1905

—China and Japan: American Tobacco Co. purchases controlling interest in local tobacco companies

1900

—Japan: health concerns voiced by military lead to law prohibiting smoking by minors and setting the legal age of smoking at 20

1904

—Japan: government nationalizes the domestic and foreign-owned tobacco industry

1906

—U.S.: federal Pure Food and Drug Act exempts cigarettes from any government scrutiny

1909

—U.S.: all tobacco sales prohibited in 17 states, but the advent of the First World War will rescind all bans

1920s

—North America, Europe, and East Asia: consumers begin steady conversion to machine-made cigarettes

—U.S.: per capita consumption of cigarettes doubles to 1,370 cigarettes

1920

—Japan: ban on smoking in all commuting vehicles and cinemas

1926

—France: state tobacco monopoly reorganized as Service de l'exploitation industrielle des tabacs (SEIT; later SEITA) to retire the public debt resulting from the First World War

1931–45

—Asia-Pacific War and Second World War disrupt the rise of heavy smoking except in the U.S., the U.K., Canada, Australia

1940–45

—U.S.: per capita smoking consumption soars 75%, reaching 3,500 cigarettes

1942

—France: ban on smoking in theaters and cinemas

1950

—U.S.: Federal Trade Commission finds cigarette advertising false and deceptive

1954–58

—France: Denoix and Schwartz smoking study sponsored by SEITA

1962–64

—U.K. and U.S.: major government reports warning of smoking hazards released

1963

—U.S.: per capita cigarette consumption peaks at 4,345 cigarettes

1965

—Japan: highest smoking rates reached: 47.1% (83.2% of adult men, 15.7% of adult women); Hirayama cohort study begins to confirm health effects of smoking on Japanese

—U.S.: smoking rates stand at 42.6% (52.5% of adult men, 33.9% of adult women)

1967–70

—U.S.: Federal Communications Commission "Fairness Doctrine" forces broadcasters to allot free airtime for anti-smoking spots

1970

—U.S.: tobacco companies withdraw all broadcast advertising; anti-smoking spots, losing free airtime, cease, and smoking rates rise; per capita rates peak at 4,000 cigarettes in early 1970s

1972

—U.S.: nonsmoking sections mandated for domestic airlines; Marlboro becomes leading brand

1975

—U.S.: Minnesota passes clear indoor air act

1976

—France: *loi Veil* is passed, and first anti-smoking media campaign runs with ads targeting secondhand smoke; Marlboro cigarettes hold 2% of domestic market

—California: Group Against Smoking Pollution (GASP) founded; will later become Americans for Nonsmokers' Rights

1977

—Japan: *Mainichi shinbun* publishes yearlong series on the hazards of smoking

—California: Berkeley clean indoor air ordinance passes

1978

—Japan: three anti-smoking groups are founded

1980
- —U.S., U.K., Canada, Australia: older men's smoking rate begins to decline precipitously

1981
- —Japan: Hirayama study of married couples indicates higher risk of lung cancer in nonsmoking wives of smokers

1984
- —Japan: Liberal Democratic Party–led coalition passes Tobacco Industry Law deeming the tobacco industry to be in the national interest

1985
- —Japan: government privatizes the tobacco monopoly, creating Japan Tobacco (JT) and allotting 100% of the stock to the Finance Ministry
- —Japan: per capita consumption peaks at 3,200 cigarettes (primarily blond tobacco)
- —France: per capita consumption peaks at 2,300 cigarettes (35% dark tobacco)
- —U.S.: Minnesota is the first state to levy a cigarette tax for tobacco control

1986
- —Japan: Sen. Jesse Helms and U.S. Trade Representative successfully pressure government to eliminate tariffs on foreign brands and open the Japanese cigarette market to U.S. and U.K. products, which will cause an explosion in tobacco advertising
- —U.S.: surgeon general's report on hazards of involuntary smoking released; 117 local clean air ordinances in effect in California, 246 nationally

1987
- —Japan: Women's Action on Smoking created; Ministry of Health and Welfare publishes White Paper; Sixth World Conference on Smoking and Health; Tokyo court dismisses lawsuits filed by nonsmokers
- —U.S.: smoking restricted in all federal buildings

1988
- —U.S.: all domestic flights 2 hours or less are mandated nonsmoking; extended to all flights 6 hours or less in 1990
- —California: Proposition 99 passes, raising cigarette taxes and mandating tobacco education program; New York City passes clean indoor air ordinance
- —France: Marlboro cigarettes grab 19.9% of domestic market

1990–2000
- —Eastern Europe: foreign manufacturers—Philip Morris in the lead—buy up former state-owned cigarette companies

1990
- —U.S.: Uptown Coalition defeats R. J. Reynolds' attempt to test market first ethnically marketed brand in Philadelphia
- —California: launch of anti-smoking media campaign based on social marketing and community empowerment principles

—U.S.: per capita consumption falls to 2,800 from 3,850 in 1980; 440,000 reported tobacco-related deaths
—France: tobacco-related deaths reach 60,000

1991
—France: Socialist government passes *loi Evin*, which bans all print and electronic tobacco advertising and introduces mandatory nonsmoking sections
—U.S.: Parmley and Glantz article in *Circulation* estimates 52,700 annual deaths from tobacco-related diseases due to secondhand smoke

1992
—France: *loi Evin* goes into effect without any media campaign; Affaire Williams-Renault; Council of State ends the price-fixing of foreign brands by the French state
—U.S.: Environmental Protection Agency (EPA) releases report on hazards of environmental tobacco smoke (ETS); Massachusetts Question 1 passes establishing highest per capita tobacco education funding
—California: Gov. Pete Wilson slashes tobacco media campaign budget
—China: cigarette consumption triples from 1972, reaching 1,900 cigarettes per capita

1993
—California: drop in smoking prevalence from 24% to 19% (1988–93)

1994
—U.S.: Henry Waxman holds hearings of the House Energy and Commerce Subcommittee on Health and the Environment in which tobacco industry executives deny under oath that cigarettes are addictive
—California: secret Brown and Williamson tobacco documents deposited at the University of California, San Francisco
—France: Ninth World Conference on Tobacco or Health (Paris)

1995
—California: the State of California extends its ban on smoking in enclosed public spaces to restaurants with the exception of bar areas; University of California, San Francisco, publishes on the Internet smoking studies suppressed by the Brown and Williamson Tobacco Corp.
—U.S.: smoking rates at 28%; per capita at around 1,800 cigarettes
—France: smoking rates at 38%; per capita at around 1,500 cigarettes; SEITA completely privatized
—Japan: per capita consumption at 3,000 cigarettes

1996
—Japan: foreign cigarette brands have 22.3% of domestic market share
—U.S.: President Bill Clinton signs a decree declaring nicotine an addictive drug and gives the federal Food and Drug Administration (FDA) broad authority to regulate cigarettes and other tobacco products (later thrown out by courts); some communities begin to ban smoking in public parks

—International: *The Global Burden of Disease*, published by the World Health Organization (WHO), World Bank, and Harvard School of Public Health, estimates that the number of annual tobacco-related deaths will reach 10 million by 2020

1997

—Japan: smoking rates decline to 34.6%
—U.S.: *Washington Post* adds the frequency of smoking to its content ratings of films; R. J. Reynolds retires Joe Camel ads in the U.S. after being accused for years of targeting teenagers and children
—China: Tenth World Conference on Tobacco or Health (Beijing)
—Europe: European Union health ministers vote to phase out tobacco advertising

1998

—U.S: California bans smoking in all bars and bar areas; federal government bans smoking on all U.S.-bound flights; the tobacco Master Settlement Agreement is signed between state attorneys general and the tobacco industry (including Japan Tobacco); tobacco industry settles class-action lawsuit filed by 60,000 former flight attendants; State of Minnesota Depository makes available 3 million confidential tobacco industry documents
—Japan: 21st Century Tobacco Policy Deliberation Committee; smoking banned on all domestic flights

1999

—International: the WHO launches Tobacco-Free Initiative; promotes the California Tobacco Control Program as a worldwide model
—International: the World Bank publishes *Curbing the Epidemic*, advocating sharp increases in cigarette taxes and regulation of the international tobacco industry
—California: ethnic tobacco education networks denounce Philip Morris' "Find Your Own Voice" print ad campaign targeting communities of color
—U.S.: American Legacy Foundation established to combat youth smoking
—Japan: WHO Conference on tobacco, women, and youth (Kobe); Ministry of Health and Welfare report *Health Japan 21* released; second national survey reports 90,000–114,000 tobacco-related deaths annually and that 52.8% of men and 13.4% of women smoke; Japan Tobacco buys R. J. Reynolds' overseas operations; Japan Airlines bans smoking on all flights
—France: 77% of cigarettes sold in France are foreign brands (45% US); SEITA buys Tabacalera (Spain) to become Altadis; Gherlain lawsuit ends with French court holding SEITA partly responsible for smoker's death

2000

—U.S.: Eleventh World Conference on Tobacco or Health held in Chicago, originally slated to be held in South Africa; democrat Gray Davis becomes governor of California but offers little support to tobacco control
—France: Air France bans smoking on all flights

2003

—Japan: Health Promotion Law, art. 25 on secondhand smoke
—France: France, the U.K., and the European Community sign the WHO's Framework Convention on Tobacco Control
—U.S.: New York City bans smoking in all restaurants and bars

2004

—U.S., Japan, and Germany: the WHO's Framework Convention on Tobacco Control signed
—International: 22.5% of population smokes in the U.S., 30.5% in France, and 16.4% in California

2005

—International: the WHO's Framework Convention on Tobacco Control treaty takes effect

Index

Page numbers in italics refer to illustrations.

Index

Index

and Research Oversight Committee (TEROC, Calif.)

An interdisciplinary scholar working at the intersection of French studies, science studies, communication, and cultural studies, **Roddey Reid** is Professor in the Department of Literature at the University of California, San Diego. A former Japan Foundation Abe Fellow, he is author of *Families in Jeopardy: Regulating the Social Body in France, 1750–1910* and co-editor (with Sharon Traweek) of *Doing Science + Culture.*

TRACKING GLOBALIZATION

Illicit Flows: States, Borders, and the
Criminal Life of Things
Edited by Willem van Schendel and Itty Abraham

Globalizing Tobacco Control:
Anti-smoking Campaigns in California, France, and Japan
Roddey Reid